Iterative Error Correction

Turbo, Low-Density Parity-Check and Repeat–Accumulate Codes

Iterative error correction codes have found widespread application in cellular communications, digital broadcasting, deep space communications, and wireless LANs. This self-contained treatment of iterative error correction presents all the key ideas needed to understand, design, implement, and analyze these powerful codes.

Turbo, low-density parity-check, and repeat–accumulate codes are given equal, detailed coverage, with precise presentations of encoding and decoding procedures. Worked examples are integrated into the text to illuminate each new idea and pseudo-code is included for important algorithms to facilitate the reader's development of the techniques described. For each subject, the treatment begins with the simplest case before generalizing. There is also coverage of advanced topics such as density-evolution and EXIT charts for those readers interested in gaining a deeper understanding of the field. This text is ideal for graduate students in electrical engineering and computer science departments, as well as practitioners in the communications industry.

Sarah J. Johnson is a Research Fellow in the School of Electrical Engineering and Computer Science at the University of Newcastle, Australia. She is a member of the IEEE Information Theory and Communications Societies.

Iterative Error Correction

Turbo, Low-Density Parity-Check and Repeat–Accumulate Codes

SARAH J. JOHNSON

University of Newcastle, New South Wales

CAMBRIDGE UNIVERSITY PRESS
Cambridge, New York, Melbourne, Madrid, Cape Town, Singapore, São Paulo, Delhi

Cambridge University Press
The Edinburgh Building, Cambridge CB2 8RU, UK

Published in the United States of America by Cambridge University Press, New York

www.cambridge.org
Information on this title: www.cambridge.org/9780521871488

First published 2010

Printed in the United Kingdom at the University Press, Cambridge

A catalog record for this publication is available from the British Library

ISBN 978-0-521-87148-8 Hardback

To Ro and Clara

Contents

Preface

The field of error correction coding was launched with Shannon's revolutionary 1948 work showing – quite counter to intuition – that it is possible to transmit digital data with arbitrarily high reliability over noise-corrupted channels, provided that the rate of transmission is below the capacity of the channel. The mechanism for achieving this reliable communication is to encode a digital message with an error correction code prior to transmission and apply a decoding algorithm at the receiver.

Classical block and convolutional error correction codes were described soon afterwards and the first iterative codes were published by Gallager in his 1962 thesis; however they received little attention until the late 1990s. In the meantime, the highly structured algebraic codes introduced by Hamming, Elias, Reed, Muller, Solomon and Golay among others dominated the field. Despite the enormous practical success of these classical codes, their performance fell well short of the theoretically achievable performances set down by Shannon in his seminal 1948 paper. By the late 1980s, despite decades of attempts, researchers were largely resigned to this seemingly insurmountable theory–practice gap.

The relative quiescence of the coding field was utterly transformed by the introduction of "turbo codes", proposed by Berrou, Glavieux and Thitimajshima in 1993, wherein all the key ingredients of successful error correction codes were replaced: turbo codes involve very little algebra, employ iterative, distributed, algorithms and focus on average (rather than worst-case) performance. This was a ground-shifting paper, forcing coding theorists to revise strongly held beliefs. Overnight, the gap to the Shannon capacity limit was all but eliminated, using decoders with manageable complexity.

As researchers worked through the 1990s to understand just why turbo codes worked as well as they did, two researchers, McKay and Neal, introduced a new class of block codes designed to possess many features of the new turbo codes. It was soon recognized that these block codes were in fact a rediscovery of the low-density parity-check (LDPC) codes developed years earlier by Gallager. Generalizations of Gallager's LDPC codes by a number of researchers, including Luby, Mitzenmacher, Shokrollahi, Spielman, Richardson and Urbanke, to irregular LDPC codes actually achieve Shannon's capacity limits on some channels and approach Shannon's capacity to within hundredths of a decibel on others.

A new class of iterative codes introduced by Divsalar, Jin and McEliece, called repeat–accumulate (RA) codes, combine aspects of both turbo and LDPC codes, arguably containing the best features of each. The combination of simplicity with excellent decoding performance has made RA codes a hot topic for the coding community. Indeed, modified RA codes have been shown to be the first capacity-achieving codes with bounded complexity per message bit for both encoding and decoding.

So rapid has progress been in this area that coding theory today is in many ways unrecognizable from its state less than two decades ago. In addition to the strong theoretical interest in iterative codes, such codes have already been adopted in deep space applications, satellite-based digital video broadcasting, long-haul optical communication standards, wireless local area networking and mobile telephony.

This book presents an introductory treatment of these extremely powerful error-correcting codes and their associated decoding algorithms. The focus of the book is almost exclusively on iterative error correction. Indeed, classical coding topics are only covered where they are necessary for the development of iterative codes, as in Chapters 1 and 4. The umbrella term *iterative decoding*, sometimes referred to as *message-passing decoding*, *probabilistic decoding*, decoding of *codes on graphs*, or simply the *turbo principle*, here encompasses the decoding algorithms associated with turbo codes, low-density parity-check codes and repeat–accumulate codes.

The aim of this book is to provide a gentle introduction to turbo, low-density parity-check and repeat–accumulate codes, with equal coverage given to all three. No prior knowledge of coding theory is required. An emphasis on examples is employed to illuminate each new idea and pseudo-code is presented to facilitate the reader's development of the techniques described. The focus of this book is on ideas and examples rather than theorems and proofs. Where results are mentioned but not proved, references are provided for the reader wishing to delve deeper. Readers interested in a thorough exposition of the theory, however, need go no further than the excellent text by Richardson and Urbanke [1].

None of the material in this text is new; all of it is available in the iterative decoding literature. However, we have decided not to attempt to attribute every idea to the location of its first publication, not least of all because many ideas in the field have been developed independently by a number of authors in different ways. Most importantly, though, we want to keep the focus of the text on explanations of ideas rather than as a historical record. We do use references, however, to point out a few main turning points in the field and to provide direction for the reader wishing to pursue topics we have not fully covered.

Chapters 2 and 3 together can be read as a stand-alone introduction to LDPC codes and Chapters 4 and 5 together provide a stand-alone introduction to turbo codes. Subsequently, however, the distinction is less sharp. The material on RA codes in Chapter 6 draws heavily on the previous material and Chapters 7 and 8 apply to all three code types.

Book outline

Chapter 1 provides a general overview of point-to-point communications systems and the role played by error correction. Firstly we define three channel models, each of which will play an important role in the remainder of the book: the binary symmetric channel (BSC), the binary erasure channel (BEC) and the binary input additive white Gaussian noise channel (BI-AWGNC). From these channel models we introduce the concept of the capacity of a communications channel, a number that precisely captures the minimum redundancy needed in a coded system to permit (essentially) zero errors following decoding. We then turn to error correction and briefly introduce classical error correction techniques and analysis.

Chapters 2 and 3 serve as a self-contained introduction to low-density parity-check codes and their iterative decoding. To begin, block error correction codes are described as a linear combination of parity-check equations and thus defined by their parity-check matrix representation. Iterative decoding algorithms are introduced using a hard decision (bit-flipping) algorithm so that the topic is first developed without reference to probability theory. Subsequently, the sum–product decoding algorithm is presented. Taken alone, Chapter 2 provides a working understanding of the encoding and decoding of LDPC codes without any material on their design and analysis. In Chapter 3 we then discuss the properties of an LDPC code which affect its iterative decoding performance and we present some common construction methods used to produce codes with desired properties.

In Chapters 4 and 5 we turn to turbo codes and their soft, iterative, decoding, which has sparked an intense level of interest in such codes over the past decade. Our starting point is convolutional codes and their trellis representation. This opens the way to the optimal decoding of trellis-based codes using the BCJR algorithm for the computation of maximum a posteriori (MAP) (symbol) probabilities. Having presented the principles of convolutional component codes in Chapter 4, we develop the turbo encoder and decoder in Chapter 5, using the convolutional encoder and MAP decoder as the primary building blocks, and discuss design principles for turbo codes.

In Chapter 6 we consider iterative codes created using serial concatenation. Having so far treated separately the two main classes of iteratively decodable codes, turbo and LDPC codes, we now consider the class of repeat–accumulate codes, which draws from both and arguably combines the best features of each.

Rather than consider the behaviour of any particular code, our focus in Chapter 7 shifts to *ensembles* of codes, and the *expected* behaviour of an iterative decoding algorithm over this ensemble. Tracking the evolution of the probability density function through successive iterations of the decoding algorithm, a process known as *density evolution* allows us to compute the thresholds of iterative code ensembles and compare their performance with the channel capacity. The chapter concludes by showing how density evolution can be usefully approximated by extrinsic information transfer (EXIT) analysis techniques to understand better the performance of iterative decoding algorithms.

In Chapter 8 we also use the concept of code ensembles but now focus on low-noise channels and consider the error floor performance of iterative codes. Except for the special case of the binary erasure channel, our analysis considers the properties of the codes independently of their respective iterative decoding algorithms. In fact we assume maximum likelihood decoding, for which the performance of a code depends only on its codeword weight distribution.

Acknowledgements

I am very grateful to the many students who have provided useful feedback and suggestions on earlier versions of this text. I would also like to thank Adrian Wills, Vladimir Trajkovic, David Hayes, Björn Rüffer and Tristan Perez for their valuable assistance.

Phil Meyler and Sarah Matthews from Cambridge University Press were very helpful and patient through the many delays. Thanks also to Anna-Marie Lovett, Karen Sawyer and the typesetters from Aptara Inc. for their assistance post production and to Susan Parkinson, who performed wonders with her copy-editing in a very tight time-frame.

I am particularly grateful to Steven Weller. As well as being responsible for starting me on this endeavor he has provided invaluable help and guidance along the way.

Most of all I owe more than I can say to my family, from whom much of the time to write this book was taken.

Notation

The following is a list of notation used across more than one chapter.

$\alpha_t(S_i)$	Forward metric of the BCJR decoder at time t for state S_i
a	Combiner parameter of an RA code
a_d	Number of codewords of weight d
$a_{w,d}$	Number of messages of weight w that produce codewords of weight d
$\mathbf{A}$	Vector of a priori LLRs
$\mathbf{A}_l^{(j)}$	Vector of a priori LLRs into the jth decoder at the lth iteration of turbo decoding
A_w	Number of messages of weight w
$A_{w,p}$	Number of messages of weight w that produce parity bits of weight p
$A(D)$	Weight enumerating function (of a_d's)
$A(W, D)$	Input–output weight enumerating function (of $a_{w,d}$'s)
$A(W, P)$	Input–redundancy weight enumerating function (of $A_{w,p}$'s)
$\mathcal{A}(S_i)$	LLR of $\alpha_t(S_i)$
$\beta_t(S_i)$	Backward metric of the BCJR decoder at time t for state S_i
$\mathcal{B}(S_i)$	LLR of $\beta_t(S_i)$
$\mathbf{c}$	Binary vector containing the codeword bits
$c_t^{(i)}$	Value of the codeword bit output at time t from the ith output of a convolutional encoder
$c(x)$	Capacity of a channel with parameter x
C	An error correction code
d	Hamming weight of a binary codeword
$\mathbf{d}$	Binary vector containing the interleaved message bits of an RA code
$d_{\min}$	Minimum distance of an error correction code

d_f	Free distance of a convolutional code
d_ef	Effective free distance of a convolutional or concatenated code
$d_\mathrm{f}^{(\mathrm{I})}$	Free distance of the inner code in a serially concatenated code system
$d_\mathrm{f}^{(\mathrm{O})}$	Free distance of the outer code in a serially concatenated code system
ε	Erasure probability of the binary erasure channel
ϵ	Crossover probability of the binary symmetric channel
$\mathbf{E}$	Vector of extrinsic LLRs
$E_{j,i}$	Extrinsic message from the jth check node to the ith bit node
$\mathbf{E}_l^{(j)}$	Vector of extrinsic LLRs from the jth decoder at the lth iteration of turbo decoding
E_b/N_0	Signal-to-noise ratio of an additive white Gaussian noise channel
$\gamma_t(S_r, S_s)$	State transition metric into the BCJR decoder for states S_r to S_s at time t
$\Gamma_t(S_r, S_s)$	LLR of $\gamma_t(S_r, S_s)$
G	Code generator matrix
$\mathbf{h}$	Row (check node) degree distribution of an irregular LDPC parity-check matrix
h_i	Fraction of Tanner graph check nodes that are degree i
H	Code parity-check matrix
$I(X;Y)$	The mutual information between X and Y
I_x	Identity matrix of size x
I_max	Maximum number of decoder iterations
k	Number of input bits to a convolutional encoder at each time instant
k_O	k for the outer code in a serially concatenated code system
k_I	k for the inner code in a serially concatenated code system
k_i	k for the ith code in a parallel concatenated code system
K	Code dimension (length of the messages)
K_O	K for the outer code in a serially concatenated code system
K_I	K for the inner code in a serially concatenated code system
K_i	K for the ith code in a parallel concatenated code system
$\lambda(x)$	The (edge perspective) bit node degree distribution an irregular LDPC code
λ_i	Fraction of Tanner graph edges connected to a degree-i bit node
l	Iteration number
$\mathbf{L}$	Vector containing the LLRs of the decoded bits
$L(x)$	Log likelihood ratio of x
log	Logarithm to the base e

$\log_2$	Logarithm to the base 2
m	Number of parity-check equations in a block code
$m_t(S_r, S_s)$	State transition metric calculated by the BCJR decoder for states S_r to S_s at time t
M_i	Message from the ith bit node
$M_{j,i}$	Message from the ith bit node to the jth check node
n	Number of output bits from a convolutional encoder at each time instant
n_{O}	n for the outer code in a serially concatenated code system
n_{I}	n for the inner code in a serially concatenated code system
n_i	n for the ith code in a parallel concatenated code system
N	Code length (length of the codewords)
N_O	N for the outer code in a serially concatenated code system
N_I	N for the inner code in a serially concatenated code system
N_i	N for the ith code in a parallel concatenated code system
$\oplus$	Modulo-2 addition
Π	Permutation sequence
π_i	ith element of permutation sequence
$\mathbf{p}$	Binary vector containing the parity bits
p	Weight of the parity bits in a binary codeword
$p(x)$	Probability of x
P	Matrix giving a puncturing pattern
q	Repetition parameter of an RA code
$Q(x)$	The error function
$\rho(x)$	The (edge perspective) check node degree distribution of an irregular LDPC code
ρ_i	Fraction of Tanner graph edges connected to a degree-i check node
r	Code rate
$\mathbf{R}$	Vector containing the LLRs of the received bits
σ^2	Variance of the additive white Gaussian noise
$s_t^{(i)}$	Value of the ith shift register element at time t in a convolutional encoder
$\mathbf{s}$	Binary vector containing the bits at the output of the RA code combiner
$\mathbf{s}$	Syndrome of a parity-check code
S	Convolutional encoder state
t	Time elapsed
T	Number of trellis segments
$u_t^{(i)}$	Value of the message bit input at time t to the ith input of the convolutional encoder

$\mathbf{u}$	Binary vector containing the message bits
$\hat{\mathbf{u}}$	Binary vector containing the hard decisions on the decoded message bits
v	Memory order of a convolutional encoder
v_i	Fraction of Tanner graph bit nodes which are degree i
$\mathbf{v}$	Binary vector containing the repeated message bits of an RA code
$\mathbf{v}$	Column (bit node) degree distribution of an irregular LDPC parity-check matrix
w	Weight of a binary message
w_{c}	Column weight (bit node degree) of a regular LDPC parity-check matrix
w_{r}	Row weight (check node degree) of a regular LDPC parity-check matrix
$\mathbf{x}$	Vector containing the transmitted bits $\in \{+1, -1\}$
$\mathbf{y}$	Vector containing the values received from the channel
$\mathbf{z}$	Vector containing the additive white Gaussian noise

Commonly used abbreviations

APP	a posteriori probability
ARA	accumulate–repeat–accumulate
BCJR	Bahl, Cock, Jelenik and Raviv (algorithm)
BER	bit error rate
BEC	binary erasure channel
DE	density evolution
EXIT	extrinsic information transfer
BI-AWGN	binary input additive white Gaussian noise
BSC	binary symmetric channel
FIR	finite impulse response
FL	finite length
IIR	infinite impulse response
IOWEF	input–output weight enumerating function
IRA	irregular repeat–accumulate
IRWEF	input–redundancy weight enumerating function
LDPC	low-density parity-check
LLR	log-likelihood ratio
MAP	maximum a posteriori probability
ML	maximum likelihood
RA	repeat–accumulate
SC	serially concatenated
SNR	signal-to-noise ratio
WER	word error rate
WEF	weight enumerating function

1
Channels, codes and capacity

In this chapter we introduce our task: communicating a digital message without error (or with as few errors as possible) despite an imperfect communications medium. Figure 1.1 shows a typical communications system. In this text we will assume that our source is producing binary data, but it could equally be an analog source followed by analog-to-digital conversion.

Through the early 1940s, engineers designing the first digital communications systems, based on pulse code modulation, worked on the assumption that information could be transmitted usefully in digital form over noise-corrupted communication channels but only in such a way that the transmission was unavoidably compromised. The effects of noise could be managed, it was believed, only by increasing the transmitted signal power enough to ensure that the received signal-to-noise ratio was sufficiently high.

Shannon's revolutionary 1948 work changed this view in a fundamental way, showing that it is possible to transmit digital data with arbitrarily high reliability, over noise-corrupted channels, by encoding the digital message with an error correction code prior to transmission and subsequently decoding it at the receiver. The error correction encoder maps each vector of K digits representing the message to longer vectors of N digits known as codewords. The redundancy implicit in the transmission of codewords, rather than the raw data alone, is the quid pro quo for achieving reliable communication over intrinsically unreliable channels. The code rate $r = K/N$ defines the amount of redundancy added by the error correction code. The transmitted bits may be corrupted in some way by the channel, and it is the function of the error correction decoder to use the added redundancy to determine the corresponding K message bits despite the imperfect reception.

In this chapter we will introduce the basic ideas behind error correction. In Section 1.1 we describe the channels considered in this text and in Section 1.2 the fundamental limits to communicating on those channels. Finally, in Section 1.3 we introduce error correction techniques.

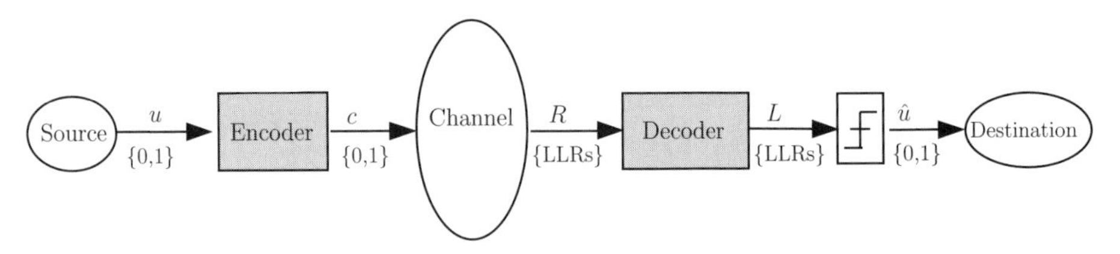

Figure 1.1 A typical communications system.

1.1 Binary input memoryless channels

A *discrete channel* is one that transmits a symbol x from a discrete set $X = \{X_1, X_1, \ldots, X_l\}$, known as the source alphabet, and returns a symbol y from another (possibly different) discrete alphabet, $Y = \{Y_1, Y_1, \ldots, Y_m\}$. A binary input channel transmits two discrete symbols, usually 0, 1, or 1, −1. The channel can also output binary symbols but may equally well output symbols from a larger discrete alphabet or a continuous range of values. Unfortunately, the channels do not always map a given transmitted symbol to the same received symbol (which is why we need error correction).

A communication channel can be modeled as a random process. For a given symbol x_i transmitted at time i, such that x_i is one of the symbols from the set X, i.e. $x_i = X_j \in \{X_1, X_1, \ldots, X_l\}$, the channel transition probability $p(y|x) = p(y = Y_j|x = X_j)$ gives the probability that the returned symbol y_i at time i is the symbol Y_i from the set Y, i.e. $y_i = Y_j \in \{Y_1, Y_1, \ldots, Y_m\}$. A channel is said to be *memoryless* if the channel output at any time instant depends only on the input at that time instant, not on previously transmitted symbols. More precisely, for a sequence of transmitted symbols $\mathbf{x} = [x_1, x_2, \ldots, x_N]$ and received symbols $\mathbf{y} = [y_1, y_2, \ldots, y_N]$:

$$p(\mathbf{y}|\mathbf{x}) = \prod_{i=1}^{N} p(y_i|x_i). \tag{1.1}$$

A memoryless channel is therefore completely described by its input and output alphabets and the conditional probability distribution $p(y|x)$ for each input–output symbol pair.

The three channels we consider in this text are the binary symmetric channel (BSC), the binary erasure channel (BEC) and the binary input additive white Gaussian noise (BI-AWGN) channel. They are all binary input memoryless channels.

Example 1.1 The *binary symmetric channel (BSC)* shown in Figure 1.2 transmits one of two symbols, the binary digits $x \in \{0, 1\}$, and returns one of two symbols, $y \in \{0, 1\}$. This channel flips the transmitted bit with probability ϵ, i.e. with

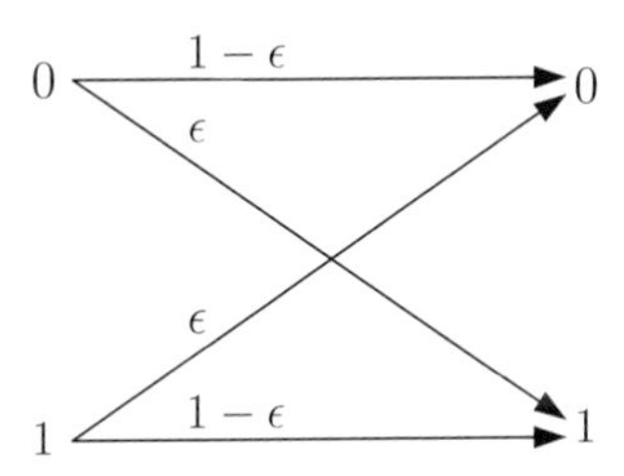

Figure 1.2 The binary symmetric channel (BSC).

probability ϵ the symbol y output by the channel is not the symbol that was sent and with probability $1 - \epsilon$ the symbol y is the symbol that was sent. The parameter ϵ is called the *crossover probability* of the channel. So, for the BSC the channel transition probabilities are:

$$\begin{aligned} p(y = 0|x = 0) &= 1 - \epsilon, \\ p(y = 0|x = 1) &= \epsilon, \\ p(y = 1|x = 0) &= \epsilon, \\ p(y = 1|x = 1) &= 1 - \epsilon. \end{aligned}$$

A binary input channel is *symmetric* if both input bits are corrupted equally by the channel. The BSC channel is easily seen to be symmetric, as $p(y = 0|x = 0) = p(y = 1|x = 1)$ and $p(y = 0|x = 1) = p(y = 1|x = 0)$. Indeed all three channels we consider in this text, i.e. the BSC, BEC and BI-AWGN channels, are symmetric.

At the decoder, the symbol y received from the channel is used to decode the symbol x that was sent. In this case we are interested in the probability $p(x|y)$, i.e. given that we have received y, how likely was it that x was sent? We will assume that each bit is equally likely to be transmitted. Thus, for the binary symmetric channel, if $y = 1$ the probability that $x = 1$ is the probability that no error occurred, i.e. $p(x = 1|y = 1) = 1 - \epsilon$, and the probability that $x = 0$ is the probability that the channel flipped the transmitted bit $p(x = 0|y = 1) = \epsilon$. Similarly, $p(x = 1|y = 0) = \epsilon$ and $p(x = 0|y = 0) = 1 - \epsilon$.

For a binary variable x it is easy to find $p(x = 1)$ given $p(x = 0)$, since $p(x = 1) = 1 - p(x = 0)$ and so we only need to store one probability value for x. Log likelihood ratios (LLRs) are used to represent the metrics for a binary variable by a single value: the LLR is given by

$$L(x) = \log \frac{p(x = 0)}{p(x = 1)}, \tag{1.2}$$

where in this text we will use log to mean log to the base e, or $\log_e$. If $p(x = 0) > p(x = 1)$ then $L(x)$ is positive and, furthermore, the greater the difference between $p(x = 0)$ and $p(x = 1)$, i.e. the more sure we are that $p(x) = 0$, the larger the positive value for $L(x)$. Conversely, if $p(x = 1) > p(x = 0)$ then $L(x)$ is negative and, furthermore, the greater the difference between $p(x = 0)$ and $p(x = 1)$, the larger the negative value for $L(x)$. Thus the sign of $L(x)$ provides a hard decision (see the text after (1.9)) on x and the magnitude $|L(x)|$ is the reliability of this decision. Translating from LLRs back to probabilities, we obtain

$$p(x = 1) = \frac{p(x = 1)/p(x = 0)}{1 + p(x = 1)/p(x = 0)} = \frac{e^{-L(x)}}{1 + e^{-L(x)}} \tag{1.3}$$

and

$$p(x = 0) = \frac{p(x = 0)/p(x = 1)}{1 + p(x = 0)/p(x = 1)} = \frac{e^{L(x)}}{1 + e^{L(x)}}. \tag{1.4}$$

A benefit of the logarithmic representation of probabilities is that when probabilities need to be multiplied, log-likelihood ratios need only be added; this can reduce the implementation complexity.

Example 1.2 Given that the probabilities $p(x|y)$ for the BSC are

$$\begin{cases} p(x_i = 1|y_i) = 1 - \epsilon \quad \text{and} \quad p(x_i = 0|y_i) = \epsilon & \text{if } y_i = 1, \\ p(x_i = 1|y_i) = \epsilon \quad \text{and} \quad p(x_i = 0|y_i) = 1 - \epsilon & \text{if } y_i = 0, \end{cases}$$

the *received* LLRs for the ith transmitted bit are

$$R_i = L(x_i|y_i) = \log \frac{p(x_i = 0|y_i)}{p(x_i = 1|y_i)} = \begin{cases} \log \epsilon/(1 - \epsilon) & \text{if } y_i = 1, \\ \log(1 - \epsilon)/\epsilon & \text{if } y_i = 0. \end{cases}$$

Example 1.3 The *binary erasure channel* (BEC) shown in Figure 1.3 transmits one of two symbols, usually the binary digits $x \in \{0, 1\}$. However, the receiver either receives the bit correctly or it receives a message "e" that the bit was not received (it was *erased*). The BEC erases a bit with probability ε, called the *erasure probability* of the channel. Thus the channel transition probabilities for the BEC are

$$\begin{aligned} p(y = 0|x = 0) &= 1 - \varepsilon, \\ p(y = e|x = 0) &= \varepsilon, \\ p(y = 1|x = 0) &= 0, \\ p(y = 0|x = 1) &= 0, \\ p(y = e|x = 1) &= \varepsilon, \\ p(y = 1|x = 1) &= 1 - \varepsilon. \end{aligned}$$

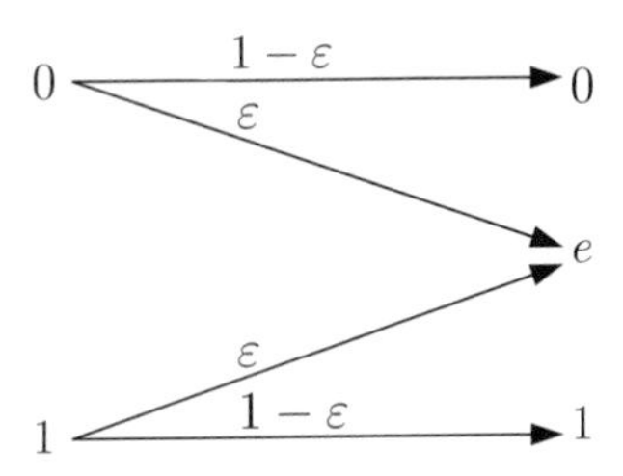

Figure 1.3 The binary erasure channel (BEC).

The BEC does not flip bits, so if y is received as a 1 or a 0 the receiver can be completely certain of the value of x:

$$\begin{aligned} p(x = 0|y = 0) &= 1, \\ p(x = 1|y = 0) &= 0, \\ p(x = 0|y = 1) &= 0, \\ p(x = 1|y = 1) &= 1. \end{aligned}$$

However, if the channel has erased the transmitted bit the receiver has no information about x and can only use the a priori probabilities of the source. If the source is equiprobable (i.e. the bits 1 and 0 are equally likely to have been sent) the receiver can only make a fifty–fifty guess:

$$\begin{aligned} p(x = 0|y = e) &= 0.5, \\ p(x = 1|y = e) &= 0.5. \end{aligned}$$

So, for this channel we have that the received LLRs for the ith transmitted bit are

$$R_i = L(x_i|y_i) = \log \frac{p(x_i = 0|y_i)}{p(x_i = 1|y_i)} = \begin{cases} \log \dfrac{0}{1} = -\infty & \text{if } y_i = 1, \\ \log \dfrac{1}{0} = \infty & \text{if } y_i = 0. \\ \log \dfrac{0.5}{0.5} = 0 & \text{if } y_i = e. \end{cases}$$

The final channel we consider, and the one most commonly used by coding theorists, is a binary input channel with additive noise modeled as samples from a Gaussian probability distribution.

Example 1.4 The *binary-input additive white Gaussian noise (BI-AWGN)* channel can be described by the equation

$$y_i = \mu x_i + z_i, \tag{1.5}$$

where $x_i \in \{-1, +1\}$ is the ith transmitted symbol, y_i is the ith received symbol and z_i is the additive noise sampled from a Gaussian random variable with mean 0 and variance σ^2. This is sometimes written $z_i = AWGN(0, \sigma)$.

The probability density function for z is

$$p(z) = \frac{1}{\sqrt{2\pi\sigma^2}} e^{-z^2/2\sigma^2}, \tag{1.6}$$

where $e^x = \exp(x)$ is the exponential function.

When transmitting a binary codeword on the BI-AWGN channel, the codeword bits $c_i \in \{0, 1\}$ can be mapped to the symbols $x_i \in \{-1, +1\}$ in one of two ways: $\{0 \to 1, 1 \to -1\}$ or $\{0 \to -1, 1 \to 1\}$. We will use the traditional convention $\{0 \to 1, 1 \to -1\}$.[1]

The received LLRs for the BI-AWGN channel are then

$$\begin{aligned} R_i = L(x_i|y_i) &= \log \frac{p(c_i = 0|y_i)}{p(c_i = 1|y_i)} \\ &= \log \frac{p(x_i = 1|y_i)}{p(x_i = -1|y_i)} \\ &= \log \frac{p(y_i|x_i = 1)p(x_i = 1)/p(y_i)}{p(y_i|x_i = -1)p(x_i = -1)/p(y_i)} \\ &= \log \frac{p(y_i|x_i = 1)p(x_i = 1)}{p(y_i|x_i = -1)p(x_i = -1)}, \end{aligned}$$

where we have used Bayes' rule

$$p(x_i|y_i) = p(x_i, y_i)/p(y_i) = p(y_i|x_i)p(x_i)/p(y_i)$$

to substitute for $p(x_i = 1|y_i)$ and $p(x_i = -1|y_i)$. If the source is equiprobable then $p(x_i = -1) = p(x_i = 1)$, and we have

$$R_i = L(x_i|y_i) = \log \frac{p(y_i|x_i = 1)}{p(y_i|x_i = -1)}.$$

For the BI-AWGN channel:

$$p(y_i|x_i = 1) = \frac{1}{\sqrt{2\pi\sigma^2}} \exp\left(-\frac{(y_i - \mu)^2}{2\sigma^2}\right), \tag{1.7}$$

$$p(y_i|x_i = -1) = \frac{1}{\sqrt{2\pi\sigma^2}} \exp\left(-\frac{(y_i + \mu)^2}{2\sigma^2}\right); \tag{1.8}$$

[1] The mapping $\{0 \to 1, 1 \to -1\}$ is used because the modulo-2 arithmetic on $\{0, 1\}$ maps directly to multiplication on $\{-1, +1\}$ when this mapping is used.

thus

$$R_i = L(x_i|y_i) = \log \frac{\frac{1}{\sqrt{2\pi\sigma^2}} \exp\left(-\frac{(y_i-\mu)^2}{2\sigma^2}\right)}{\frac{1}{\sqrt{2\pi\sigma^2}} \exp\left(-\frac{(y_i+\mu)^2}{2\sigma^2}\right)}$$
$$= \log \exp\left(-\frac{(y_i-\mu)^2}{2\sigma^2} + \frac{(y_i+\mu)^2}{2\sigma^2}\right)$$
$$= \frac{1}{2\sigma^2}(-(y_i^2 - 2\mu y_i + \mu^2) + (y_i^2 + 2\mu y_i + \mu^2))$$
$$= \frac{2\mu}{\sigma^2} y_i. \tag{1.9}$$

The LLR value for a bit c_i is sometimes called a *soft decision* for c_i. A *hard decision* for c_i will return $c_i = 0$, equivalently $x_i = 1$, if R_i is positive and $c_i = 1$, equivalently $x_i = -1$, if R_i is negative.

When considering the relative noise level of a BI-AWGN channel, it is convenient to assume that $\mu = 1$ and adjust σ to reflect the noise quality of the channel. In this case R_i can be written as

$$R_i = \frac{2}{\sigma^2} y_i.$$

Often the noise level is expressed as the ratio of the energy per transmitted symbol, E_s, and the noise power spectral density N_0:

$$\frac{E_s}{N_0} = \frac{\mu^2}{2\sigma^2},$$

and (1.5) is sometimes written in the form

$$y_i = \sqrt{E_s} x_i + z_i.$$

When using error correction coding on a BI-AWGN channel, a fraction r of the transmitted bits correspond to bits in the message and the remainder are extra, redundant, bits added by the code. For channels using error correction the noise level is often expressed as the ratio of energy per message bit, E_b, and N_0, the *signal-to-noise ratio (SNR)*:

$$\frac{E_b}{N_0} = \frac{1}{r}\frac{E_s}{N_0} = \frac{1}{r}\frac{\mu^2}{2\sigma^2},$$

and the received LLR is often given as

$$R_i = L(x_i|y_i) = 4\frac{\sqrt{E_s}}{N_0} y_i = 4\frac{\sqrt{rE_b}}{N_0} y_i,$$

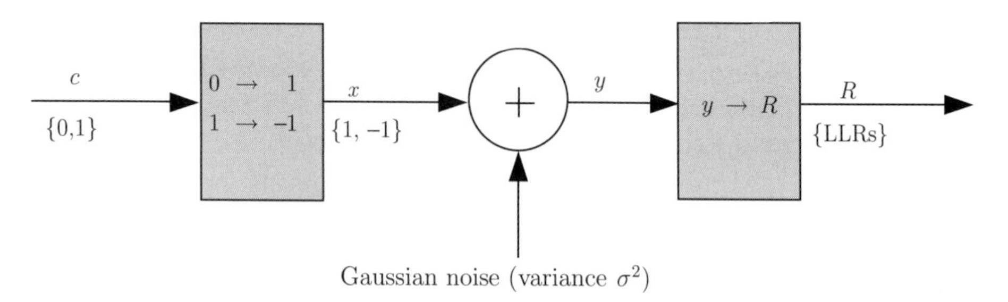

Figure 1.4 The BI-AWGN channel; $R = 4y/\sigma^2$.

or, when μ is assumed to be 1,

$$R_i = \frac{4}{N_0} y_i.$$

The signal-to-noise ratio can be also expressed in dB:

$$\frac{E_b}{N_0}(\text{dB}) = 10 \log_{10} \frac{E_b}{N_0} = 10 \log_{10} \frac{\mu^2}{2r\sigma^2}.$$

Figure 1.4 shows a block diagram for the BI-AWGN channel that we will consider in this text.

1.2 Entropy, mutual information and capacity

In the previous section we mentioned that a communications channel can be modeled as a random process and we described three common such models. In this section we will define some useful properties of random variables and use them to define limits on how well we can communicate over our three channels.

A discrete random variable X has a symbol alphabet $\{X_1, X_2, \ldots, X_q\}$ and probability distribution $p = \{p_1, p_2, \ldots, p_q\}$, where $p_j = p(x = X_j)$ gives the probability that a random sample x from X will return the symbol X_i.

A continuous random variable X can take any value in an uncountable set A_x. For example, A_x could define all real numbers between 0 and 1. A continuous random variable has a probability density function $p(x)$.

1.2.1 A measure of information

In order to motivate the concept of information we will consider a simple contest. There are two competitors and each has a "black box" that emits symbols x

from a random variable X with symbol alphabet $\{X_1, X_2\}$ and probabilities $p_1 = p(X_1) = \frac{99}{100}$, $p_2 = p(X_2) = \frac{1}{100}$. Each competitor receives only symbols from their own black box. The winner of the contest is the first to correctly name both symbols, that is, the first to have complete information about the symbol set of X. Now, suppose on the first round the first contestant receives the symbol X_1 while the second receives X_2. At this point it is clear that the second contestant is more likely to win (as in order to know both X_1 and X_2 the second contestant now only needs to receive the symbol X_1, which is much more likely to occur). In a sense the second contestant has received more *information* about X than the first, and this is reflected in the way information is measured. The lower the probability that a symbol occurs, the more information that is obtained from an occurrence of that symbol.

We denote as $I(p_j)$ the information obtained from receiving a symbol X_j, because the information is not a function of the symbol itself but of the symbol's probability of occurrence. There are three properties that we may expect the function $I(p)$ to have:

I1 $I(p) \geq 0$, that is, the information we gain by receiving a symbol cannot be negative (i.e. even though we may learn nothing from receiving a symbol we cannot lose information we already have).

I2 $I(p)$ is continuous in p.

I3 $I(p_1 p_2) = I(p_1) + I(p_2)$, that is, the information obtained from the knowledge that both X_1 and X_2 occurred is equal to the information obtained from the knowledge that X_1 occurred plus the information obtained from the knowledge that X_2 occurred.

The only function that satisfies all three assumptions is a logarithm:

$$I(p) = A \log_2 \frac{1}{p} \qquad \text{for some constant } A > 0,$$

and that is why this function was proposed by Hartley in 1928 as the measure of the information produced when a symbol with probability of occurrence p is received. Although any logarithm would work, base 2 logarithms are used most commonly.

The unit of measurement of information is the binary unit, and the constant A is chosen so as to equate one binary unit to the information received from one symbol of a binary source when both symbols are equally probable:

$$I(p) = A \log_2 \frac{1}{p} = A \log_2 \frac{1}{0.5} = A,$$

so $I(p) = 1$ when $A = 1$. Binary units (bits) are the measure of information regardless of whether X is binary, and it is important not to confuse them with binary digits (also shortened to bits).

1.2.2 Entropy

In the previous section we saw how to measure the information of a sample from a binary random variable; now we look at how to measure the amount of information in the variable itself. The information content of a random variable X is the average information over all its symbols and is called its *entropy* $H(X)$:

$$H(X) = \mathrm{E}[I(p(x))] = \sum_{j=1}^{q} p_j I(p_j) = \sum_{j=1}^{q} p_j \log_2 \frac{1}{p_j} = -\sum_{j=1}^{q} p_j \log_2 p_j.$$

We have assumed that the emission of a symbol is independent of time, i.e. the fact that a given symbol is emitted at one instant has no effect on which source symbol will be emitted at any other instant. Since the entropy is the average information per symbol its units are bits per symbol.

Example 1.5 Consider a discrete random variable X for which each symbol is equally likely to occur, that is $p_j = 1/q$ for all $j = 1, \ldots, q$. Then

$$H(X) = \sum_{j=1}^{q} p_j \log_2 \frac{1}{p_j} = \sum_{i=1}^{q} \frac{1}{q} \log_2 q = \log_2 q.$$

If $q = 3$ then

$$H(X) = \log_2 3 = 1.585 \text{ bits per symbol.}$$

However, for a discrete random variable $X = \{X_1, X_2, X_3\}$ with $p = \{\frac{1}{4}, \frac{1}{4}, \frac{1}{2}\}$, we have

$$H(X) = \sum_{j=1}^{q} p_j \log_2 \frac{1}{p_j} = \tfrac{1}{4} \log_2 4 + \tfrac{1}{4} \log_2 4 + \tfrac{1}{2} \log_2 2 = 1.5 \text{ bits per symbol.}$$

Thus the equiprobable random variable has a higher entropy.

To obtain an intuitive feel for the above result we return to our contest but this time each competitor has a different random variable. The winner of the contest is still the first to correctly name both symbols, that is, the first to have complete information about their random variable.

Suppose the first competitor has a binary random variable that is equiprobable. In two samples (x_1 and x_2) the probability that both symbols have been sampled is

$$P(\text{both } X_1, X_2 \text{ sampled}) = p(x_1 = X_1, x_2 = X_2) + p(x_1 = X_2, x_2 = X_1)$$
$$= 0.5 \times 0.5 + 0.5 \times 0.5 = 0.5.$$

Suppose that the second competitor has a random binary variable with $p_1 = 0.1$ and $p_2 = 0.9$. In two trials the probability that both symbols have been sampled is

$$P(\text{both } X_1, X_2 \text{ sampled}) = p(x_1 = X_1, x_2 = X_2) + p(x_1 = X_2, x_2 = X_1)$$
$$= 0.1 \times 0.9 + 0.9 \times 0.1 = 0.18.$$

What about after three trials? For the first competitor, the probability that both X_1 and X_2 are sampled is

$$1 - [p(x_1 = X_1, x_2 = X_1, x_3 = X_1) + p(x_1 = X_2, x_2 = X_2, x_3 = X_2)]$$
$$= 1 - [0.5 \times 0.5 \times 0.5 + 0.5 \times 0.5 \times 0.5] = 0.75.$$

For the second competitor, the probability that both X_1 and X_2 are sampled is

$$1 - [p(x_1 = X_1, x_2 = X_1, x_3 = X_1)|p(x_1 = X_2, x_2 = X_2, x_3 = X_2)]$$
$$= 1 - [0.1 \times 0.1 \times 0.1 + 0.9 \times 0.9 \times 0.9] = 0.27.$$

Thus, both after two turns and after three turns, the first competitor is much more likely to know the symbol set of their random variable than the second competitor (and this continues for any number of turns). We can say that the first competitor gets more information per sample of the random variable than the second. The entropy, which is the average information per sample, reflects this fact: the first random variable has a higher entropy than the second.

Entropy can also be thought of as the amount of uncertainty we have about the random variable X. For the first source both symbols are equally likely so we are completely uncertain about which symbol will be emitted next. However, with the second source, if we are asked to guess the next symbol we would choose the second and be correct nine out of ten times. We thus have less uncertainty about the second source and correspondingly it has a lower entropy.

For a continuous, rather than discrete, random variable X we define the *differential entropy*:

$$H(X) = \mathrm{E}[I(p(x))] = -\int p(x) \log_2 p(x)\, dx.$$

Example 1.6 For a random variable Z with a continuous Gaussian distribution, the (differential) entropy of Z is

$$\begin{aligned}
H(Z) &= \mathrm{E}[I(p(z))] \\
&= -\mathrm{E}\left(\log_2 \frac{1}{\sqrt{2\pi\sigma^2}} e^{-z^2/2\sigma^2}\right) \\
&= -\mathrm{E}\left(\log_2 \frac{1}{\sqrt{2\pi\sigma^2}} + (\log_2 e)\left(-\frac{z^2}{2\sigma^2}\right)\right) \\
&= -\log_2 \frac{1}{\sqrt{2\pi\sigma^2}} - (\log_2 e)\left(\frac{-1}{2\sigma^2}\right)\mathrm{E}(z^2) \\
&= \frac{1}{2}\log_2 2\pi\sigma^2 + (\log_2 e)\left(\frac{1}{2\sigma^2}\right)\sigma^2 \\
&= \frac{1}{2}\log_2 2\pi\sigma^2 + \frac{1}{2}\log_2 e \\
&= \frac{1}{2}\log_2 2\pi e\sigma^2.
\end{aligned}$$

We can also define the joint entropy of two random discrete variables, X and Y:

$$H(X, Y) = \sum p(x, y) \log_2 \frac{1}{p(x, y)}.$$

The conditional entropy

$$H(X|Y) = \sum p(x, y) \log_2 \frac{1}{p(x|y)}$$

can be thought of as the amount of uncertainty about X that remains after Y is known.

Mutual information

The mutual information $I(X; Y)$ between the random variables X and Y quantifies how much information knowing one provides about the other. For two discrete random variables X and Y, the mutual information is defined as

$$I(X; Y) = H(X) - H(X|Y) = H(Y) - H(Y|X).$$

Thus we can say that $I(X; Y)$ gives the amount of uncertainty in X that is removed by knowing Y and vice versa.

Mutual information will play an important role in the EXIT chart analysis in Chapter 7 but here we will use it in a far more fundamental communications role. When Y is the output of a channel with input X, the mutual information $I(X; Y)$

measures the amount of uncertainty about X that is removed by knowing Y, or, put another way, the amount of information about X that can be transmitted through the channel.

1.2.3 Capacity

A communications channel with input random variable X having probability distribution $p(x)$ and output random variable Y with probability distribution $p(y)$ is defined to have a *capacity* c given by

$$c = \max_{p(x)} I(X;Y).$$

Thus, the capacity of a channel defines the maximum amount of information that can be conveyed through the channel per transmitted symbol over all possible input distributions.

Example 1.7 For the capacity of a binary symmetric channel (BSC) we have

$$I(X;Y) = H(Y) - H(Y|X) \tag{1.10}$$
$$= H(Y) - \sum_{x\in\{0,1\}} p(x)H(Y|X=x). \tag{1.11}$$

Consider

$$H(Y|X=0) = -\sum_y p(y|0)\log_2 p(y|0) \tag{1.12}$$
$$= -(p(0|0)\log_2 p(0|0) + p(1|0)\log_2 p(1|0)) \tag{1.13}$$
$$= -(\epsilon\log_2\epsilon + (1-\epsilon)\log_2(1-\epsilon)) \tag{1.14}$$
$$\triangleq H_2(\epsilon), \tag{1.15}$$

where in the last line the symbol $\triangleq$ means that $H_2(\epsilon)$ is defined as a shorthand for the right-hand side of (1.14).

Likewise $H(Y|X=1) = H_2(\epsilon)$, and so

$$I(X;Y) = H(Y) - \sum_x p(x)H_2(\epsilon) \tag{1.16}$$
$$= H(Y) - H_2(\epsilon) \tag{1.17}$$
$$\le 1 - H_2(\epsilon). \tag{1.18}$$

The upper bound requires $H(Y) = 1$, which occurs when the distribution is uniform (i.e. every symbol is equally probable). Setting $p(x)$ to be the uniform distribution on $\{0, 1\}$ gives the capacity for the binary symmetric channel:

$$c(\epsilon) = 1 - H_2(\epsilon) \qquad \text{bits per channel use.} \tag{1.19}$$

Example 1.8 For the BI-AWGN channel with $\mu = 1$,

$$I(X;Y) = H(Y) - H(Y|X)$$
$$= H(Y) - H(Z).$$

From Example 1.6 we know that

$$H(Z) = \frac{1}{2}\log_2 2\pi e\sigma^2.$$

Furthermore

$$H(Y) = -\int_{-\infty}^{\infty} p(y)\log_2 p(y)dy.$$

If the source is equiprobable then

$$p(y) = p(y|x=+1)p(x=+1) + p(y|x=-1)p(x=-1)$$
$$= \tfrac{1}{2}p(y|x=+1) + \tfrac{1}{2}p(y|x=-1), \tag{1.20}$$

where

$$p(y|x=\pm 1) = \frac{1}{\sqrt{2\pi\sigma^2}}\exp\left(-\frac{(y\pm 1)^2}{2\sigma^2}\right).$$

Thus

$$p(y) = \frac{1}{2}\frac{1}{\sqrt{2\pi\sigma^2}}\exp\left(-\frac{(y-1)^2}{2\sigma^2}\right) + \frac{1}{2}\frac{1}{\sqrt{2\pi\sigma^2}}\exp\left(-\frac{(y+1)^2}{2\sigma^2}\right) \tag{1.21}$$

$$= \frac{1}{\sqrt{8\pi\sigma^2}}\left(\exp\left(-\frac{(y-1)^2}{2\sigma^2}\right) + \exp\left(-\frac{(y+1)^2}{2\sigma^2}\right)\right) \tag{1.22}$$

$$\triangleq \phi_\sigma(y). \tag{1.23}$$

Therefore, for equiprobable $x \in \{1, -1\}$ we have

$$c(\sigma) = H(Y) - H(Z) \tag{1.24}$$

$$= -\int_{-\infty}^{\infty} \phi_\sigma(y)\log_2 \phi_\sigma(y)dy - \tfrac{1}{2}\log_2 2\pi e\sigma^2, \tag{1.25}$$

where

$$\phi_\sigma(y) = \frac{1}{\sqrt{8\pi\sigma^2}}\left(\exp\left(-\frac{(y+1)^2}{2\sigma^2}\right) + \exp\left(-\frac{(y-1)^2}{2\sigma^2}\right)\right). \tag{1.26}$$

We leave it as an exercise for the reader to prove that the capacity of the BEC with erasure probability ε (see Example 1.3) is

$$c(\varepsilon) = 1 - \varepsilon. \tag{1.27}$$

The importance of the capacity c in coding theory is due to Shannon's remarkable noisy channel coding theorem. Shannon proved that, provided the coded rate of transmission r is less that the channel's capacity, there exists an error correction code that will achieve an arbitrarily low probability of error despite the noise added by the channel.

Put another way, for a channel with noise level parameter x and an error correction code rate r, the noise level x_{Sh}, such that $r = c(x_{\text{Sh}})$, is a threshold for error correction codes with that rate. The noise level x_{Sh} is often called the *Shannon limit*. Shannon's noisy coding theorem says that for any noise level x below x_{Sh} there exists a rate-r code that can achieve an arbitrarily low probability of error, while for any noise level above x_{Sh}, no rate-r code can achieve an arbitrarily low probability of error.

To find the Shannon limit of error correction coding with rate r on a channel with parameter x requires that we find a value for x_{Sh} such that $c(x_{\text{Sh}}) = r$. Pseudo-code for computing the Shannon limit on a BI-AWGN channel within a tolerance δ is given in Algorithm 1.1. The algorithm will search between σ_{L}

Algorithm 1.1 Shannon limit of a BI-AWGN channel

1: **procedure** SHANNONLIMIT(r,δ,σ_L,σ_H)
2:
3:     **repeat**
4:         $\sigma = \frac{1}{2}(\sigma_{\text{L}} + \sigma_{\text{H}})$
5:         $c(\sigma) = -\int_{-\infty}^{\infty} \phi_\sigma(y) \log_2 \phi_\sigma(y) dy$ ▷ Numerical integration
6:             where $\phi_\sigma(y) := \dfrac{1}{\sqrt{8\pi\sigma^2}}(e^{-(y+1)^2/2\sigma^2} + e^{-(y-1)^2/2\sigma^2})$
7:         $c(\sigma) = c(\sigma) - \frac{1}{2}\log_2 2\pi e\sigma^2$
8:         **if** $c(\sigma) > r$ **then**
9:             $\sigma_L = \sigma$
10:        **else**
11:            $\sigma_H = \sigma$
12:        **end if**
13:    **until** $\sigma_H - \sigma_L < \delta$
14:
15:    $E_b/N_0 = 10\log_{10}\dfrac{1}{2r\sigma^2}$ ▷ Shannon limit in dB
16:    **return** E_b/N_0
17: **end procedure**

and σ_H to find σ_{Sh}, so these need to be set to appropriately small and large values respectively. The integral $c(\sigma)$ can be solved using any desired numerical integration technique (Matlab users might like to use the QUAD function). On a BI-AWGN channel the Shannon limit can also be expressed in dB:

$$\left(\frac{E_b}{N_0}\right)_{Sh} = 10\log_{10}\left(\frac{1}{2r\sigma_{Sh}^2}\right).$$

Example 1.9 The BI-AWGN capacity thresholds for various code rates are as shown in the table below:

Rate	σ_{Sh}
0.1	2.5926
0.2	1.7665
0.3	1.3859
0.4	1.1488
0.5	0.9786
0.6	0.8439
0.7	0.7298
0.8	0.6247
0.9	0.5154
0.999	0.2859

Unfortunately Shannon's proof (see e.g. [2, Chapter 8]) showed that such a code exists, not how to construct it, and certainly not how to encode and decode it with reasonable complexity. This is the task that has faced coding theorists since 1948 and is only now being achieved for some channels by the new iterative codes which are the topic of this book.

1.3 Codes and decoding

Figure 1.1 shows our communications system. The error correction encoder processes the message bits from the source in blocks of K bits at a time and the code is called a block code. Redundancy is added so that $N > K$ bits are actually transmitted through the channel for every K bits from the source. The number K is called the *message length* or *dimension index* of the code, and N is the *code length*. The code rate $r = K/N$ defines the amount of redundancy added. The transmitted bits may be corrupted in some way by the channel, and the error correction decoder uses the added redundancy to determine the corresponding K message bits despite the imperfect reception.

A block code is linear if its codewords form a vector space over the Galois field GF(q). In this text we will be interested in binary codes so will only consider the field GF(2), which is simply the set of binary elements 0, 1 with modulo-2 arithmetic as shown in the following table:

Multiplication	Addition
$0 \cdot 0 = 0$	$0 \oplus 0 = 0$
$1 \cdot 0 = 0$	$1 \oplus 0 = 1$
$0 \cdot 1 = 0$	$0 \oplus 1 = 1$
$1 \cdot 1 = 1$	$1 \oplus 1 = 0$

The addition of elements a and b in GF(2) is simply an exclusive OR operation, which we will write as $a \oplus b$ or $(a + b \bmod 2)$.

An (N, K) binary linear block code C consists of a set of 2^K length-N binary codewords with the property that the linear combination (using modulo-2 arithmetic) of any set of codewords is always a codeword.

Example 1.10 The (5, 3) code C has $2^3 = 8$ codewords:

$$C = \{00000, 00101, 01010, 01110, 10010, 10111, 11001, 11100\},$$

and its rate r is $3/5$.

Each codeword can be expressed as a row vector $\mathbf{c}$. Thus the codeword 00101 in Example 1.10 can be expressed as the row vector $\mathbf{c} = [0\ 0\ 1\ 0\ 1]$. Each codeword vector can be mapped to one of the possible 2^K length-K message vectors $\mathbf{u}$. In what follows we shall use "codeword" and "vector" interchangeably for "codeword vector". We distinguish between *codes*, which define the set of codewords, and *encoders*, which define the mapping of messages to those codewords.

Example 1.11 An encoder for the code C in Example 1.10 maps messages to codewords as follows:

$$000 \to 00000,\ 001 \to 00101,\ 010 \to 01010,\ 011 \to 01110,$$
$$100 \to 10010,\ 101 \to 10111,\ 110 \to 11001,\ 111 \to 11100.$$

The receiver performs an inverse mapping from codewords back to messages to determine the particular message communicated. This task is made simpler if the message bits make up part of the codeword. Such an encoder is called *systematic*. The encoder in Example 1.11 is systematic since the first three bits of every codeword are the three message bits.

If the channel adds noise to the codewords then the receiver will first need to determine which codeword was actually sent. This task is called *decoding*.

1.3.1 Decoding

Suppose that a codeword has been sent down a binary symmetric channel and one or more of the codeword bits may have been flipped. The task of the decoder is to detect any flipped bits and, if possible, to correct them.

Example 1.12 The codeword $\mathbf{c} = [1\,0\,1\,1\,1\,0]$ from the code

$$C = \{000000, 001011, 010111, 011100, 100101, 101110, 110010, 111001\}$$

was transmitted through a binary symmetric channel (by the mapping $\mathbf{x} = \mathbf{c}$) and the vector $\mathbf{y} = [1\,0\,1\,0\,1\,0]$ was received. Comparing $\mathbf{y}$ with all possible codewords shows that $\mathbf{y}$ is not a codeword of this code. We therefore conclude that bit flipping errors must have occurred during transmission.

Example 1.12 demonstrates the use of an error correction code to detect transmission errors, but suppose that the channel is even noisier and that three bits are flipped to produce the vector $\mathbf{y} = [0\,0\,1\,0\,1\,1]$. Now, $\mathbf{y}$ is a valid codeword and so we cannot detect the transmission errors that have occurred. In general, a decoder can only detect a set of bit errors if the errors do not change one codeword into another.

The minimum number of bit-flipping errors required before one codeword is changed into another depends on the number of bit positions that differ between the two codewords. For example the codewords $[1\,0\,1\,0\,0\,1\,1\,0]$ and $[1\,0\,0\,0\,0\,1\,1\,1]$ differ in two positions, the third and eighth codeword bits. So two appropriately located bit-flipping errors could change one codeword into the other.

The number of bit positions in which two codewords differ is called the *Hamming distance* between them. The *minimum Hamming distance* or just *minimum distance* of a code, $d_{\min}$, is defined as the smallest Hamming distance between any pair of codewords in the code. For the code in Example 1.12, $d_{\min} = 3$, so the corruption of three or more bits in a codeword could result in another valid codeword. In general, a code with minimum distance $d_{\min}$ can detect t errors whenever

$$t < d_{\min}. \tag{1.28}$$

The smaller the code rate r, the smaller the subset of 2^N binary vectors that are codewords and so the better the minimum distance that can be achieved by a code with length N. The importance of the code's minimum distance in determining its performance is reflected in the description of error correction codes by the three parameters $N, K, d_{\min}$.

We have seen how bit-flipping errors can be detected; to go further and correct those errors requires an error correction decoder.

Maximum likelihood (ML) decoding

The first *error correction* decoder we consider is called the *maximum likelihood* (ML) decoder, as it will always choose the codeword that is most likely to have produced $\mathbf{y}$. Specifically, given a received vector $\mathbf{y}$ the ML decoder will choose the codeword $\mathbf{c}$ that maximizes the probability $p(\mathbf{y}|\mathbf{c})$. The ML decoder returns the decoded codeword $\hat{\mathbf{c}}$ according to the rule

$$\hat{\mathbf{c}} = \text{argmax}_{\mathbf{c} \in C}\, p(\mathbf{y}|\mathbf{c}),$$

where, assuming a memoryless channel,

$$p(\mathbf{y}|\mathbf{c}) = \prod_{i=1}^{N} p(y_i|c_i).$$

For a binary symmetric channel with crossover probability less than 0.5, the most likely codeword is the one that requires the fewest number of flipped bits to produce $\mathbf{y}$, since a bit is more likely to be received correctly than flipped. Then the ML decoder is equivalent to choosing the codeword closest in Hamming distance to $\mathbf{y}$.

Example 1.13 In Example 1.12 we detected that the received vector $\mathbf{y} = [1\,0\,1\,0\,1\,0]$ was not a codeword of the code. By comparing $\mathbf{y}$ with each codeword in this code the ML decoder will choose $\mathbf{c} = [1\,0\,1\,1\,1\,0]$ as the closest codeword, as it is the only codeword with Hamming distance 1 from $\mathbf{y}$.

The minimum distance of the code in Example 1.12 is 3, so the flipping of a single bit always results in a vector $\mathbf{y}$ closer to the codeword that was sent than to any other codeword and thus can always be corrected by the ML decoder. However, if two bits are flipped in $\mathbf{y}$ then there may be a different codeword that is closer to $\mathbf{y}$ than the one which was sent, in which case the decoder will choose an incorrect codeword.

Example 1.14 The codeword $\mathbf{c} = [1\ 0\ 1\ 1\ 1\ 0]$ from the code in Example 1.12 was transmitted through a BSC channel that introduced two flipped bits, producing the vector $\mathbf{y} = [0\ 0\ 1\ 0\ 1\ 0]$. By comparison of $\mathbf{y}$ with each codeword of this code, the ML decoder will choose $\mathbf{c} = [0\ 0\ 1\ 0\ 1\ 1]$ as the closest codeword as it has Hamming distance 1 from $\mathbf{y}$. In this case the ML decoder has actually added errors rather than corrected them.

In general, for a code with minimum distance $d_{\min}$, t bit flips can always be corrected by choosing the closest codeword whenever

$$t \leq \lfloor (d_{\min} - 1)/2 \rfloor, \tag{1.29}$$

where $\lfloor x \rfloor$ is the largest integer less than or equal to x.

For a channel with non-binary output symbols the Hamming distance metric is replaced by the Euclidean distance metric

$$\mathbf{y} - \mathbf{x} = \sum_i |y_i - x_i|.$$

Example 1.15 A rate-3/5 code is defined by the codeword set:

$$\mathcal{C} = \{00000, 00011, 01110, 01101, 11000, 11011, 10101, 10110\}.$$

The codewords $\mathbf{c} = [c_1 \cdots c_N]$ are transmitted by mapping the codeword bits $c_i \in \{0, 1\}$ to the symbols $x_i \in \{1, -1\}$:

$$0 \mapsto +1,$$
$$1 \mapsto -1.$$

The x_i are then transmitted over a BI-AWGN channel with variance σ^2 and $\mu = 1$:

$$y_i = x_i + z_i.$$

Suppose that a codeword $\mathbf{c}$ is transmitted ($\mathbf{x} = -2\mathbf{c} + \mathbf{1}$) and that the channel measurement $\mathbf{y} = [y_1 \cdots y_5]$ is

$$\mathbf{y} = [-0.9604\ -0.2225\ -1.4173\ -0.5932\ -1.0430].$$

The Euclidean distance from $\mathbf{y}$ to

$$\mathbf{x} = [-1\ -1\ -1\ -1\ -1]$$

is

$$\begin{aligned}\sum_i |y_i - x_i| &= |-0.9604 - (-1)| + |-0.2225 - (-1)| \\ &\quad + |-1.4173 - (-1)| + |-0.5932 - (-1)| + |-1.0430| \\ &= 9.2364.\end{aligned}$$

Calculating the Euclidean distance from $\mathbf{y}$ to the transmitted vector for every codeword in C gives the following table:

$\mathbf{c}$	$\mathbf{x}$	$\lvert\mathbf{y}-\mathbf{x}\rvert$
[0 0 0 0 0]	[1 1 1 1 1]	9.2364
[0 0 0 1 1]	[1 1 1 −1 −1]	6.0500
[0 1 1 1 0]	[1 −1 −1 −1 1]	5.6050
[0 1 1 0 1]	[1 −1 −1 1 −1]	4.7914
[1 1 0 0 0]	[−1 −1 1 1 1]	6.8706
[1 1 0 1 1]	[−1 −1 1 −1 −1]	3.6842
[1 0 1 0 1]	[−1 1 −1 1 −1]	3.3156
[1 0 1 1 0]	[−1 1 −1 −1 1]	4.1292

The decoded codeword using ML decoding is thus $\mathbf{c} = [1\ 0\ 1\ 0\ 1]$, which has the closest Euclidean distance to $\mathbf{y}$.

In Example 1.15 we performed ML decoding using brute force methods. A much more efficient algorithm for performing ML decoding, the Viterbi algorithm, is presented in Chapter 4.

If instead a hard decision (see the text after (1.9)) decoder had been employed in Example 1.15, a hard decision on $\mathbf{y}$, giving $\bar{\mathbf{y}} = [1\ 1\ 1\ 1\ 1]$, would have been compared with every codeword $\mathbf{c}$ in C. In this case

$$\hat{\mathbf{c}} = [1\ 1\ 0\ 1\ 1]$$

would have been chosen as this codeword is closest in Hamming distance to $\bar{\mathbf{y}}$, only differing in one bit position. The hard decision decoder has made a different decision from the soft decision ML decoder because it does not have the information from the channel conveyed by the magnitude of the LLRs (see (1.2)). The magnitude of the LLR for y_3 is high, suggesting a high probability that c_3 is 1, while the magnitudes for the LLRs of y_2 and y_4 are much smaller, suggesting a much lower probability that c_3 and c_4 are 1. Thus

$$\hat{\mathbf{c}} = [1\ 0\ 1\ 0\ 1]$$

is more likely to be the codeword sent.

If no extra information is known about $\mathbf{c}$ other than that received from the channel, an ML decoder is the type most likely to choose the correct codeword successfully. However, if the decoder has a priori information about $\mathbf{c}$ – for example, it knows that the all-zeros message is used often, so that the all-zeros codeword is much more likely than any other – then it is desirable that this extra information is taken into account. In such a case a more sophisticated decoder is needed.

Maximum a posteriori (MAP) decoding

A *maximum a posteriori (MAP)* or *block-MAP decoder* chooses the codeword $\mathbf{c}$ that maximizes $p(\mathbf{c}|\mathbf{y})$, i.e. $\hat{\mathbf{c}}$ is chosen according to the rule

$$\hat{\mathbf{c}} = \text{argmax}_{\mathbf{c}\in C}\, p(\mathbf{c}|\mathbf{y}).$$

The probability $p(\mathbf{c}|\mathbf{y})$ is called the *a posteriori probability (APP)* for $\mathbf{c}$, so a decoder that selects the codeword with the maximum APP probability is a maximum a posteriori decoder.

The probabilities $p(\mathbf{c}|\mathbf{y})$ and $p(\mathbf{y}|\mathbf{c})$ are closely related. By Bayes' rule,

$$p(\mathbf{y}|\mathbf{c}) = p(\mathbf{y}, \mathbf{c})/p(\mathbf{c}) = p(\mathbf{c}|\mathbf{y})p(\mathbf{y})/p(\mathbf{c}),$$

where $p(\mathbf{y})$ can be treated as a normalizing constant:

$$p(\mathbf{y}) = \sum_{\mathbf{c}\in C} p(\mathbf{y}|\mathbf{c})p(\mathbf{c}).$$

If each codeword is equally likely to have been sent then $p(\mathbf{c}|\mathbf{y}) = p(\mathbf{y}|\mathbf{c})$, so ML and MAP decoding will return an identical result. However, if the decoder has a priori information about $\mathbf{c}$ then the MAP decoder will choose the most probable codeword after taking into account this extra information.

Example 1.16 The rate-3/5 code from Example 1.15 is used for transmission over a BI-AWGN channel with $\mu = 1$ and $\sigma = 1$, and the values

$$\mathbf{y} = [-0.9604\ -0.2225\ -1.4173\ -0.5932\ -1.0430]$$

are received.

For the BI-AWGN channel with $\mu = 1$ and $\sigma = 1$ we have, from (1.6),

$$p(y_i|c_i = 1) = \frac{1}{\sqrt{2\pi}} \exp\left(-\frac{(y_i - 1)^2}{2}\right), \tag{1.30}$$

$$p(y_i|c_i = 0) = \frac{1}{\sqrt{2\pi}} \exp\left(-\frac{(y_i + 1)^2}{2}\right). \tag{1.31}$$

For the codeword $\mathbf{c} = [0\ 0\ 0\ 0\ 0]$,

$$\begin{aligned} p(\mathbf{y}|\mathbf{c}) &= \prod_{i=1}^{5} p(y_i|c_i) \\ &= 5.8395e^{-2} \times 1.8897e^{-1} \times 2.1481e^{-2} \times 1.1213e^{-1} \times 4.9496e^{-2} \\ &= 1.3155e^{-6}. \end{aligned}$$

Repeating this calculation for the remaining seven codewords gives the values for $p(\mathbf{y}|\mathbf{c})$ in the following table:

$\mathbf{c}$	$p(\mathbf{y}\|\mathbf{c})$	$p(\mathbf{c}\|\mathbf{y})$
[0 0 0 0 0]	$1.3155e^{-6}$	$5.1644e^{-4}$
[0 0 0 1 1]	$3.4697e^{-5}$	$1.3621e^{-2}$
[0 1 1 1 0]	$1.1446e^{-4}$	$4.4935e^{-2}$
[0 1 1 0 1]	$2.8142e^{-4}$	$1.1048e^{-1}$
[1 1 0 0 0]	$1.4014e^{-5}$	$5.5014e^{-3}$
[1 1 0 1 1]	$3.6961e^{-4}$	$1.4510e^{-1}$
[1 0 1 0 1]	$1.2311e^{-3}$	$4.8328e^{-1}$
[1 0 1 1 0]	$5.0072e^{-4}$	$1.9657e^{-1}$

If all codewords are equally likely then

$$p(\mathbf{c}) = \frac{1}{|C|} = \frac{1}{8} \quad \forall \mathbf{c}.$$

Furthermore

$$\begin{aligned} p(\mathbf{y}) &= \sum_{\mathbf{c} \in C} p(\mathbf{y}|\mathbf{c})p(\mathbf{c}) \\ &= \tfrac{1}{8}(1.3155e^{-6} + 3.4697e^{-5}1.1446e^{-4} + 2.8142e^{-4} \\ &\quad + 1.4014e^{-5} + 3.6961e^{-4} + 1.2311e^{-3} + 5.0072e^{-4}) \\ &= 3.1841e^{-4}, \end{aligned}$$

so that

$$p(\mathbf{c}|\mathbf{y}) = p(\mathbf{y}, \mathbf{c})/p(\mathbf{y}) = p(\mathbf{y}|\mathbf{c})p(\mathbf{c})/p(\mathbf{y}) = \tfrac{1}{8}p(\mathbf{y}|\mathbf{c})/3.1841e^{-4}.$$

Substituting the values already calculated for $p(\mathbf{y}|\mathbf{c})$ gives the $p(\mathbf{c}|\mathbf{y})$ values in the table. The block-MAP decoder thus selects the codeword 10101 as being the most probable.

Maximum a posteriori decoding can also be done on a symbol by symbol basis. The symbol-MAP decoder will choose the most probable symbol, in our case bit, for each transmitted symbol (even if the set of chosen bits does not make up a valid codeword). A symbol-MAP decoder chooses the symbol $\hat{\mathbf{c}}_\mathbf{i}$ according to the rule

$$\hat{c}_i = \text{argmax}_{c_i \in \{0,1\}} \, p(c_i|\mathbf{y}).$$

An efficient algorithm for performing symbol-MAP decoding for a particular class of codes, the BCJR algorithm, is presented in Chapter 4.

In general, though, ML and MAP decoders must compute $\mathbf{y} - \mathbf{c}$ or $p(\mathbf{c}|\mathbf{y})$ for all the 2^K codewords in the code. The value for K does not have to be very large for this to become completely impossible.

Unlike the ML and MAP algorithms defined above, the defining feature of the decoding algorithms of iterative error correction codes is that decoding proceeds in an *iterative* manner to produce accurate *estimates* of $p(c_i|y_i)$ using repeated low-complexity processes.

Except in a few special cases, the iterative algorithms are not optimal (in the sense of ML or MAP decoders). Nevertheless iterative decoding algorithms can come very close to ML or MAP performance, and the properties that make a "good" code for ML or MAP decoding also play a significant role in the performance of codes for iterative decoding algorithms. Both factors make an understanding of the performance of ML or MAP decoders a useful tool for designing codes for iterative decoding.

1.3.2 Performance measures

Suppose that a codeword $\mathbf{c}$ is transmitted through a binary input memoryless channel and a vector $\mathbf{y}$ is received. The decoder will return a codeword $\hat{\mathbf{c}}$ that it has determined to be the codeword sent. Unless the channel is completely noise-free, however, there will always be some probability that the decoder chooses a codeword $\hat{\mathbf{c}}$ that differs from the codeword transmitted.

The *word error rate (WER)* or *block error rate* for a given decoder on a given channel gives the number of times the decoder chooses the wrong codeword as a fraction of the total number of codewords decoded. Since each codeword is mapped to a different message, decoding to an incorrect codeword will result in the selection of a message that is not the message transmitted. Thus the word error rate also gives the number of times the receiver chooses the wrong message as a fraction of the total number of messages transmitted.

For some applications the fraction of bits in error, or *bit error rate (BER)*, is also important. The number of incorrect codeword bits depends on the Hamming distance between the codewords $\mathbf{c}$ and $\hat{\mathbf{c}}$, while the number of incorrect message bits depends on the Hamming distance between their corresponding messages.

Example 1.17 If the code in Example 1.11 is used to send 10 messages and one of those messages is decoded incorrectly then the word error rate is $1/10$. If the codeword $\mathbf{c} = [0\ 1\ 0\ 1\ 0]$ for the message 010 was the one incorrectly decoded, because the decoder chose $\hat{\mathbf{c}} = [1\ 1\ 0\ 0\ 1]$, three codeword bits are in error and so the *codeword* bit error rate is $3/10.5 = 3/50$. The codeword $\hat{\mathbf{c}}$ corresponds to the message 110 so the number of message bits in error is 1. The *message* bit error rate is thus $1/3.10 = 1/30$.

Usually, the message BER is the most relevant BER measure since it is the messages that we ultimately wish to use. However, the codeword BER can sometimes be easier to measure and, depending on the encoder, can be essentially the same as the message BER so may be reported instead.

The WER (and codeword BER) will depend on the choice of code and decoder. The message BER depends, however, on the number of message bits in error for each erroneous codeword and so is determined by the mapping of messages to codewords (in the encoder) as well as by the code and decoder.

One way to analyze the performance of a given error correction code is simply to simulate it. In a process called Monte-Carlo simulation, a large number of messages are randomly generated, encoded, transmitted and decoded. The channel is simulated by randomly generating noise from the appropriate distribution (e.g. $p(z)$ in (1.6)) and using it to corrupt the codeword according to the channel model. The decoded message is compared with the transmitted message and the number of errors counted. The BER or WER performance of the code is then the error rate averaged over all the transmitted messages. For a chosen code and a chosen channel noise level, a common rule of thumb is that around 500 word errors should be observed to obtain confidence in the calculated BER or WER value. A BER or WER curve is found by repeating the simulation for a range of different channel noise levels using the same code. For example, in Figure 2.5 the BI-AWGN channel model is used to generate BER and WER curves for three different codes.

Another common way to analyze the performance of an error correction code is using union bound techniques.

Union bounds

If every message is equally likely or if the a priori message probabilities are not known then the best decoding strategy, maximum likelihood decoding, is to pick the codeword closest to the received sequence as the decoded codeword. The ML decoder will thus decode incorrectly if the received sequence $\mathbf{y}$ is closer to a codeword other than the codeword sent. The probability $Pe(\mathbf{c}_1, \mathbf{c}_2)$ that a codeword $\mathbf{c}_1$ is incorrectly decoded to a codeword $\mathbf{c}_2$ is called the *pairwise error probability* and will depend on the level of noise in the channel and the number of codeword bit locations in which the two codewords differ.

The *union bound* states that the total probability of error $Pe(\mathbf{c}_1)$ for the codeword $\mathbf{c}_1$ can be upper bounded by the sum of the probabilities that it will be decoded incorrectly to any other codeword in the code C:

$$Pe(\mathbf{c}_1) \leq \sum_{\mathbf{c}\in C, \mathbf{c}\neq\mathbf{c}_1} Pe(\mathbf{c}_1, \mathbf{c}).$$

Assuming, as throughout this text, a binary input memoryless symmetric channel, the probability that $\mathbf{c}_1$ will be selected instead of $\mathbf{c}_2$ depends only on the number of bit locations d that differ between $\mathbf{c}_1$ and $\mathbf{c}_2$ and not on the location of those differing entries. When the first codeword $\mathbf{c}_1$ is the all-zeros codeword

$\mathbf{0}$ then d is equal to the number of ones in $\mathbf{c}_2$; this number is called its *Hamming weight*.

In a linear code, every codeword has the same number of other codewords differing from it in d bit locations for every possible value of d. Thus, for a linear code, and a memoryless binary symmetric channel, the probability of error for every codeword is the same as the probability of error for the all-zeros codeword $\mathbf{0}$. Consequently the word error rate for the code C is

$$Pe(C) = Pe(\mathbf{0}) \leq \sum_{\mathbf{c} \in C, \mathbf{c} \neq \mathbf{0}} Pe(\mathbf{0}, \mathbf{c}). \tag{1.32}$$

For a BI-AWGN channel, the transmitted sequence $\mathbf{x}_0$ corresponding to the all-zeros codeword is all ones, while the transmitted sequence $\mathbf{x}_d$ corresponding to a weight-d codeword is -1 in d locations and 1 elsewhere. Thus, the probability that $\mathbf{c}_d$ will be selected instead of the all-zeros codeword is the probability that the sum of those d locations in the received vector, $\mathbf{y}$, is negative:

$$Pe(\mathbf{0}, \mathbf{c}_d) = P_d = p\left(\sum_{i \in \mathcal{D}} y_i \leq 0\right) \tag{1.33}$$

where $\mathcal{D}$ is the set of bit locations of the non-zero bits in $\mathbf{c}_d$.

For a BI-AWGN channel, $y_i = \mu x_i + z_i$ where z_i is the additive white Gaussian noise with variance σ^2. Thus Pe is the sum of d independent identically distributed Gaussian random variables with mean μ and variance σ^2, so Pe is itself a Gaussian random variable with mean μd and variance $\sigma^2 d$. For the BI-AWGN channel, P_d is given by the error function

$$P_d = Q(\sqrt{2\mu^2 d/\sigma^2}) \tag{1.34}$$

or, equivalently,

$$P_d = Q(\sqrt{2rdE_b/N_0}) \tag{1.35}$$

where

$$Q(x) = \frac{1}{\sqrt{2\pi}} \int_x^{\infty} e^{-y^2/2} dy.$$

From the above discussion, we can see that the important factor in the ML decoding of a linear code C is the Hamming distance d between all the codewords in the code or, equivalently, the Hamming weight d of all the non-zero codewords in the code. The number of codewords of each weight in C is called the *weight distribution* of C and the weight distribution of a given code is described by its *weight enumerating function (WEF)*

$$A(D) = \sum_{i=1}^{N} a_i D^i \tag{1.36}$$

where D is an indeterminate parameter that allows us to keep track of the codeword weights in a calculation; a_i is the number of codewords of weight i in C. Then, using (1.33) and (1.36), (1.32) becomes

$$Pe(C) \leq \sum_d a_d P_d, \tag{1.37}$$

where the sum runs over the Hamming distance d. For channels with very little noise, it becomes very unlikely that a codeword would be incorrectly decoded to another codeword far away from it in Hamming distance, and so the terms with minimum d dominate the sum in (1.37). Using just the smallest value of d for a code C, i.e. its minimum distance $d_{\min}$, $Pe(C)$ can be approximated as follows:

$$Pe(C) \approx a_{d_{\min}} P_{d_{\min}}. \tag{1.38}$$

Example 1.18 The rate-3/5 code from Example 1.10 has the codeword set

$$C = \{00000, 00101, 01010, 01110, 10010, 10111, 11001, 11100\},$$

with weights 0, 2, 2, 3, 2, 4, 3 and 3 respectively. Thus the minimum distance of C is 2, the weight distribution of C is given by $a_0 = 1$, $a_2 = 3$, $a_3 = 3$ and $a_4 = 1$ and the weight enumerating function for C is

$$A(D) = 1 + 3D^2 + 3D^3 + D^4.$$

An upper bound on the ML decoding word error probability due to using C in a BI-AWGN channel with SNR $\mu^2/(2\sigma^2)$ is, from (1.37),

$$Pe(C) \leq 3Q(\sqrt{4\mu^2/\sigma^2}) + 3Q(\sqrt{6\mu^2/\sigma^2}) + Q(\sqrt{8\mu^2/\sigma^2}),$$

and the word error probability can be approximated by, using (1.38),

$$Pc(C) \approx 3Q(\sqrt{4\mu^2/\sigma^2}).$$

An upper bound on the *codeword* bit error rate $Pe(C)$ of the code can be derived by noting that d of the N codeword bits will be in error when the all-zero codeword is decoded incorrectly to a weight-d codeword:

$$Pc(C) \leq \sum_{d=d_{\min}}^{N} \frac{d}{N} a_d P_d,$$

where N is the code length.

To calculate the *message* bit error rate requires the *input–output weight enumerating function (IOWEF)* of the encoder, which is defined as

$$A(W, D) = \sum_{w=1}^{K} \sum_{d=d_{\min}}^{N} a_{w,d} W^w D^d, \tag{1.39}$$

where W performs the same function for the message weights as D performs for the codeword weights; K is the message length and $a_{w,d}$ is the number of codewords of weight d generated by a message of weight w.

For each codeword weight d, we have

$$B_d = \sum_{w=1}^{K} \frac{w}{K} a_{w,d} \tag{1.40}$$

and so the message bit error rate $Pb(C)$ can be bounded as follows:

$$\begin{aligned} Pb(C) &\le \sum_{d=d_{\min}}^{d_{\max}} B_d P_d, \\ &= \sum_{d=d_{\min}}^{d_{\max}} \sum_{w=1}^{K} \frac{w}{K} a_{w,d} P_d, \\ &\approx B_{d_{\min}} P_{d_{\min}}, \\ &= \sum_{w=1}^{K} \frac{w}{K} a_{w,d_{\min}} P_{d_{\min}}. \end{aligned} \tag{1.41}$$

For systematic codes we define $A_{w,p}$ as the number of sequences of redundant codeword bits of weight p generated by a message of weight w. The overall codeword weight is $d = w + p$. The *input–redundancy weight enumerating function (IRWEF)* is then

$$A(W, P) = \sum_{w=1}^{K} \sum_{p=d_{\min}-w}^{N-K} A_{w,p} W^w P^p, \tag{1.42}$$

and the message bit error rate can be written as

$$Pb(C) \le \sum_{w=1}^{K} \sum_{p=d_{\min}-w}^{N-K} \frac{w}{K} A_{w,p} P_{w+p}. \tag{1.43}$$

Example 1.19 The rate-3/5 encoder from Example 1.11 maps messages to codewords as follows:

$$\begin{aligned} &000 \to 00000,\ 001 \to 00101,\ 010 \to 01010,\ 011 \to 01110, \\ &100 \to 10010,\ 101 \to 10111,\ 110 \to 11001,\ 111 \to 11100. \end{aligned}$$

Thus the first message–codeword pair maps a weight-0 message to a weight-0 codeword, the second and third message–codeword pairs map a weight-1 message to a weight-2 codeword, the fourth message–codeword pair maps a weight-2 message to a weight-3 codeword, etc. Counting the number of codewords of weight d mapping to a message of weight w gives $a_{w,d}$. Thus $a_{0,0} = 1$, $a_{1,2} = 3$, $a_{2,3} = 2$, $a_{3,3} = 1$ and $a_{2,4} = 1$, and the input–output weight enumerating function for this encoder is

$$A(W, D) = 1 + 3WD^2 + 2W^2D^3 + W^3D^3 + W^2D^4.$$

The upper bound on the ML bit error probability due to using C in a BI-AWGN channel with SNR $\mu^2/2\sigma^2$ is given by

$$\begin{aligned} Pb(C) \le\ & \tfrac{1}{3}3Q(\sqrt{4\mu^2/\sigma^2}) + \tfrac{2}{3}2Q(\sqrt{6\mu^2/\sigma^2}) + \tfrac{3}{3}Q(\sqrt{6\mu^2/\sigma^2}) \\ & + \tfrac{2}{3}Q(\sqrt{8\mu^2/\sigma^2}), \end{aligned}$$

and the bit error probability can be approximated by (1.41):

$$Pb(C) \approx Q(\sqrt{4\mu^2/\sigma^2}).$$

In classical error correction coding the primary goal of the code designer has been to improve $Pb(C)$ by maximizing $d_{\min}$. Iterative codes, however, focus on reducing $a_{w,d}$ and so making the multiplicities of P_d small when d is small.

1.4 Bibliographic notes

In this chapter we have presented the three channel models that we will use in this book, introduced the concept of error correction coding and presented a little bit of information theory. Information theory is a whole field in its own right and there are many excellent texts on it such as [2] , [3] and [4]. We have modeled our presentation style on [4].

We have also introduced some concepts in error control coding, which we will apply to iterative error correction codes in the remainder of this book. The error correction codes used as examples in this chapter were all block codes, in which the messages are broken up into blocks to be encoded at the transmitter and decoded as separate blocks at the receiver. Convolutional error correction codes, which encode a continuous stream of input bits into a continuous stream of output bits, are considered in Chapter 4.

There is a great deal of interesting material on classical block and convolutional codes that we have not even attempted to cover in this book. For more detail on classical error correction codes, see for example [5], [6] and [7] or, for those interested in a more mathematical treatment, [8] and [9].

1.5 Exercises

1.1 The uncoded binary vector [1 1] is transmitted through a binary symmetric channel (BSC) with crossover probability $\epsilon = 0.1$. What is the probability that
(a) $\mathbf{y} = [0\ 1]$ will be received,
(b) $\mathbf{y} = [1\ 0]$ will be received,
(c) if two bits are transmitted they will both be received error-free?

1.2 The vector $\mathbf{y} = [1\ 0]$ is received from a BSC with crossover probability $\epsilon = 0.1$. Assuming that the transmitted symbols are equiprobable, what is the probability that
(a) $\mathbf{x} = [1\ 1]$ was transmitted,
(b) $\mathbf{x} = [0\ 0]$ was transmitted?

1.3 Repeat Exercise 1.2 but assume that you have a priori information that the source symbol 1 is three times more likely to have been transmitted than the symbol 0.

1.4 For the source–channel combinations in Exercises 1.2 and 1.3, give the LLRs corresponding to the probabilities $p(x_i|y_i)$.

1.5 The uncoded binary vector [1 1] is transmitted through a binary erasure channel (BEC) with erasure probability $\varepsilon = 0.1$. What is the probability that
(a) $\mathbf{y} = [1\ 1]$ will be received,
(b) $\mathbf{y} = [1\ 0]$ will be received,
(c) if two bits are transmitted they will both be erased,
(d) if two bits are transmitted they will both be received error free?

1.6 The vector $\mathbf{y} = [1\ e]$ is received from a BEC with erasure probability $\varepsilon = 0.1$. Assuming that the transmitted symbols are equiprobable, what is the probability that
(a) $\mathbf{x} = [1\ 1]$ was transmitted,
(b) $\mathbf{x} = [1\ 0]$ was transmitted,
(c) $\mathbf{x} = [0\ 1]$ was transmitted,
(d) $\mathbf{x} = [0\ 0]$ was transmitted?

1.7 Repeat Exercise 1.6 but assume that you have a priori information that the symbol 1 is four times more likely to have been transmitted than the symbol 0.

1.8 For the source–channel combinations in Exercises 1.6 and 1.7, give the LLRs corresponding to the probabilities $p(x_i|y_i)$.

1.9 The uncoded binary vector [1 0] is to be transmitted through a BI-AWGN channel with $\mu = 1$ and $\sigma = 0.1$.
(a) Give $\mathbf{x}$ for this message, and
(b) give the signal-to-noise ratio of this channel (in dB).

1.10 The vector $\mathbf{y} = [-1.1\ 1.3]$ was received at the output of a BI-AWGN channel with $\mu = 1$ and $\sigma = 1$. Assuming that the transmitted symbols are equiprobable, give the probability that
(a) the binary vector [1 1] was transmitted,
(b) the binary vector [1 0] was transmitted.

1.11 Repeat Exercise 1.10 but assume that you have a priori information that the symbol 1 is three times as likely to have been transmitted as the symbol 0.

1.12 For the source–channel combinations in Exercises 1.10 and 1.11 give the LLRs corresponding to the probabilities $p(x_i|y_i)$.

1.13 A binary input channel is completely described by the following channel transition probabilities:

$$
\begin{aligned}
p(y = 0|x = 0) &= 0.2, \\
p(y = 0|x = 1) &= 0.3, \\
p(y = 1|x = 0) &= 0.8, \\
p(y = 1|x = 1) &= 0.7.
\end{aligned}
$$

(a) Is the channel memoryless?
(b) Is the channel symmetric?
(c) Calculate the probabilities $p(x|y)$ for every possible x and y pair. Assume that x is equiprobable.

1.14 Calculate the entropy of a discrete random variable $X = \{X_1, X_2, X_3\}$ with
(a) $p = \{1/5, 2/5, 2/5\}$,
(b) $p = \{1/10, 1/10, 8/10\}$,
(c) equiprobable symbols.

1.15 The random variable $X = \{0, 1\}$ has distribution $p = \{1/3, 2/3\}$. The random variable $Y = \{0, 1\}$ is related to X by

$$
\begin{aligned}
P(y = 0|x = 0) &= 0.8, \\
P(y = 0|x = 1) &= 0.2, \\
P(y = 1|x = 0) &= 0.2, \\
P(y = 1|x = 1) &= 0.8.
\end{aligned}
$$

Use Bayes' rule to calculate the distributions $p(y)$ and $p(x, y)$.

1.16 Calculate the mutual information between the variables X and Y in Exercise 1.15.

1.17 Calculate the mutual information between the variables X and Y, where X is an equiprobable binary source and Y is the received symbol when X is transmitted through a BSC with crossover probability $\epsilon = 0.1$.

1.18 Calculate the mutual information between the variables X and Y, where X is a binary source with distribution $p(x) = \{0.2, 0.8\}$ and Y is the received symbol when X is transmitted through a BSC with crossover probability $\epsilon = 0.1$.

1.19 What is the capacity of the BSC in Exercise 1.17?

1.20 What is the capacity of the BEC in Exercise 1.5?

1.21 What is the capacity of the BI-AWGN channel in Exercise 1.9?

1.22 For each of the Exercises 1.19–1.21, what is the maximum code rate below which an error correction code is proved to exist that will achieve an arbitrarily low probability of error despite the noise added by the channel?

1.23 Give Shannon's capacity limit for a rate-$1/4$ code on
(a) a BSC,
(b) a BEC,
(c) a BI-AWGN channel.

1.24 Plot Shannon's capacity limit for a rate-r code on a BI-AWGN channel as a function of r.

1.25 Show that the capacity of the binary erasure channel can be given as

$$c(\varepsilon) = 1 - \varepsilon.$$

1.26 The encoder for a rate-$1/q$ repetition code repeats each message bit q times. For example, if $q = 2$ then the message $\mathbf{u} = [u_1\ u_2]$ will be encoded into the codeword $\mathbf{c} = [u_1\ u_1\ u_2\ u_2]$. If a codeword from a length $N = 6$, rate-$1/3$, repetition code is transmitted through a BSC and $\mathbf{y} = [1\ 1\ 0\ 0\ 0\ 1]$ is received, use ML decoding to find the most likely transmitted message, assuming that the message was generated by a binary equiprobable source,
(a) when $\epsilon = 0.1$,
(b) when $\epsilon = 0.4$,
(c) when $\epsilon = 0.6$.

1.27 Repeat Exercise 1.26 but assume that the message was generated by a binary source that is three times as likely to generate a 1 symbol as a 0 symbol.

1.28 Repeat Exercise 1.26 but use MAP decoding.

1.29 Repeat Exercise 1.27 but use MAP decoding.

1.30 For the code in Example 1.12,
(a) calculate its input–output weight enumerating function,
(b) determine its minimum distance,
(c) determine its probability of ML decoding error on a BI-AWGN channel with signal-to-noise ratio 2 dB.

1.31 For the code in Exercise 1.26,
(a) calculate its input–output weight enumerating function,
(b) determine its minimum distance,
(c) determine its probability of ML decoding error on a BI-AWGN channel with signal-to-noise ratio 2 dB.

1.32 For a length $N = 6$, rate-$1/2$, repetition code,
(a) calculate its input–output weight enumerating function,
(b) determine its minimum distance,
(c) give an upper bound for the probability of ML decoding error using this code on a BI-AWGN channel with a signal-to-noise ratio of 3 dB.
(d) estimate the probability of ML decoding error using this code on a BI-AWGN channel with a high signal-to-noise ratio, 10 dB.

1.33 Two length-5 rate $3/5$ codes are given in Examples 1.12 and 1.15. Which of the two codes would you choose to use over a very low noise channel if the decoder was implementing ML decoding? Explain your choice.

2
Low-density parity-check codes

2.1 Introduction

In this chapter we introduce *low-density parity-check (LDPC)* codes, a class of error correction codes proposed by Gallager in his 1962 PhD thesis 12 years after error correction codes were first introduced by Hamming (published in 1950). Both Hamming codes and LDPC codes are block codes: the messages are broken up into blocks to be encoded at the transmitter and similarly decoded as separate blocks at the receiver. While Hamming codes are short and very structured with a known, fixed, error correction ability, LDPC codes are the opposite, usually long and often constructed pseudo-randomly with only a probabilistic notion of their expected error correction performance.

The chapter begins by presenting parity bits as a means to detect and, when more than one is employed, to correct errors in digital data. Block error correction codes are described as a linear combination of parity-check equations and thus defined by their parity-check matrix representation. The graphical representation of codes by Tanner graphs is presented and the necessary graph theoretic concepts introduced.

In Section 2.4 iterative decoding algorithms are introduced using a hard decision algorithm (bit flipping), so that the topic is developed first without reference to probability theory. Subsequently the sum–product decoding algorithm is presented.

This chapter serves as a self-contained introduction to LDPC codes and their decoding. It is intended that the material presented here will enable the reader to implement LDPC encoders and iterative decoders. A deeper understanding of how and why LDPC codes work will be developed in later chapters.

2.2 Error correction using parity checks

The essential idea of error correction coding is to augment the *message* bits to be communicated with deliberately introduced redundancy in the form of extra bits in order to produce a *codeword* for the message. These extra bits are added in such a way that the codewords are sufficiently distinct from one another to allow the transmitted message to be correctly inferred at the receiver even when some bits in the codeword are corrupted during transmission over the channel.

The simplest coding scheme is the single parity-check code. This code involves the addition of a single extra bit, called a *parity-check bit*, to the binary message; the value of this bit depends on the bits in the message. In an even-parity code the additional bit added to each message ensures an even number of 1s in every codeword.

Example 2.1 The 7-bit ASCII[1] vector for the letter S is [1 0 1 0 0 1 1]. A parity bit is to be added as the eighth bit. The vector for S already has an even number of 1s (namely four) and so the value of the parity bit must be 0. The codeword for S is thus [1 0 1 0 0 1 1 0].

More formally, for the 7-bit ASCII plus even-parity code we define a codeword c to have the following structure:

$$\mathbf{c} = [c_1 \; c_2 \; c_3 \; c_4 \; c_5 \; c_6 \; c_7 \; c_8]$$

where each c_i is either 0 or 1 and every codeword satisfies the constraint

$$c_1 \oplus c_2 \oplus c_3 \oplus c_4 \oplus c_5 \oplus c_6 \oplus c_7 \oplus c_8 = 0. \tag{2.1}$$

Equation (2.1) is called a *parity-check equation*. The symbol $\oplus$ represents modulo-2 addition.

Example 2.2 A 7-bit ASCII letter is encoded with the single-parity check code from Example 2.1. The resulting codeword is sent through a noisy channel and the vector $\mathbf{y} = [1\;0\;0\;1\;0\;0\;1\;0]$ is received. To check whether $\mathbf{y}$ is a valid codeword we test it with (2.1):

$$y_1 \oplus y_2 \oplus y_3 \oplus y_4 \oplus y_5 \oplus y_6 \oplus y_7 \oplus y_8 = 1 \oplus 0 \oplus 0 \oplus 1 \oplus 0 \oplus 0 \oplus 1 \oplus 0 = 1.$$

Since the sum is 1, the parity-check equation is not satisfied and $\mathbf{y}$ is not a valid codeword. Thus we have detected that at least one error occurred during transmission.

While the inversion of a single bit due to channel noise can easily be detected with a single-parity check code, this code is not sufficiently powerful to indicate which bit, or perhaps bits, were inverted. Moreover, since any even number of

[1] American Standard Code for Information Interchange.

bit inversions produces a vector satisfying the constraint (2.1), patterns of even numbers of errors go undetected by this simple code. Detecting more than a single bit error calls for increased redundancy in the form of additional parity bits. These more sophisticated codes contain multiple parity-check equations, every one of which must be satisfied by every codeword in the code.

Example 2.3 A code C consists of all length-6 vectors

$$\mathbf{c} = [c_1\ c_2\ c_3\ c_4\ c_5\ c_6]$$

that satisfy the three parity-check equations

$$\begin{aligned} c_1 \oplus c_2 \oplus c_4 &= 0, \\ c_1 \oplus c_2 \oplus c_3 \oplus c_6 &= 0, \\ c_2 \oplus c_3 \oplus c_5 &= 0. \end{aligned} \tag{2.2}$$

Checking the vector $\tilde{\mathbf{c}} = [1\ 1\ 0\ 0\ 0\ 0]$ we see that

$$\begin{aligned} 1 \oplus 1 \oplus 0 &= 0, \\ 1 \oplus 1 \oplus 0 \oplus 0 &= 0, \\ 1 \oplus 0 \oplus 0 &= 1, \end{aligned}$$

so $\tilde{\mathbf{c}}$ is not a valid codeword for this code.

Codeword constraints are often written in matrix form, and so the constraints (2.2) become

$$\underbrace{\begin{bmatrix} 1 & 1 & 0 & 1 & 0 & 0 \\ 1 & 1 & 1 & 0 & 0 & 1 \\ 0 & 1 & 1 & 0 & 1 & 0 \end{bmatrix}}_{H} \begin{bmatrix} c_1 \\ c_2 \\ c_3 \\ c_4 \\ c_5 \\ c_6 \end{bmatrix} = \begin{bmatrix} 0 \\ 0 \\ 0 \end{bmatrix}. \tag{2.3}$$

The matrix H is called a *parity-check matrix*. Each row of H corresponds to a parity-check equation and each column of H corresponds to a bit in the codeword. The (j, i)th entry of H is 1 if the ith codeword bit is included in the jth parity-check equation. Thus for a binary code with m parity-check constraints and length-N codewords the parity-check matrix is an $m \times N$ binary matrix. In matrix form a vector $\tilde{\mathbf{c}} = [c_1\ c_2\ c_3\ c_4\ c_5\ c_6]$ is a valid codeword for the code with parity-check matrix H if and only if it satisfies the matrix equation

$$H\tilde{\mathbf{c}}^T = \mathbf{0} \pmod 2, \tag{2.4}$$

where $\mathbf{0}$ is the length-m all-zeros vector and $\tilde{\mathbf{c}}^T$ denotes the transpose of $\tilde{\mathbf{c}}$.

More than one parity-check matrix can describe a particular code; two parity-check matrices for the same code do not even have to have the same number of rows, but they must satisfy (2.4) for every codeword in the code.

Example 2.4 The code C in Example 2.3 can also be described by four parity-check equations:

$$\begin{aligned} c_1 \oplus c_2 \oplus c_4 &= 0, \\ c_1 \oplus c_2 \oplus c_3 \oplus c_6 &= 0, \\ c_2 \oplus c_3 \oplus c_5 &= 0, \\ c_3 \oplus c_4 \oplus c_6 &= 0. \end{aligned} \tag{2.5}$$

The extra equation is the linear combination of the first and second parity-check equations and so is *linearly dependent* on the existing parity-check equations. (In modulo-2 arithmetic a linear combination of parity-check equations is simply the modulo-2 sum of two or more equations.) It is easy to see that any codeword for which $c_1 \oplus c_2 \oplus c_4 = 0$ and $c_1 \oplus c_2 \oplus c_3 \oplus c_6 = 0$ will also satisfy $c_3 \oplus c_4 \oplus c_6 = 0$.

In Examples 2.3 and 2.4 there are three linearly independent equations in six variables, so there are three dependent variables (one for each equation). These are the variables corresponding to the parity-check bits. The remaining independent variables correspond to the message bits. So, for the code in Example 2.3, the number of codeword bits $N = 6$, the number of message bits $K = 3$ and so we have a rate $r = 1/2$ code.

It seems plausible that the more parity-check constraints there are, the fewer the number of codewords that can satisfy them. Indeed, this is true as long as the parity-check equations are linearly independent. An extra equation that is linearly dependent on the existing equations adds no extra constraint to the codeword set so does not reduce the number of independent variables.

In general, a code can have any number of parity-check constraints but only $N - K$ of them will be linearly independent, where K is the number of code message bits. In matrix notation $N - K$ is the rank of H over GF(2):

$$N - K = \text{rank}_2 H, \tag{2.6}$$

where $\text{rank}_2 H$ is the number of rows in H that are linearly independent over GF(2).

2.2.1 Low-density parity-check codes

As their name suggests, LDPC codes are block codes with parity-check matrices that contain only a very small number of non-zero entries. This sparseness of

H is essential for an iterative decoding complexity that increases only linearly with the code length.

Aside from the requirement that H be sparse, an LDPC code itself is no different from any other block code. Indeed, classical block codes will work well with iterative decoding algorithms if they can be represented by a sparse parity-check matrix. Generally, however, finding such a matrix for an existing code is not practical. Instead LDPC codes are designed by constructing a suitable sparse parity-check matrix first and then determining an encoder for the code afterwards.

The biggest difference between LDPC codes and classical block codes is how they are decoded. Classical block codes are generally decoded with ML-like decoding algorithms and so are usually short and designed algebraically to make this task less complex. However, LDPC codes are decoded iteratively using a graphical representation of their parity-check matrix and so are much longer, less structured, and designed with the properties of H as a focus.

An LDPC code parity-check matrix is called *(w_c, w_r)-regular* if each code bit is contained in a fixed number w_c of parity checks and each parity-check equation contains a fixed number w_r of code bits.

Example 2.5 A regular parity-check matrix, with $w_c = 2$ and $w_r = 3$, for the code in Example 2.3 is

$$H = \begin{bmatrix} 1 & 1 & 0 & 1 & 0 & 0 \\ 0 & 1 & 1 & 0 & 1 & 0 \\ 1 & 0 & 0 & 0 & 1 & 1 \\ 0 & 0 & 1 & 1 & 0 & 1 \end{bmatrix}. \tag{2.7}$$

For an *irregular* parity-check matrix we designate the fraction of columns having weight i by v_i and the fraction of rows having weight i by h_i. Collectively the pair $\mathbf{v}$, $\mathbf{h}$, where $\mathbf{v} = [v_1\ v_2\ v_3 \cdots]$ and $\mathbf{h} = [h_1\ h_2\ h_3 \cdots]$, is called the *degree distribution* of the code.

Example 2.6 The parity-check matrix in (2.3) is irregular with degree distribution $v_1 = 1/2$, $v_2 = 1/3$, $v_3 = 1/6$, $h_3 = 2/3$ and $h_4 = 1/3$.

A regular LDPC code will have

$$m w_r = N w_c \tag{2.8}$$

1s in its parity-check matrix. Similarly, for an irregular code,

$$m\left(\sum_i h_i i\right) = N\left(\sum_i v_i i\right). \tag{2.9}$$

Recall that the rate of a block code is given by, taking into account (2.6),

$$r = \frac{K}{N} = \frac{N - \mathrm{rank}_2 H}{N}.$$

When the parity-check matrix is full rank, i.e.

$$\mathrm{rank}_2 H = m,$$

the rate can be written as

$$r = \frac{K}{N} = \frac{N-m}{N} = \frac{N - Nw_\mathrm{c}/w_\mathrm{r}}{N} = 1 - \frac{w_\mathrm{c}}{w_\mathrm{r}}, \tag{2.10}$$

for a regular code and

$$r = \frac{K}{N} = 1 - \frac{\sum_i v_i i}{\sum_i h_i i}, \tag{2.11}$$

for an irregular code.

In all but a few specially designed cases (some of which we will see in Section 3.3.2), parity-check matrices are full rank, or very close to it, and so the rate of an LDPC code is often given as

$$r \approx 1 - \frac{w_\mathrm{c}}{w_\mathrm{r}}$$

or

$$r \approx 1 - \frac{\sum_i v_i i}{\sum_i h_i i},$$

even when the rank of H has not been explicitly measured. In this case r is called the *design rate* of the code.

Note that parity-check matrices for LDPC codes are generally not systematic (see Example 1.11); the exception is RA codes, which we will consider later. Thus it is not generally possible to distinguish between message bits and parity bits in LDPC codewords.

Often LDPC parity-check matrices are represented in graphical form by a *Tanner graph*. The Tanner graph consists of two sets of nodes: N nodes for the codeword bits (called *bit nodes*), and m nodes for the parity-check equations (called *check nodes*). An edge joins a bit node to a check node if that bit is included in the corresponding parity-check equation and so the number of edges in the Tanner graph is equal to the number of 1s in the parity-check matrix.

Example 2.7 The Tanner graph of the parity-check matrix in Example 2.5 is shown in Figure 2.1. The bit nodes are represented by the circular vertices and the check nodes by the square vertices.

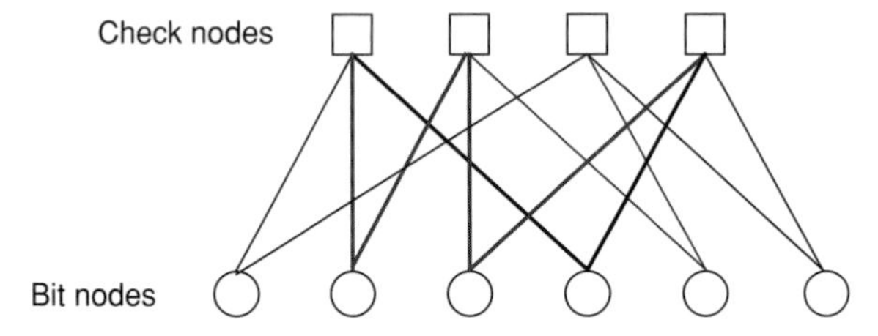

Figure 2.1 The Tanner graph representation of the parity-check matrix (2.7). A 6-cycle is shown in bold.

A *cycle* in a Tanner graph is a sequence of connected nodes which starts and ends at the same node in the graph and which contains other nodes no more than once. The length of a cycle is the number of edges it contains, and the *girth* of a graph is the size of its smallest cycle.

Example 2.8 A cycle of size 6 is shown in bold in Figure 2.1. There are no smaller cycles in the Tanner graph so its girth is also 6.

2.3 Encoding

The parity-check matrix defines the set of valid codewords (those that satisfy $\mathbf{c}H^T = \mathbf{0}$), but we have not yet seen how to map messages to those codewords.

Example 2.9 To distinguish between the message bits and parity bits in the codeword in Example 2.3 we can rewrite the code parity-check constraints so that each one solves for a different codeword bit. The codeword constraints from Example 2.3 can be rewritten as

$$\begin{aligned} c_4 &= c_1 \oplus c_2, \\ c_5 &= c_2 \oplus c_3, \\ c_6 &= c_1 \oplus c_2 \oplus c_3. \end{aligned} \tag{2.12}$$

The codeword bits c_4, c_5 and c_6 contain the three parity-check bits while the codeword bits c_1, c_2 and c_3 contain the 3-bit message. The message bits are conventionally labeled $u_1, u_2, \ldots, u_K$, where the vector $\mathbf{u} = [u_1\ u_2\ \cdots\ u_K]$ is the message to be communicated. This code is systematic because the first K codeword bits contain the message bits.

Written as in (2.12), the codeword constraints show how to encode the message.

Example 2.10 Using the constraints in (2.12) the message $\mathbf{u} = [1\ 1\ 0]$ produces the parity-check bits

$$\begin{aligned} c_4 &= 1 \oplus 1 = 0, \\ c_5 &= 1 \oplus 0 = 1, \\ c_6 &= 1 \oplus 1 \oplus 0 = 0, \end{aligned}$$

and so the codeword for this message is $\mathbf{c} = [1\ 1\ 0\ 0\ 1\ 0]$.

Again these constraints can be written in matrix form:

$$[c_1\ c_2\ c_3\ c_4\ c_5\ c_6] = [u_1\ u_2\ u_3] \underbrace{\begin{bmatrix} 1 & 0 & 0 & 1 & 0 & 1 \\ 0 & 1 & 0 & 1 & 1 & 1 \\ 0 & 0 & 1 & 0 & 1 & 1 \end{bmatrix}}_{G} \tag{2.13}$$

or

$$\mathbf{c} = \mathbf{u}G, \tag{2.14}$$

where the matrix G is called the *generator matrix* of the code. The (i, j)th entry of G is 1 if the ith message bit plays a role in determining the jth codeword bit. For example, the fourth column of G is 1 in the first and second rows so it follows that

$$c_4 = u_1 \oplus u_2.$$

For systematic codes the generator matrix contains the $K \times K$ *identity matrix* I_K as its first K columns, since the ith message bit is directly mapped to the ith codeword bit. (The identity matrix I_K is a $K \times K$ square binary matrix with entries 1 on the diagonal from the top left corner to the bottom right corner and entries 0 everywhere else.)

From (2.14) the set of codewords for the code with generator G is the set of all possible linear combinations of the rows of G. Thus, if G is the generator matrix for a code with parity-check matrix H then

$$GH^T = \mathbf{0} \pmod 2,$$

where $\mathbf{0}$ is a $K \times m$ all-zeros matrix.

For a binary code with K message bits and length-N codewords, the generator matrix G is a $K \times N$ binary matrix. A code with K message bits contains 2^K codewords, which are a subset of the total possible 2^N binary vectors of length N.

Example 2.11 Substituting each of the $2^3 = 8$ distinct messages $u_1u_2u_3 = 000, 001, \ldots, 111$ into (2.14) yields the following set of codeword vectors for the code from Example 2.3:

$$\begin{array}{llll} [0\,0\,0\,0\,0\,0], & [0\,0\,1\,0\,1\,1], & [0\,1\,0\,1\,1\,1], & [0\,1\,1\,1\,0\,0], \\ [1\,0\,0\,1\,0\,1], & [1\,0\,1\,1\,1\,0], & [1\,1\,0\,0\,1\,0], & [1\,1\,1\,0\,0\,1]. \end{array} \tag{2.15}$$

In general, a generator matrix for a code with parity-check matrix H can be found by performing Gauss–Jordan elimination on H to obtain it in the form

$$H = [A \quad I_{N-K}], \tag{2.16}$$

where A is an $(N-K) \times K$ binary matrix and I_{N-K} is the identity matrix of order $N-K$. The generator matrix is then

$$G = [I_K \quad A^T]. \tag{2.17}$$

Example 2.12 We wish to encode the length-10 rate-1/2 LDPC code given by the parity-check matrix

$$H = \begin{bmatrix} 1 & 1 & 0 & 1 & 1 & 0 & 0 & 1 & 0 & 0 \\ 0 & 1 & 1 & 0 & 1 & 1 & 1 & 0 & 0 & 0 \\ 0 & 0 & 0 & 1 & 0 & 0 & 0 & 1 & 1 & 1 \\ 1 & 1 & 0 & 0 & 0 & 1 & 1 & 0 & 1 & 0 \\ 0 & 0 & 1 & 0 & 0 & 1 & 0 & 1 & 0 & 1 \end{bmatrix}.$$

Firstly, we put H into *row-echelon form* (i.e. in any two successive rows that do not consist entirely of 0s, the leading 1 in the lower row must occur further to the right than the leading 1 in the higher row).

The matrix H is put into this form by applying *elementary row operations* in $GF(2)$. The elementary row operations in $GF(2)$ are interchanging two rows or adding one row to another modulo 2. From linear algebra we know that after the use of only elementary row operations the modified parity-check matrix will have the same codeword set as the original (as the new system of linear equations will have an unchanged solution set).

The first and second rows of H already satisfy the requirements for row-echelon form, and the entries in the columns below the diagonal are removed by replacing the fourth row with the modulo-2 sum of the first and fourth rows. The fifth row of H has a 1 entry to the left of the leading 1 entry above it, but this can be remedied by swapping the third and fifth rows. Then replacing the fifth row with the modulo-2 sum of the fifth and fourth rows gives H_{r}, the matrix H in row-echelon form:

$$H_{\mathrm{r}} = \begin{bmatrix} 1 & 1 & 0 & 1 & 1 & 0 & 0 & 1 & 0 & 0 \\ 0 & 1 & 1 & 0 & 1 & 1 & 1 & 0 & 0 & 0 \\ 0 & 0 & 1 & 0 & 0 & 1 & 0 & 1 & 0 & 1 \\ 0 & 0 & 0 & 1 & 1 & 1 & 1 & 1 & 1 & 0 \\ 0 & 0 & 0 & 0 & 1 & 1 & 1 & 0 & 0 & 1 \end{bmatrix}.$$

Next the parity-check matrix is put into *reduced* row-echelon form (i.e. any column that contains a leading 1 has 0s everywhere else). The first column is already correct, and the entry in the second column above the diagonal is removed by replacing the first row with the modulo-2 sum of the first and second rows. Similarly the entry in the third column above the diagonal is removed by replacing the second row with the modulo-2 sum of the second and third rows. To clear the fourth column the first row is replaced with the modulo-2 sum of the first and fourth rows. To clear the fifth column involves adding the fifth row to the first, second and fourth rows, which gives the parity-check matrix in reduced row-echelon form:

$$H_{\mathrm{rr}} = \begin{bmatrix} 1 & 0 & 0 & 0 & 0 & 0 & 1 & 1 & 1 & 0 \\ 0 & 1 & 0 & 0 & 0 & 1 & 0 & 1 & 0 & 0 \\ 0 & 0 & 1 & 0 & 0 & 1 & 0 & 1 & 0 & 1 \\ 0 & 0 & 0 & 1 & 0 & 0 & 0 & 1 & 1 & 1 \\ 0 & 0 & 0 & 0 & 1 & 1 & 1 & 0 & 0 & 1 \end{bmatrix}.$$

Lastly, using column permutations we put the parity-check matrix into standard form (where the last m columns of H_{std} are the m columns of H_{rr} that contain the leading 1s):

$$H_{\mathrm{std}} = \begin{bmatrix} 0 & 1 & 1 & 1 & 0 & 1 & 0 & 0 & 0 & 0 \\ 1 & 0 & 1 & 0 & 0 & 0 & 1 & 0 & 0 & 0 \\ 1 & 0 & 1 & 0 & 1 & 0 & 0 & 1 & 0 & 0 \\ 0 & 0 & 1 & 1 & 1 & 0 & 0 & 0 & 1 & 0 \\ 1 & 1 & 0 & 0 & 1 & 0 & 0 & 0 & 0 & 1 \end{bmatrix}.$$

In this final step column permutations have been used and so the codewords of H_{std} will be permuted versions of the codewords corresponding to H. A solution is to keep track of the column permutation Π used to create H_{std}, which in this case is

$$\Pi = [6, 7, 8, 9, 10, 1, 2, 3, 4, 5],$$

and apply the inverse permutation to each H_{std} codeword before it is transmitted.

Alternatively, if the channel is memoryless, so that the order of the codeword bits is unimportant, a far easier option is to apply Π to the original H to give a parity-check matrix

$$H' = \begin{bmatrix} 1 & 1 & 0 & 1 & 1 & 0 & 0 & 1 & 0 & 0 \\ 0 & 1 & 1 & 0 & 1 & 1 & 1 & 0 & 0 & 0 \\ 0 & 0 & 0 & 1 & 0 & 0 & 0 & 1 & 1 & 1 \\ 1 & 1 & 0 & 0 & 0 & 1 & 1 & 0 & 1 & 0 \\ 0 & 0 & 1 & 0 & 0 & 1 & 0 & 1 & 0 & 1 \end{bmatrix},$$

which has the same properties as H but which shares the same codeword bit ordering as H_{std}.

Finally, the generator G for the code with parity-check matrices H_{std} and H' is given by

$$G = \begin{bmatrix} 1 & 0 & 0 & 0 & 0 & 0 & 1 & 1 & 0 & 1 \\ 0 & 1 & 0 & 0 & 0 & 1 & 0 & 0 & 0 & 1 \\ 0 & 0 & 1 & 0 & 0 & 1 & 1 & 1 & 1 & 0 \\ 0 & 0 & 0 & 1 & 0 & 1 & 0 & 0 & 1 & 0 \\ 0 & 0 & 0 & 0 & 1 & 0 & 0 & 1 & 1 & 1 \end{bmatrix}.$$

All this processing can be done off-line and just the matrices G and H' provided to the encoder and decoder respectively. However, the drawback of this approach is that, unlike H, the matrix G will most likely not be sparse and so the matrix multiplication

$$\mathbf{c} = \mathbf{u}G$$

at the encoder will have a complexity of order N^2 operations, where N is the number of bits in a codeword. As N is large for LDPC codes, from thousands to hundreds of thousands, the encoder can become prohibitively complex. Later we will see that structured parity-check matrices can be used to lower this implementation complexity significantly. For arbitrary parity-check matrices, a good approach is to avoid constructing G altogether and, instead, to encode using back substitution with H, as will be demonstrated in the following subsection.

2.3.1 (Almost) linear-time encoding for LDPC codes

Rather than finding a generator matrix for H, an LDPC code can be encoded using the parity-check matrix directly by transforming it into lower triangular form and using back substitution. The idea is to do as much of the transformation as possible using only row and column permutations so as to keep H as sparse as possible.

Firstly, using only row and column permutations the parity-check matrix is put into the *approximate lower triangular form*

$$H_t = \begin{bmatrix} A & B & T \\ C & D & E \end{bmatrix},$$

where the matrix T is a lower triangular matrix of size $(m-g)\times(m-g)$ (that is, T has 1s on the diagonal from the upper left corner to the lower right corner, and all entries above the diagonal are 0) . If H_t is full rank then the matrix B is of size $(m-g)\times g$ and A is of size $(m-g)\times K$. The g rows of H left in C, D and E are called the *gap* of the approximate representation and the smaller g, the lower the encoding complexity for the code.

Example 2.13 We wish to encode the message $\mathbf{u} = [1\ 1\ 0\ 0\ 1]$ with the same length-10 rate-1/2 LDPC code as in Example 2.12:

$$H = \begin{bmatrix} 1 & 1 & 0 & 1 & 1 & 0 & 0 & 1 & 0 & 0 \\ 0 & 1 & 1 & 0 & 1 & 1 & 1 & 0 & 0 & 0 \\ 0 & 0 & 0 & 1 & 0 & 0 & 0 & 1 & 1 & 1 \\ 1 & 1 & 0 & 0 & 0 & 1 & 1 & 0 & 1 & 0 \\ 0 & 0 & 1 & 0 & 0 & 1 & 0 & 1 & 0 & 1 \end{bmatrix}.$$

Instead of putting H into reduced row-echelon form we put it into approximate lower triangular form using only row and column swaps. Thus we swap the second and third rows and sixth and 10th columns to obtain

$$H_t = \left[\begin{array}{ccccc|cc|ccc} 1 & 1 & 0 & 1 & 1 & 0 & 0 & 1 & 0 & 0 \\ 0 & 0 & 0 & 1 & 0 & 1 & 0 & 1 & 1 & 0 \\ 0 & 1 & 1 & 0 & 1 & 0 & 1 & 0 & 0 & 1 \\ \hline 1 & 1 & 0 & 0 & 0 & 0 & 1 & 0 & 1 & 1 \\ 0 & 0 & 1 & 0 & 0 & 1 & 0 & 1 & 0 & 1 \end{array}\right],$$

with gap 2.

Once in approximate lower triangular form, Gauss–Jordan elimination is applied to clear E. This is equivalent to multiplying H_t by

$$\begin{bmatrix} I_{m-g} & \mathbf{0} \\ -ET^{-1} & I_g \end{bmatrix},$$

to give

$$\tilde{H} = \begin{bmatrix} I_{m-g} & \mathbf{0} \\ -ET^{-1} & I_g \end{bmatrix} H_t = \begin{bmatrix} A & B & T \\ \tilde{C} & \tilde{D} & \mathbf{0} \end{bmatrix}$$

where

$$\tilde{C} = -ET^{-1}A + C,$$
$$\tilde{D} = -ET^{-1}B + D$$

and

$$\tilde{E} = -ET^{-1}T + E = \mathbf{0}.$$

Example 2.14 Continuing from Example 2.13 and using the above discussion, we have

$$T^{-1} = \begin{bmatrix} 1 & 0 & 0 \\ 1 & 1 & 0 \\ 0 & 0 & 1 \end{bmatrix}$$

and

$$\begin{bmatrix} I_{m-g} & 0 \\ -ET^{-1} & I_g \end{bmatrix} = \begin{bmatrix} 1 & 0 & 0 & 0 & 0 \\ 0 & 1 & 0 & 0 & 0 \\ 0 & 0 & 1 & 0 & 0 \\ 1 & 1 & 1 & 1 & 0 \\ 1 & 0 & 1 & 0 & 1 \end{bmatrix},$$

to give

$$\tilde{H} = \left[\begin{array}{ccccc|cc|ccc} 1 & 1 & 0 & 1 & 1 & 0 & 0 & 1 & 0 & 0 \\ 0 & 0 & 0 & 1 & 0 & 1 & 0 & 1 & 1 & 0 \\ 0 & 1 & 1 & 0 & 1 & 0 & 1 & 0 & 0 & 1 \\ \hline 0 & 1 & 1 & 0 & 0 & 1 & 0 & 0 & 0 & 0 \\ 1 & 0 & 0 & 1 & 0 & 1 & 1 & 0 & 0 & 0 \end{array}\right].$$

Since T is lower triangular, only $\tilde{C}$ and $\tilde{D}$ are affected when applying Gauss–Jordan elimination to clear E; the rest of the parity-check matrix remains sparse.

Finally, to perform encoding using $\tilde{H}$, the codeword $\mathbf{c} = [c_1\ c_2\ \cdots\ c_N]$ is divided into three parts, so that $\mathbf{c} = [\mathbf{u}\ \mathbf{p}^{(1)}\ \mathbf{p}^{(2)}]$ where $\mathbf{u} = [u_1\ u_2\ \cdots\ u_K]$ is the K-bit message, $\mathbf{p}^{(1)} = [p_1^{(1)}\ p_2^{(1)}\ \cdots\ p_g^{(1)}]$ holds the first g parity bits and $\mathbf{p}^{(2)} = [p_1^{(2)}\ p_2^{(2)}\ \cdots\ p_{m-g}^{(2)}]$ holds the remaining parity bits.

The codeword $\mathbf{c} = [\mathbf{u}\ \mathbf{p}^{(1)}\ \mathbf{p}^{(2)}]$ must satisfy the parity-check equation $\mathbf{c}\tilde{H}^T = \mathbf{0}$; thus

$$A\mathbf{u} + B\mathbf{p}^{(1)} + T\mathbf{p}^{(2)} = \mathbf{0} \tag{2.18}$$

and

$$\tilde{C}\mathbf{u} + \tilde{D}\mathbf{p}^{(1)} + \mathbf{0}\mathbf{p}^{(2)} = \mathbf{0}. \tag{2.19}$$

Since E has been cleared, the parity bits in $\mathbf{p}^{(1)}$ depend only on the message bits and so can be calculated independently of the parity bits in $\mathbf{p}^{(2)}$. If $\tilde{D}$ is invertible, $\mathbf{p}^{(1)}$ can be found from (2.19):

$$\mathbf{p}^{(1)} = \tilde{D}^{-1}\tilde{C}\mathbf{u}. \tag{2.20}$$

If $\tilde{D}$ is not invertible then the columns of $\tilde{H}$ can be permuted until it is. By keeping g as small as possible the added complexity burden of the matrix multiplication in (2.20) is kept low.

Once $\mathbf{p}^{(1)}$ is known, $\mathbf{p}^{(2)}$ can be found from (2.18):

$$\mathbf{p}^{(2)} = -T^{-1}(A\mathbf{u} + B\mathbf{p}^{(1)}), \tag{2.21}$$

where the sparseness of A, B and T can be employed to keep the complexity of this operation low and, as T is lower triangular, $\mathbf{p}^{(2)}$ can in fact be found using back substitution.

Example 2.15 Continuing from Example 2.14 we partition the length-10 codeword $\mathbf{c} = [c_1\ c_2\ \cdots\ c_{10}]$ as $\mathbf{c} = [\mathbf{u}\ \mathbf{p}^{(1)}\ \mathbf{p}^{(2)}]$, where $\mathbf{p}^{(1)} = [c_6\ c_7]$ and $\mathbf{p}^{(2)} = [c_8\ c_9\ c_{10}]$. The parity bits in $\mathbf{p}^{(1)}$ are calculated from the message using (2.20):

$$\mathbf{p}^{(1)} = \tilde{D}^{-1}\tilde{C}\mathbf{u} = \begin{bmatrix} 1 & 0 \\ 1 & 1 \end{bmatrix} \begin{bmatrix} 0 & 1 & 1 & 0 & 0 \\ 1 & 0 & 0 & 1 & 0 \end{bmatrix} \begin{bmatrix} 1 \\ 1 \\ 0 \\ 0 \\ 1 \end{bmatrix} = [1 \quad 0].$$

As T is lower triangular the bits in $\mathbf{p}^{(2)}$ can then be calculated using back substitution.

$$\begin{aligned} p_1^{(2)} &= u_1 \oplus u_2 \oplus u_4 \oplus u_5 = 1 \oplus 1 \oplus 0 \oplus 1 = 1, \\ p_2^{(2)} &= u_4 \oplus p_1^{(1)} \oplus p_1^{(1)} = 0 \oplus 1 \oplus 1 = 0, \\ p_3^{(2)} &= u_2 \oplus u_3 \oplus u_5 \oplus p_2^{(1)} = 1 \oplus 0 \oplus 1 \oplus 0 = 0, \end{aligned}$$

and the codeword is $\mathbf{c} = [1\ 1\ 0\ 0\ 1\ 1\ 0\ 1\ 0\ 0]$.

Again column permutations have been used to obtain H_t from H, and so either H_t or H with the same column permutation applied will be used at the decoder. Note that since the algorithm used to compute G in Example 2.12 used a different column permutation, the set of codewords generated by G will be permuted versions of the codewords generated using $\tilde{H}$.

2.3.2 Repeat–accumulate codes

Another type of LDPC codes, called *repeat–accumulate codes* (see Section 6.2), has weight-2 columns in a step pattern for the last m columns of H. This structure makes repeat–accumulate codes systematic and allows them to be easily encoded.

Example 2.16 A length-12 rate-1/4 repeat–accumulate code is

$$H = \begin{bmatrix} 1 & 0 & 0 & 1 & 0 & 0 & 0 & 0 & 0 & 0 & 0 & 0 \\ 1 & 0 & 0 & 1 & 1 & 0 & 0 & 0 & 0 & 0 & 0 & 0 \\ 0 & 1 & 0 & 0 & 1 & 1 & 0 & 0 & 0 & 0 & 0 & 0 \\ 0 & 0 & 1 & 0 & 0 & 1 & 1 & 0 & 0 & 0 & 0 & 0 \\ 0 & 0 & 1 & 0 & 0 & 0 & 1 & 1 & 0 & 0 & 0 & 0 \\ 0 & 1 & 0 & 0 & 0 & 0 & 0 & 1 & 1 & 0 & 0 & 0 \\ 1 & 0 & 0 & 0 & 0 & 0 & 0 & 0 & 1 & 1 & 0 & 0 \\ 0 & 1 & 0 & 0 & 0 & 0 & 0 & 0 & 0 & 1 & 1 & 0 \\ 0 & 0 & 1 & 0 & 0 & 0 & 0 & 0 & 0 & 0 & 1 & 1 \end{bmatrix}.$$

The first three columns of H correspond to the message bits. The first parity bit (the fourth column of H) can be encoded as $c_4 = c_1$, the second as $c_5 = c_4 \oplus c_1$ and the next as $c_6 = c_5 \oplus c_2$ and so on. In this way the parity bits can be computed one at a time using only the message bits and the previously calculated parity bit.

We will consider repeat–accumulate codes in more detail in Chapter 6.

2.4 Decoding

Suppose that a codeword has been sent down a binary symmetric channel and one or more of the codeword bits may have been flipped. The task of the decoder is to detect any flipped bits and, if possible, to correct them.

Every codeword in the code must satisfy (2.4), and so any received word that does not satisfy this equation must have errors. The parity-check matrix H thus considerably reduces the complexity of detecting errors.

Example 2.17 The codeword $\mathbf{c} = [1\ 0\ 1\ 1\ 1\ 0]$ from the code in Example 2.3 was sent through a binary symmetric channel and the vector $\mathbf{y} = [1\ 0\ 1\ 0\ 1\ 0]$

was received. Substitution into the left-hand side of (2.4) gives

$$Hy^T = \begin{bmatrix} 1 & 1 & 0 & 1 & 0 & 0 \\ 1 & 1 & 1 & 0 & 0 & 1 \\ 0 & 1 & 1 & 0 & 1 & 0 \end{bmatrix} \begin{bmatrix} 1 \\ 0 \\ 1 \\ 0 \\ 1 \\ 0 \end{bmatrix} = \begin{bmatrix} 1 \\ 0 \\ 0 \end{bmatrix}. \tag{2.22}$$

The result is non-zero and so the vector $\mathbf{y}$ is not a codeword of this code. We therefore conclude that bit-flipping errors must have occurred during transmission.

The vector

$$\mathbf{s} = H\mathbf{y}^T,$$

is called the syndrome of $\mathbf{y}$. The syndrome indicates which parity-check constraints are not satisfied by $\mathbf{y}$.

Example 2.18 The result of (2.22), i.e. the syndrome $\mathbf{s}$, indicates that the first parity-check equation represented in H is not satisfied by $\mathbf{y}$. Since this parity-check equation involves the first, second and fourth codeword bits we can conclude that at least one of these three bits has been inverted by the channel.

As we saw in Chapter 1, error correction by direct comparison of the received vector with every other codeword in the code is guaranteed to return the most likely codeword. However, such an exhaustive search is feasible only when the number of message bits K is small. For codes with K in the thousands it becomes far too computationally expensive to compare the received vector directly with every one of the 2^K codewords in the code. Numerous ingenious solutions have been proposed to make this task less complex; they include choosing algebraic codes, and exploiting their structure, so as to speed up the decoding or, as for LDPC codes, devising decoding methods which are not maximum likelihood but which can perform very well with a much reduced complexity.

The class of algorithms used to decode LDPC codes are collectively termed *message-passing* algorithms, since their operation can be explained by the passing of messages along the edges of a Tanner graph. Each Tanner graph node works in isolation, having access only to the messages on the edges connected to it. Message-passing algorithms are a type of *iterative decoding* algorithm where the messages pass back and forward between the bit and check nodes iteratively until a result is achieved (or the process is halted); they are named for the type of message passed or for the type of operation performed at the nodes.

In some algorithms, such as *bit-flipping* decoding, the messages are binary and in others, such as *belief-propagation* or *sum–product* decoding, the messages are probabilities (or their log likelihood ratios) that represent a level of belief about the value of the codeword bits.

2.4.1 Message passing on the binary erasure channel

On a binary erasure channel (BEC) a transmitted bit is either received correctly or completely erased with some probability ε. Since the bits that are received are always completely correct the task of the decoder is to determine the value of the unknown bits.

If there exists a parity-check equation that includes only one erased bit the correct value for the erased bit can be determined by choosing the value that satisfies the requirement of even parity.

Example 2.19 The code in Example 2.3 includes the parity-check equation

$$c_1 \oplus c_2 \oplus c_4 = 0.$$

If the value of bit c_1 is known to be 0 and the value of bit c_2 is known to be 1 then the value of bit c_4 must be 1 if c_1, c_2 and c_4 are to be part of a valid codeword for this code.

In the message-passing decoder, each check node can determine the value of an erased bit if it is the only erased bit in its parity-check equation. The messages passed along the Tanner graph edges are straightforward: a bit node sends the same outgoing message to each of its connected check nodes. This message, labeled M_i for the ith bit node, declares the value of the bit as 1 or 0 if the value is known or e if it is erased. If a check node receives only one e message, it can calculate the value of the unknown bit by choosing the value that satisfies parity.

The check nodes send back different messages to each of their connected bit nodes. The message $E_{j,i}$ from the jth check node to the ith bit node declares the value of the ith bit to be 1, 0 or e as determined by the jth check node. If the bit node of an erased bit receives an incoming message that is 1 or 0 the bit node changes its value to the value of the incoming message. In one iteration of message-passing decoding the bit node to check node messages are sent once and the return check node to bit node messages are sent once. This process is repeated until all the bit values are known, or until some maximum number of decoder iterations has passed and the decoder halts.[2]

We use the notation B_j to represent the set of bits in the jth parity-check equation of H. So for the parity-check matrix in Example 2.5 we have

$$B_1 = \{1, 2, 4\}, \quad B_2 = \{2, 3, 5\}, \quad B_3 = \{1, 5, 6\}, \quad B_4 = \{3, 4, 6\}.$$

Similarly, we use the notation A_i to represent the parity-check equations for the ith bit of the code. For the code in Example 2.5 we have

$$A_1 = \{1, 3\}, \quad A_2 = \{1, 2\}, \quad A_3 = \{2, 4\},$$
$$A_4 = \{1, 4\}, \quad A_5 = \{2, 3\}, \quad A_6 = \{3, 4\}.$$

Algorithm 2.1 outlines message-passing decoding on the BEC. The input is the received vector of values from the detector, $\mathbf{y} = [y_1 \ \cdots \ y_N]$; those can be 1, 0 or e. The output is $\hat{\mathbf{c}} = [c_1 \ \cdots \ c_N]$; these values can also be 1, 0 or e. The decoder continues until all the erased bits are corrected or the maximum allowed number of iterations, $I_{\max}$, has been reached.

Example 2.20 The LDPC code from Example 2.5 is used to encode the codeword

$$\mathbf{c} = [0\ 0\ 1\ 0\ 1\ 1].$$

The vector $\mathbf{c}$ is sent through an erasure channel and the vector

$$\mathbf{y} = [0\ 0\ 1\ e\ e\ e]$$

is received. Message-passing decoding is used to recover the erased bits. Figure 2.2 shows graphically the messages passed in the message-passing decoder.

At initialization $M_i = y_i$, so

$$\mathbf{M} = [0\ 0\ 1\ e\ e\ e].$$

For *Step 1* of the algorithm the check node messages are calculated. The first check node is joined to the first, second and fourth bit nodes and so has incoming messages 0, 0 and e. Since this check node has one incoming e message, from the fourth bit node, it can calculate the value of this bit and transmit the result as the outgoing message $E_{1,4}$ on the edge from the first check node to the fourth bit node:

$$\begin{aligned} E_{1,4} &= M_1 \oplus M_2 \\ &= 0 \oplus 0 \\ &= 0. \end{aligned}$$

The second check node is joined to the second, third and fifth bits and so has incoming messages 0, 1 and e. Since this check node has one incoming e

[2] In practice the decoder can also halt once a single iteration has found no erasures to correct as no new single-erasure parity-check equations will have been created for the following iteration.

Algorithm 2.1 Erasure decoding

1: **procedure** DECODE($\mathbf{y}$,$I_{\max}$)
2:
3:     **for** $i = 1 : N$ **do** ▷ Initialization
4:         $M_i = y_i$
5:     **end for**
6:     $l = 0$ ▷ Iteration counter
7:     **repeat**
8:
9:         **for** $j = 1 : m$ **do** ▷ Step 1: Check messages
10:             **for** all $i \in B_j$ **do**
11:                 **if** all messages into check j other than M_i are known **then**
12:                     $E_{j,i} = \sum_{i' \in B_j, i' \neq i} (M_{i'} \mod 2)$
13:                 **else**
14:                     $E_{j,i} = e$
15:                 **end if**
16:             **end for**
17:         **end for**
18:
19:         **for** $i = 1 : N$ **do** ▷ Step 2: Bit messages
20:             **if** $M_i = e$ **then**
21:                 **if** there exists a $j \in A_i$ s.t. $E_{j,i} \neq e$ **then**
22:                     $M_i = E_{j,i}$
23:                 **end if**
24:             **end if**
25:         **end for**
26:
27:         **if** all M_i known or $l = I_{\max}$ **then** ▷ Stopping criteria
28:             Finished
29:         **else**
30:             $l = l + 1$
31:         **end if**
32:     **until** Finished
33: **end procedure**

message, from the fifth bit node, its outgoing message $E_{2,5}$ on this edge will be the value of the fifth codeword bit:

$$\begin{aligned} E_{2,5} &= M_2 \oplus M_3 \\ &= 0 \oplus 1 \\ &= 1. \end{aligned}$$

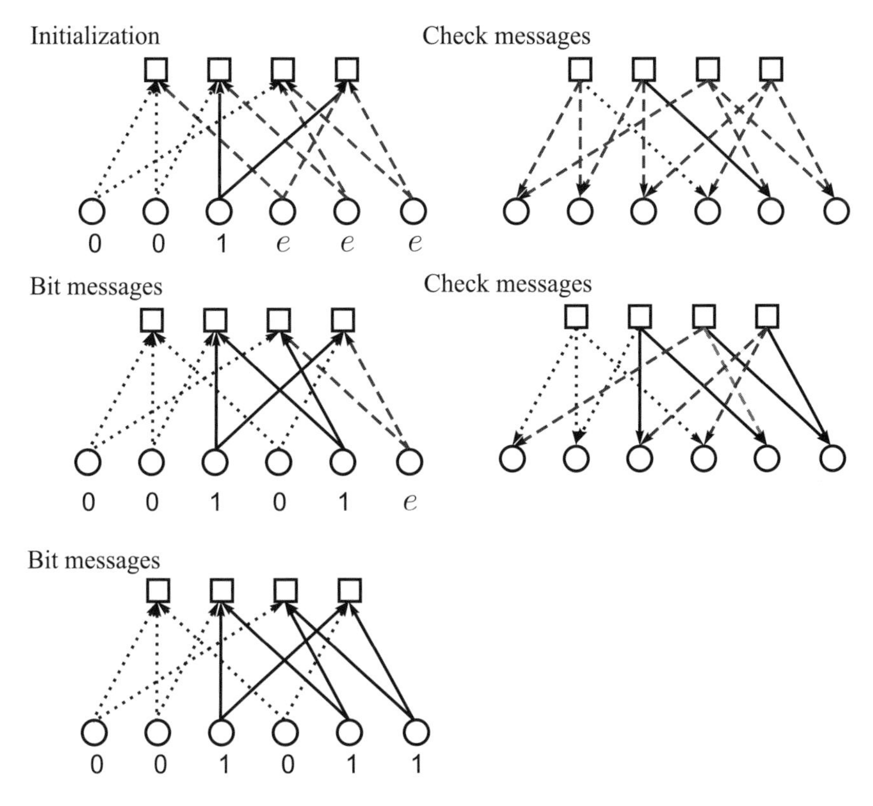

Figure 2.2 Message-passing decoding of the received vector $y = [0\ 0\ 1\ e\ e\ e]$. Each sub-figure indicates the decision made at a step of the decoding algorithm on the basis of the messages from the previous step. The dotted arrows correspond to messages "bit = 0" while the solid arrows correspond to "bit = 1" and the broken arrows correspond to "bit $= e$".

The third check node is joined to the first, fifth and sixth bits and so has incoming messages 0, e and e. Since this check node receives two e messages, it cannot be used to determine the value of any bit. In this case, therefore, the outgoing messages from the check node are all e. Similarly, the fourth check node includes the third, fourth and sixth bits and so receives two e messages, and thus it also cannot be used to determine the value of any bit.

In *Step 2* each bit node that has an unknown value uses its incoming messages to update its value if possible. The fourth bit is unknown and has incoming messages 0 ($E_{1,4}$) and e ($E_{4,4}$) and so it changes its value to 0. The fifth bit is also unknown and has incoming messages 1 ($E_{2,5}$) and e ($E_{3,5}$) and so it changes its value to 1. The sixth bit is also unknown but it has incoming messages e ($E_{3,6}$) and e ($E_{4,6}$) so it cannot change its value. At the end of Step 2 we thus have

$$\mathbf{M} = [0\ 0\ 1\ 0\ 1\ e].$$

There is a remaining unknown bit (the sixth bit) and so the algorithm continues.

Repeating Step 1, the third check node is joined to the first, fifth and sixth bit nodes, and so this check node has one incoming e message, M_6. The outgoing message from this check to the sixth bit node, $E_{3,6}$, is the value of the 6th codeword bit:

$$\begin{aligned} E_{3,6} &= M_1 \oplus M_5 \\ &= 1 \oplus 0 \\ &= 1. \end{aligned}$$

The fourth check node is joined to the third, fourth and sixth bit nodes, and so this check node has one incoming e message, M_6. The outgoing message from this check to the sixth bit node, $E_{4,6}$, is the value of the 6th codeword bit:

$$\begin{aligned} E_{3,6} &= M_3 \oplus M_4 \\ &= 0 \oplus 1 \\ &= 1. \end{aligned}$$

In the repeat of Step 2 the sixth bit is unknown but has incoming messages $E_{3,6}$ and $E_{4,6}$ with value 1, and so it changes its value to 1. Note that, since the received bits from the BEC are always correct, the messages from the check nodes will always agree. (In the bit-flipping algorithm we will see a strategy that can be adopted when this is not the case.) This time, at the test there are no unknown codeword bits and so the algorithm halts and returns

$$\hat{\mathbf{c}} = \mathbf{M} = [0\ 0\ 1\ 0\ 1\ 1]$$

as the decoded codeword. The received vector has therefore been correctly determined although half the codeword bits were erased.

This example also demonstrates how erasure decoding can be used to encode an LDPC code by treating the parity bits as erased bits.

Since the received bits in an erasure channel are either correct or unknown (no errors are introduced by the channel) the messages passed between nodes are always either the correct bit values or e. When the channel introduces errors into the received word, as in the binary symmetric or BI-AWGN channels, the messages in message-passing decoding are instead the best guesses of the codeword bit values, based on the current information available to each node.

2.4.2 Bit-flipping decoding

A *bit-flipping* algorithm is the name given to hard decision message-passing algorithms for LDPC codes. A binary (hard) decision about each received bit is made by the detector and this is passed to the decoder. For bit-flipping algorithms the messages passed along the Tanner graph edges are also binary: a bit node sends a message declaring whether it is a 1 or a 0, and each check node sends a message to each connected bit node declaring its decision on the value of that bit (on the basis of the information available to the check node) or a message indicating whether its corresponding parity-check equation is satisfied.

For the bit-flipping algorithm we will describe here, the jth check node determines its decision on the ith bit node by assuming that the ith bit has been erased and choosing the value 1 or 0 that satisfies the jth parity-check equation. (That is, the decision that check node j makes about bit node i is based only on the values of the other bits included in the jth parity-check equation.) The jth check node thus determines a value for the ith bit that is completely independent of the value for the ith bit just received by it (it will, however, use that value when calculating values for the other bits in the parity-check equation). The check node is said to be creating extra, *extrinsic*, information about the ith bit. Note, however, that, unlike for the erasure correction message-passing decoding algorithm, the bit values provided to the check node and thus the bit values calculated by the check node are not necessarily correct.

At the bit node, all the extrinsic information about a bit is compared with the information received from the channel to determine the most likely bit value. If the majority of the messages received by a bit node are different from its received value then the bit node changes (flips) its current value. This process is repeated until all the parity-check equations are satisfied or until some maximum number of decoder iterations has passed and the decoder halts.

The bit-flipping decoder immediately stops whenever a valid codeword has been found by a check of whether the parity-check equations are satisfied (i.e. $\hat{\mathbf{c}}H^T = \mathbf{s} = \mathbf{0}$ is a *stopping criterion* for bit-flipping decoding). This is true of all message-passing decoding of LDPC codes and has two important benefits: firstly, additional iterations are avoided once a solution has been found and, secondly, failure to converge to a codeword is always detected.

The bit-flipping algorithm is based on the principle that an incorrect codeword bit will be contained in a large number of parity-check equations with no other incorrect bits, each of which will be able to calculate the correct value for that bit. The sparseness of H helps spread out the bits into checks so that two parity-check equations are unlikely to contain the same set of codeword bits. In Example 2.22 below we will show the detrimental effect that occurs if parity-check equations overlap.

Algorithm 2.2 presents the bit-flipping algorithm. The inputs are the received hard decision $\mathbf{y} = [y_1 \cdots y_N]$ and the maximum number of iterations allowed, $I_{\max}$, and the output is $\hat{\mathbf{c}} = [\hat{c_1} \cdots \hat{c_N}]$.

Algorithm 2.2 Bit-flipping decoding

1: **procedure** DECODE($\mathbf{y}$,I_{max})
2:
3:     **for** $i = 1 : N$ **do** ▷ Initialization
4:         $M_i = y_i$
5:     **end for**
6:
7:     $l = 0$ ▷ Iteration count
8:     **repeat**
9:         **for** $j = 1 : m$ **do** ▷ Step 1: Check messages
10:             **for** $i = 1 : N$ **do**
11:                 $E_{j,i} = \sum_{i' \in B_j, i' \neq i} (M_{i'} \mod 2)$
12:             **end for**
13:         **end for**
14:
15:         **for** $i = 1 : N$ **do** ▷ Step 2: Bit messages
16:             **if** the majority of messages $E_{j,i}$ disagree with y_i **then**
17:                 $M_i = (y_i + 1 \mod 2)$
18:             **end if**
19:         **end for**
20:
21:         **for** $j = 1 : m$ **do** ▷ Stopping criteria: are the
22:             $s_j = \sum_{i \in B_j} (M_i \mod 2)$ ▷ parity-check equations satisfied?
23:         **end for**
24:         **if** all $s_j = 0$ or $l = I_{\text{max}}$ **then**
25:             Finished
26:         **else**
27:             $l = l + 1$
28:         **end if**
29:     **until** Finished
30: **end procedure**

Example 2.21 The LDPC code from Example 2.5 is used to encode the codeword

$$\mathbf{c} = [0\ 0\ 1\ 0\ 1\ 1].$$

The vector $\mathbf{c}$ is sent through a BSC with crossover probability $\epsilon = 0.2$, and the received signal is

$$\mathbf{y} = [0\ 1\ 1\ 0\ 1\ 1].$$

The decoding steps are shown graphically in Figure 2.3.

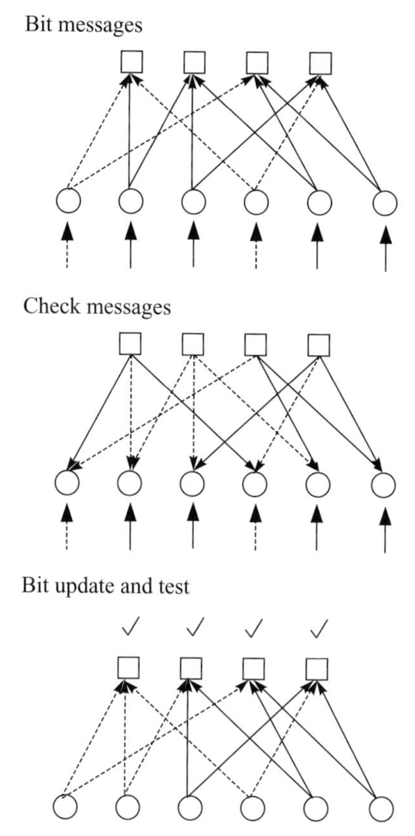

Figure 2.3 Bit-flipping decoding (see Algorithm 2.2) of the received vector $y = [0\ 1\ 1\ 0\ 1\ 1]$. Each sub-figure indicates the decision made at a step of the decoding algorithm on the basis of the messages from the previous step. A tick (✓) indicates that the parity check is satisfied. The broken arrows correspond to the messages "bit = 0" while the solid arrows correspond to "bit = 1".

An initialization $M_i = y_i$, so

$$\mathbf{M} = [0\ 1\ 1\ 0\ 1\ 1].$$

For *Step 1* of Algorithm 2.2, the check node messages are calculated. The first check node is joined to the first, second and fourth bit nodes, $B_1 = \{1, 2, 4\}$ (see Section 2.4.1), and so the messages from the first check node are

$$\begin{aligned} E_{1,1} &= M_2 \oplus M_4 \\ &= 1 \oplus 0 \\ &= 1, \end{aligned}$$

$$\begin{aligned} E_{1,2} &= M_1 \oplus M_4 \\ &= 0 \oplus 0 \\ &= 0, \end{aligned}$$

$$\begin{aligned} E_{1,4} &= M_1 \oplus M_2 \\ &= 0 \oplus 1 \\ &= 1. \end{aligned}$$

The second check includes the second, third and fifth bits, $B_2 = \{2, 3, 5\}$, and so the messages for the second check are

$$\begin{aligned} E_{2,2} &= M_3 \oplus M_5 \\ &= 1 \oplus 1 \\ &= 0, \\ E_{2,3} &= M_2 \oplus M_5 \\ &= 1 \oplus 1 \\ &= 0, \\ E_{2,5} &= M_2 \oplus M_3 \\ &= 1 \oplus 1 \\ &= 0. \end{aligned}$$

Repeating for the remaining check nodes gives

$$\begin{aligned} E_{3,1} &= 0, & E_{3,5} &= 1, & E_{3,6} &= 1, \\ E_{4,3} &= 1, & E_{4,4} &= 0, & E_{4,6} &= 1. \end{aligned}$$

In *Step 2* the first bit has messages 1 and 0 from the first and third checks respectively and 0 from the channel, so it retains its received value. The second bit has messages 0 and 0 from the first and second checks respectively and 1 from the channel. Thus the majority of the messages into the second bit node indicate a value different from the received value and so the second bit node flips its value. None of the remaining bit nodes, however, have enough check node to bit node messages that differ from their current values to change the latter. The new bit node to check node messages are thus given by

$$\mathbf{M} = [0\ 0\ 1\ 0\ 1\ 1].$$

For the test, the parity checks are calculated. For the first check node,

$$\begin{aligned} s_1 &= M_1 \oplus M_2 \oplus M_4 \\ &= 0 \oplus 0 \oplus 0 \\ &= 0. \end{aligned}$$

For the second check node,

$$\begin{aligned} s_2 &= M_2 \oplus M_3 \oplus M_5 \\ &= 0 \oplus 1 \oplus 1 \\ &= 0, \end{aligned}$$

and, similarly, for the third and fourth check nodes

$$\begin{aligned} s_3 &= 0, \\ s_4 &= 0. \end{aligned}$$

There are thus no unsatisfied parity-check equations and so the algorithm halts and returns

$$\hat{\mathbf{c}} = \mathbf{M} = [0\ 0\ 1\ 0\ 1\ 1]$$

as the decoded codeword. The received vector has therefore been correctly decoded without requiring an explicit search over all possible codewords.

The existence of cycles in the Tanner graph of a code reduces the effectiveness of the iterative decoding process. To illustrate the detrimental effect of a 4-cycle we will repeat Example 2.2.1 using a new LDPC code with Tanner graph shown in Figure 2.4. For this Tanner graph there is a 4-cycle between the first two bit nodes and the first two check nodes.

Example 2.22 A valid codeword for the code with the Tanner graph shown in Figure 2.4 is

$$\mathbf{c} = [0\ 0\ 1\ 0\ 0\ 1].$$

This codeword is sent through a binary input additive white Gaussian noise channel and

$$[-1.1\ 1.5\ -0.5\ 1\ 1.8\ -2]$$

is received. The detector makes a hard decision on each codeword bit and returns

$$\mathbf{y} = [1\ 0\ 1\ 0\ 0\ 1].$$

The steps of the bit-flipping algorithm for this received vector are shown in Figure 2.4. The initial bit values are 1, 0, 1, 0, 0 and 1 respectively, and messages are sent to the check nodes indicating these values. *Step 1* reveals that the first and second parity-check equations are not satisfied and so at the test the algorithm continues. In *Step 2*, the majority of the messages to both the first and second bits indicate that the received value is incorrect and so both flip their bit values. When Step 1 is repeated we see that again the first and second parity-check equations are not satisfied. In further iterations the first two bits continue to flip their values together, so that one of them is always incorrect and the algorithm fails to converge. As a result of the 4-cycle, the first two codeword bits are both involved in the same two parity-check equations and so when both these parity-check equations are unsatisfied it is not possible to determine which bit is in error.

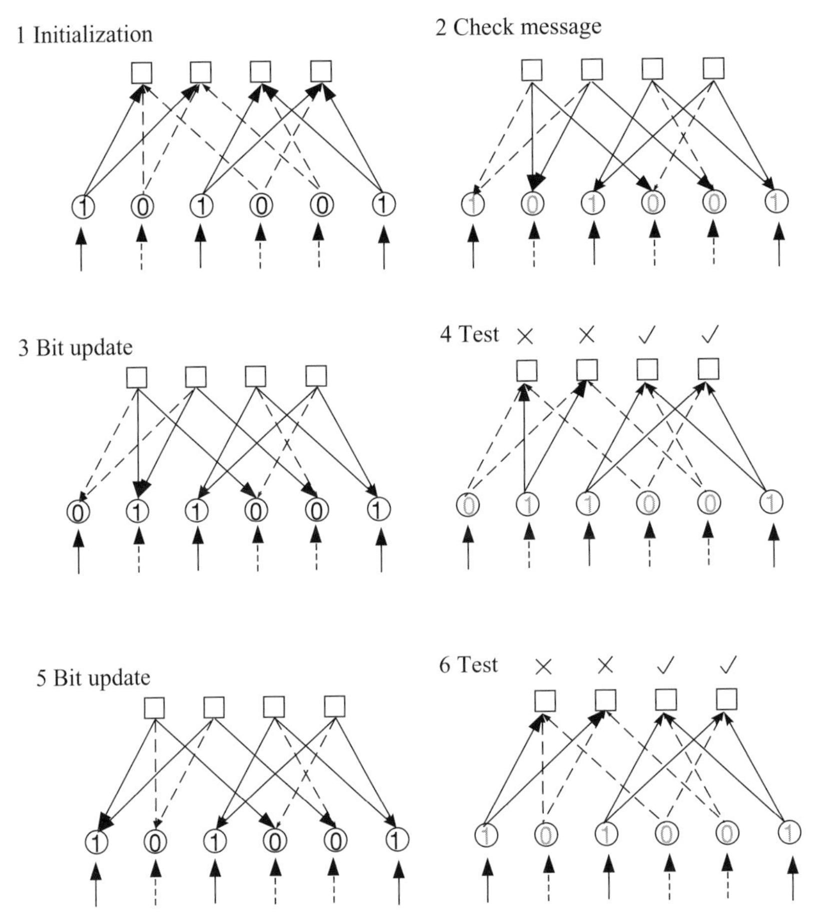

Figure 2.4 Bit-flipping decoding of $y = [1\ 0\ 1\ 0\ 0\ 1]$. Each sub-figure indicates the decision made at each step of the decoding algorithm on the basis of the messages from the previous step. A cross ($\times$) indicates that the parity check is not satisfied while a tick ($\checkmark$) indicates that it is satisfied. The broken arrows correspond to messages "bit $= 0$" while the solid arrows correspond to "bit $= 1$". The number inside a bit node gives the decoded value of the bit at that stage in the algorithm.

The bit-flipping algorithm described above was chosen to extend naturally to the sum–product algorithm discussed below. However, a lower-complexity bit-flipping algorithm will pass the same message on each outgoing edge of a given check node. This message indicates whether the check node is satisfied on the basis of all the incoming bit messages. Thus a bit node is not receiving messages about its own value directly; rather, it can be programmed to flip its bit if some predetermined number of connected check nodes are found to be unsatisfied.

2.4.3 Sum–product decoding

The sum–product algorithm is a soft decision (see the text after (1.9)) message-passing algorithm. It is similar to the bit-flipping algorithm described in the previous section but the messages representing each decision (whether the bit value is 1 or 0) are now probabilities. Whereas bit-flipping decoding accepts an initial hard decision on the received bits as input, the sum–product algorithm is a soft decision algorithm that accepts the probability for each received bit as input. The input *channel*, or *received*, bit probabilities are also called the *a priori* probabilities for the received bits, because they were known in advance before the LDPC decoder was operated.

For the sum–product decoder, the extrinsic information passed between nodes is also given as probabilities rather than hard decisions. The extrinsic message $E_{j,i}$ from check node j to a bit node i to which it is connected is check node j's opinion on the probability that $c_i = 1$, based on the information available to check node j. That is, $E_{j,i}$ gives the probability that $c_i = 1$ will cause parity-check equation j to be satisfied. (Note that $E_{j,i}$ is not defined if bit i is not included in check j since no extrinsic information is passed between nodes i and j in this case.) The probability that a parity-check equation is satisfied if $c_i = 1$ is the probability that an odd number of the bits in that parity-check equation are 1s:

$$P_{j,i}^{\text{ext}} = \frac{1}{2} - \frac{1}{2} \prod_{i' \in B_j, i' \neq i} (1 - 2P_{j,i'}), \tag{2.23}$$

where $P_{j,i'}$ is the current estimate available to check node j of the probability that $c_{i'} = 1$. The probability that the parity-check equation is satisfied if $c_i = 0$ is thus $1 - P_{j,i}^{\text{ext}}$.

For a binary variable x it is easy to find $p(x = 1)$ given $p(x = 0)$, since $p(x = 1) = 1 - p(x = 0)$, and so we need only store one probability value for x. Log likelihood ratios are used to represent the metrics for a binary variable by a single value (1.2):

$$L(x) = \log \frac{p(x = 0)}{p(x = 1)},$$

where by log we mean $\log_e$. The sign of $L(x)$ provides a hard decision on x and the magnitude $|L(x)|$ is the reliability of this decision. Translating from log likelihood ratios back to probabilities, (1.3), (1.4),

$$p(x = 1) = \frac{e^{-L(x)}}{1 + e^{-L(x)}},$$

$$p(x = 0) = \frac{e^{L(x)}}{1 + e^{L(x)}}.$$

The benefit of the logarithmic representation of probabilities is that, when probabilities need to be multiplied, log likelihood ratios need only be added; this can reduce the complexity of the sum–product decoder.

Expressed as a log likelihood ratio, the extrinsic information from check node j to bit node i is

$$E_{j,i} = L(P^{\text{ext}}_{j,i}) = \log \frac{1 - P^{\text{ext}}_{j,i}}{P^{\text{ext}}_{j,i}}. \tag{2.24}$$

Now substituting (2.23) into (2.24) gives

$$\begin{aligned} E_{j,i} &= \log \frac{\frac{1}{2} + \frac{1}{2}\prod_{i' \in B_j, i' \neq i}(1 - 2P_{j,i'})}{\frac{1}{2} - \frac{1}{2}\prod_{i' \in B_j, i' \neq i}(1 - 2P_{j,i'})} \\ &= \log \frac{1 + \prod_{i' \in B_j, i' \neq i}\left(1 - 2\dfrac{e^{-M_{j,i'}}}{1 + e^{-M_{j,i'}}}\right)}{1 - \prod_{i' \in B_j, i' \neq i}\left(1 - 2\dfrac{e^{-M_{j,i'}}}{1 + e^{-M_{j,i'}}}\right)} \\ &= \log \frac{1 + \prod_{i' \in B_j, i' \neq i}\dfrac{1 - e^{-M_{j,i'}}}{1 + e^{-M_{j,i'}}}}{1 - \prod_{i' \in B_j, i' \neq i}\dfrac{1 - e^{-M_{j,i'}}}{1 + e^{-M_{j,i'}}}}, \end{aligned} \tag{2.25}$$

where

$$M_{j,i'} \triangleq L(P_{j,i'}) = \log \frac{1 - P_{j,i'}}{P_{j,i'}}.$$

Using the relationship

$$\tanh \frac{1}{2} \log \left(\frac{1-p}{p}\right) = 1 - 2p$$

gives

$$E_{j,i} = \log \frac{1 + \prod_{i' \in B_j, i' \neq i} \tanh(M_{j,i'}/2)}{1 - \prod_{i' \in B_j,\, i' \neq i} \tanh(M_{j,i'}/2)}. \tag{2.26}$$

Alternatively, using the relationship

$$2 \tanh^{-1} p = \log \frac{1+p}{1-p},$$

(2.26) can be equivalently written as

$$E_{j,i} = 2 \tanh^{-1} \prod_{i' \in B_j, i' \neq i} \tanh(M_{j,i'}/2). \tag{2.27}$$

Each bit node has access to the input LLR, R_i, and to the LLRs from every connected check node. The total LLR of the ith bit is the sum of these LLRs:

$$L_i = L(P_i) = R_i + \sum_{j \in A_i} E_{j,i}. \tag{2.28}$$

However, the messages sent from the bit nodes to the check nodes, $M_{j,i}$, are not the complete LLR value for each bit. To avoid sending back to each check node information that it already has, the message sent from the ith bit node to the jth check node is the sum in (2.28) without the component $E_{j,i}$ just received from the jth check node:

$$M_{j,i} = \sum_{j' \in A_i,\ j' \neq j} E_{j',i} + R_i. \tag{2.29}$$

The aim of sum–product decoding is (a) to compute the a posteriori probability (APP) for each codeword bit, $p_i = p\{c_i = 1|\mathbf{s} = \mathbf{0}\}$, which is the probability that the ith codeword bit is 1-conditional on the event $\mathbf{s} = \mathbf{0}$ (i.e. that all parity-check constraints are satisfied), and (b) to select the decoded value for each bit as the value with the maximum APP probability. The sum–product algorithm iteratively computes an approximation of the MAP value for each code bit. However, the a posteriori probabilities returned by the sum–product decoder are only exact MAP probabilities if the Tanner graph is cycle-free. Briefly, the extrinsic information obtained from a parity-check constraint in the first iteration is independent of the a priori probability information for that bit (it does of course depend on the a priori probabilities of the other codeword bits). The extrinsic information provided to bit node i in subsequent iterations remains independent of the original a priori probability for c_i until this probability is returned back to bit node i via a cycle in the Tanner graph. The correlation of the extrinsic information with the original a priori bit probability is what prevents the resulting probabilities from being exact.

The sum–product algorithm is shown in Algorithm 2.3. The inputs are the log likelihood ratios for the a priori message probabilities from each channel,

$$R_i = \log \frac{p(\mathbf{c}_i = 0|y_i)}{p(\mathbf{c}_i = 1|y_i)},$$

the parity-check matrix H and the maximum number of allowed iterations, $I_{\max}$. The algorithm outputs the estimated a posteriori bit probabilities of the received bits as log likelihood ratios.

The sum–product decoder immediately stops whenever a valid codeword has been found by a checking of whether the parity-check equations are satisfied (thus $\hat{\mathbf{c}}H^T = \mathbf{0}$ is also a *stopping criterion* for sum–product decoding).

Algorithm 2.3 Sum–product decoding

```
1: procedure DECODE(R, I_max)
2:
3:     I = 0                                              ▷ Initialization
4:     for i = 1 : N do
5:         for j = 1 : m do
6:             M_{j,i} = R_i
7:         end for
8:     end for
9:
10:    repeat
11:        for j = 1 : m do                               ▷ Step 1: Check messages
12:            for i ∈ B_j do
13:                E_{j,i} = log (1 + ∏_{i'∈B_j, i'≠i} tanh(M_{j,i'}/2)) / (1 − ∏_{i'∈B_j, i'≠i} tanh(M_{j,i'}/2))
14:            end for
15:        end for
16:
17:        for i = 1 : N do                               ▷ Test
18:            L_i = ∑_{j∈A_i} E_{j,i} + R_i
19:            ĉ_i = { 1, L_i ≤ 0,
                       0, L_i > 0.
20:        end for
21:        if I = I_max or H ĉ^T = 0 then
22:            Finished
23:        else
24:            for i = 1 : N do                           ▷ Step 2: Bit messages
25:                for j ∈ A_i do
26:                    M_{j,i} = ∑_{j'∈A_i, j'≠j} E_{j',i} + R_i
27:                end for
28:            end for
29:            I = I + 1
30:        end if
31:    until Finished
32: end procedure
```

Line 1: $\text{DECODE}(\mathbf{R}, I_{\max})$

Line 13: $$E_{j,i} = \log \frac{1 + \prod_{i' \in B_j, i' \neq i} \tanh(M_{j,i'}/2)}{1 - \prod_{i' \in B_j,\, i' \neq i} \tanh(M_{j,i'}/2)}$$

Line 18: $L_i = \sum_{j \in A_i} E_{j,i} + R_i$

Line 19: $$\hat{c}_i = \begin{cases} 1, & L_i \leq 0, \\ 0, & L_i > 0. \end{cases}$$

Line 21: if $I = I_{\max}$ or $H\hat{\mathbf{c}}^T = 0$ then

Line 26: $M_{j,i} = \sum_{j' \in A_i,\, j' \neq j} E_{j',i} + R_i$

Example 2.23 The LDPC code from Example 2.5 is used to encode the codeword

$$\mathbf{c} = [0\ 0\ 1\ 0\ 1\ 1].$$

The vector $\mathbf{c}$ is sent through a BSC with crossover probability $\epsilon = 0.2$, and the received signal is

$$\mathbf{y} = [1\ 0\ 1\ 0\ 1\ 1].$$

Since the channel is binary symmetric the probability that 0 was sent if 1 is received is the probability ϵ that a crossover occurred, while the probability that 1 was sent if 1 is received is the probability $1 - \epsilon$ that no crossover has occurred. Similarly, the probability that 1 was sent if 0 is received is the probability that a crossover occurred, while the probability that 0 was sent if 0 is received is the probability that no crossover occurred. Thus the a priori LLRs for the BSC are, see Example 1.2,

$$R_i = \begin{cases} \log \dfrac{\epsilon}{1-\epsilon} & \text{if } y_i = 1, \\ \log \dfrac{1-\epsilon}{\epsilon} & \text{if } y_i = 0. \end{cases}$$

For this channel we have

$$\log \frac{\epsilon}{1-\epsilon} = \log \frac{0.2}{0.8} = -1.3863,$$
$$\log \frac{1-\epsilon}{\epsilon} = \log \frac{0.8}{0.2} = 1.3863,$$

and so

$$\mathbf{R} = [-1.3863\ 1.3863\ -1.3863\ 1.3863\ -1.3863\ -1.3863].$$

To begin decoding we set the maximum number of iterations to 3 and feed in H and $\mathbf{R}$. At initialization,

$$M_{j,i} = R_i.$$

The first bit is included in the first and third checks and so $M_{1,1}$ and $M_{3,1}$ are initialized to R_1:

$$M_{1,1} = R_1 = -1.3863 \quad \text{and} \quad M_{3,1} = R_1 = -1.3863.$$

Repeating this for the remaining bits gives:

$$\begin{array}{lll} \text{for } i = 2, & M_{1,2} = R_2 = 1.3863, & M_{2,2} = R_2 = 1.3863; \\ \text{for } i = 3, & M_{2,3} = R_3 = -1.3863, & M_{4,3} = R_3 = -1.3863; \\ \text{for } i = 4, & M_{1,4} = R_4 = 1.3863, & M_{4,4} = R_4 = 1.3863; \\ \text{for } i = 5, & M_{2,5} = R_5 = -1.3863, & M_{3,5} = R_5 = -1.3863; \\ \text{for } i = 6, & M_{3,6} = R_6 = -1.3863, & M_{4,6} = R_6 = -1.3863. \end{array}$$

Now the extrinsic probabilities are calculated for the check-to-bit messages. The first parity check includes the first, second and fourth bits and so the extrinsic probability from the first check node to the first bit node depends on the probabilities of the second and fourth bits:

$$\begin{aligned} E_{1,1} &= \log \frac{1 + \tanh(M_{1,2}/2)\tanh(M_{1,4}/2)}{1 - \tanh(M_{1,2}/2)\tanh(M_{1,4}/2)} \\ &= \log \frac{1 + \tanh(1.3863/2)\tanh(1.3863/2)}{1 - \tanh(1.3863/2)\tanh(1.3863/2)} \\ &= \log \frac{1 + 0.6 \times 0.6}{1 - 0.6 \times 0.6} = 0.7538. \end{aligned}$$

Similarly, the extrinsic probability from the first check node to the second bit node depends on the probabilities of the first and fourth bits:

$$\begin{aligned} E_{1,2} &= \log \frac{1 + \tanh(M_{1,1}/2)\tanh(M_{1,4}/2)}{1 - \tanh(M_{1,1}/2)\tanh(M_{1,4}/2)} \\ &= \log \frac{1 + (-0.6 \times 0.6)}{1 - (-0.6 \times 0.6)} = -0.7538, \end{aligned}$$

and the extrinsic probability from the first check node to the 4th bit node depends on the LLRs sent from the first and second bit nodes to the first check node:

$$\begin{aligned} E_{1,4} &= \log \frac{1 + \tanh(M_{1,1}/2)\tanh(M_{1,2}/2)}{1 - \tanh(M_{1,1}/2)\tanh(M_{1,2}/2)} \\ &= \log \frac{1 + (-0.6 \times 0.6)}{1 - (-0.6 \times 0.6)} = -0.7538. \end{aligned}$$

Next, the second check node connects to the second, third and fifth bit nodes and so the extrinsic LLRs are

$$\begin{aligned} E_{2,2} &= \log \frac{1 + \tanh(M_{2,3}/2)\tanh(M_{2,5}/2)}{1 - \tanh(M_{2,3}/2)\tanh(M_{2,5}/2)} \\ &= \log \frac{1 + (-0.6) \times (-0.6)}{1 - (-0.6) \times (-0.6)} = 0.7538, \\ E_{2,3} &= \log \frac{1 + \tanh(M_{2,2}/2)\tanh(M_{2,5}/2)}{1 - \tanh(M_{2,2}/2)\tanh(M_{2,5}/2)} \\ &= \log \frac{1 + 0.6 \times (-0.6)}{1 - 0.6 \times (-0.6)} = -0.7538, \\ E_{2,5} &= \log \frac{1 + \tanh(M_{2,2}/2)\tanh(M_{2,3}/2)}{1 - \tanh(M_{2,2}/2)\tanh(M_{2,3}/2)} \\ &= \log \frac{1 + 0.6 \times (-0.6)}{1 - 0.6 \times (-0.6)} = -0.7538. \end{aligned}$$

Repeating for all checks gives the extrinsic LLRs:

$$E = \begin{bmatrix} 0.7538 & -0.7538 & . & -0.7538 & . & . \\ . & 0.7538 & -0.7538 & . & -0.7538 & . \\ 0.7538 & . & . & . & 0.7538 & 0.7538 \\ . & . & -0.7538 & 0.7538 & . & -0.7538 \end{bmatrix}.$$

To save space the extrinsic LLRs are given in matrix form, where the (j, i)th entry of E holds $E_{j,i}$. A dot entry indicates that an LLR does not exist for that i and j.

To check for a valid codeword, we calculate the estimated a posteriori probabilities for each bit, make a hard decision and check the syndrome $\mathbf{s}$. The first bit has extrinsic LLRs from the first and third checks and an intrinsic LLR from the channel. The total LLR is their sum:

$$L_1 = R_1 + E_{1,1} + E_{3,1} = -1.3863 + 0.7538 + 0.7538 = 0.1213.$$

Thus even though the LLR from the channel is negative, indicating that the first bit is a 1, both the extrinsic LLRs are positive, indicating that the bit is 0. The extrinsic LLRs are large enough to make the total LLR positive and so the decision on the first bit has effectively been changed. Repeating for the second to sixth bits gives:

$$\begin{aligned} L_2 &= R_2 + E_{1,2} + E_{2,2} = 1.3863, \\ L_3 &= R_3 + E_{2,3} + E_{4,3} = -2.8938, \\ L_4 &= R_4 + E_{1,4} + E_{4,4} = 1.3863, \\ L_5 &= R_5 + E_{2,5} + E_{3,5} = -1.3863, \\ L_6 &= R_6 + E_{3,6} + E_{4,6} = -1.3863. \end{aligned}$$

The hard decision on the received bits is simply given by the signs of the LLRs; we obtain

$$\hat{\mathbf{c}} = [0\ 0\ 1\ 0\ 1\ 1].$$

To check whether $\hat{\mathbf{c}}$ is a valid codeword, consider

$$\mathbf{s} = \hat{\mathbf{c}}H' = [0\ 0\ 1\ 0\ 1\ 1] \begin{bmatrix} 1 & 0 & 1 & 0 \\ 1 & 1 & 0 & 0 \\ 0 & 1 & 0 & 1 \\ 1 & 0 & 0 & 1 \\ 0 & 1 & 1 & 0 \\ 0 & 0 & 1 & 1 \end{bmatrix} = [0\ 0\ 0\ 0].$$

Since $\mathbf{s}$ is zero, $\hat{\mathbf{c}}$ is a valid codeword and the decoding stops, returning $\hat{\mathbf{c}}$ as the decoded word.

The sum–product algorithm can be modified to reduce the implementation complexity of the decoder. This can be done by altering (2.27),

$$E_{j,i} = 2\tanh^{-1} \textstyle\prod_{i' \in B_j, i' \neq i} \tanh(M_{j,i'}/2),$$

in such a way as to replace the product term by a sum. For simplicity we will write

$$\prod_{i'} \triangleq \prod_{i' \in B_j, i' \neq i}$$

in the remainder of this section.

Firstly, $M_{j,i'}$ can be factored as follows:

$$M_{j,i'} = \alpha_{j,i'} \beta_{j,i'},$$

where

$$\begin{aligned} \alpha_{j,i'} &= \operatorname{sign} M_{j,i'}, \\ \beta_{j,i'} &= |M_{j,i'}|. \end{aligned} \tag{2.30}$$

Using this notation we have that

$$\prod_{i'} \tanh(M_{j,i'}/2) = \prod_{i'} \alpha_{j,i'} \prod_{i'} \tanh(\beta_{j,i'}/2).$$

Then (2.27) becomes

$$\begin{aligned} E_{j,i} &= 2\tanh^{-1} \left(\textstyle\prod_{i'} \alpha_{j,i'} \prod_{i'} \tanh(\beta_{j,i'}/2)\right) \\ &= \left(\textstyle\prod_{i'} \alpha_{j,i'}\right) 2\tanh^{-1} \textstyle\prod_{i'} \tanh(\beta_{j,i'}/2). \end{aligned} \tag{2.31}$$

Equation (2.31) can now be rearranged to replace the product by a sum:

$$\begin{aligned} E_{j,i} &= \left(\textstyle\prod_{i'} \alpha_{j,i'}\right) 2\tanh^{-1} \log^{-1} \log \textstyle\prod_{i'} \tanh(\beta_{j,i'}/2) \\ &= \left(\textstyle\prod_{i'} \alpha_{j,i'}\right) 2\tanh^{-1} \log^{-1} \textstyle\sum_{i'} \log \tanh(\beta_{j,i'}/2). \end{aligned} \tag{2.32}$$

Next, we define

$$\phi(x) = -\log \tanh \frac{x}{2} = \log \frac{e^x + 1}{e^x - 1}$$

and note that since

$$\phi(\phi(x)) = \log \frac{e^{\phi(x)} + 1}{e^{\phi(x)} - 1} = x$$

we have $\phi^{-1} = \phi$. Finally, (2.32) becomes

$$E_{j,i} = \left(\textstyle\prod_{i'} \alpha_{j,i'}\right) \phi\left(\textstyle\sum_{i'} \phi(\beta_{j,i'})\right). \tag{2.33}$$

The product of the signs can be calculated by using modulo-2 addition of the hard decisions on each $M_{j,i'}$, while the function ϕ can be implemented easily using a lookup table.

Alternatively, the *min–sum* algorithm simplifies the calculation of (2.31) even further by recognizing that the term corresponding to the smallest $M_{j,i'}$ dominates the product term and so the product can be approximated by a minimum:

$$E_{j,i} \approx \prod_{i'} \operatorname{sign} M_{j,i'} \underbrace{\min}_{i'} \left| M_{j,i'} \right|. \tag{2.34}$$

Again, the product of the signs can be calculated by using modulo-2 addition of the hard decisions on each $M_{j,i'}$, and so the resulting min–sum algorithm thus requires only additions and the calculation of minimums.

Example 2.24 In this example we consider again the performance of sum–product decoding, but this time we will look at the average performance of a $(3,6)$-regular code over a number of signal-to-noise ratios on a BI-AWGN channel using sum–product decoding with a maximum of 100 iterations. Using a randomly generated length $N = 200$ $(K = 100)$ code, and varying the signal-to-noise ratio from 1 to 5 dB, the average performance of the decoder is calculated by simulating enough codewords that the decoder fails at least 500 times. The average bit error rates of the decoder are plotted in Figure 2.5. Next we repeat this process for longer code lengths, up to $N = 10\,000$. The average bit error and word error rates of the decoder for these codeword lengths are also plotted in Figure 2.5. We can see that increasing the codeword length has an enormous impact on the error correction performance of the code.

To investigate the role of the maximum number of allowed iterations, the performance of the $(3,6)$-regular code with $N = 1000$ $(K = 500)$ was calculated for varying values of $I_{\max}$ and the results plotted in Figure 2.6. We see that a dramatic improvement in performance can be achieved by increasing $I_{\max}$ from 5 to 10, but performance returns diminish as $I_{\max}$ is increased much beyond this.

Figures 2.5 and 2.6 show the characteristic BER curve for LDPC codes, which can be divided into three regions. The first is the *non-convergence region*, where the BER is high and remains relatively flat as the SNR is increased. The second region shows a rapid decrease in BER for small increments in SNR and is commonly called the *waterfall region*. Lastly, in the *error floor region* the slope of the curve flattens out and the BER reductions are small as the SNR increases.

Figure 2.5 shows two trends common to LDPC codes in general. Firstly, increasing the codeword length N improves the code BER and WER performances by increasing the slope of the waterfall region. (The value to

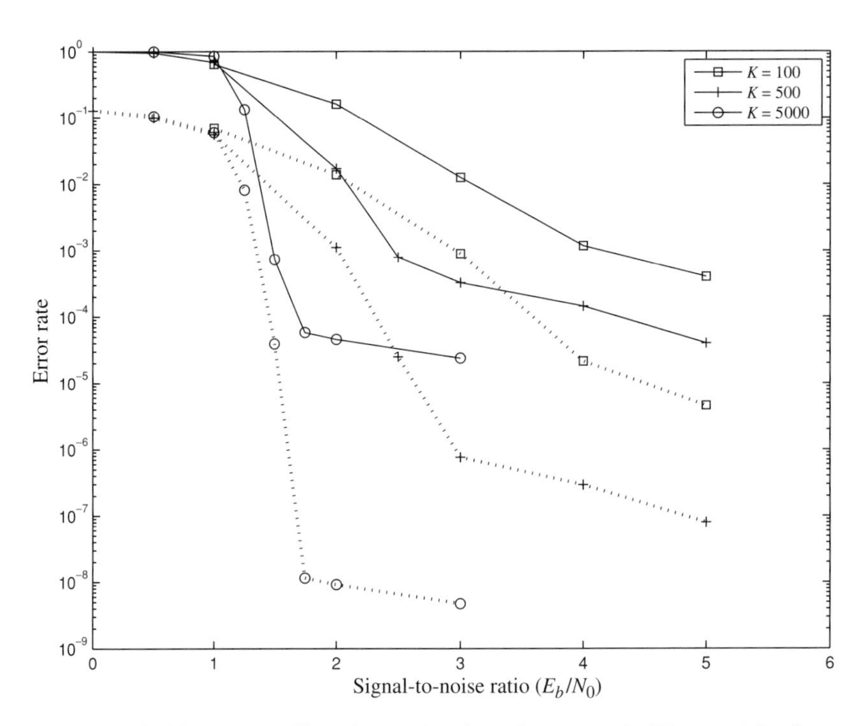

Figure 2.5 The bit error rate (dotted curves) and word error rate (solid curves) for the sum–product decoding of (3, 6)-regular LDPC codes versus the signal-to-noise ratio, for message lengths $K = 100$ ($I_{\max} = 100$), $K = 500$ ($I_{\max} = 100$) and $K = 5000$ ($I_{\max} = 1000$).

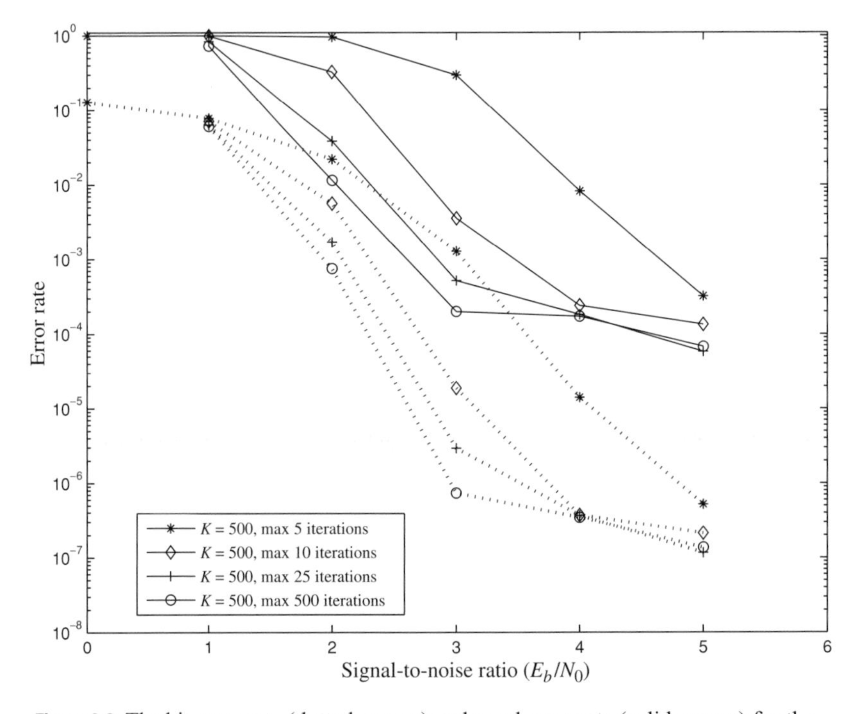

Figure 2.6 The bit error rate (dotted curves) and word error rate (solid curves) for the sum–product decoding of (3, 6)-regular LDPC codes versus the signal-to-noise ratio and for varying values of maximum number of iterations.

which the SNR of the waterfall region can be decreased simply by increasing N reaches a limit, the *threshold* SNR, which depends on the degree distribution of the code; this will be discussed in Chapter 7.) Secondly, increasing the number of decoder iterations improves decoder performance. This improvement reaches a limit, the error floor, after which extra iterations provide little benefit. The error floor region is dominated by the codeword weight distribution, which is determined by both the node degrees and the structure of the parity-check matrix. We will consider each of these properties, and their impact on the decoding performance, in the following chapter. The error floor performance of the sum–product decoder, regardless of the number of iterations used, can be bounded by considering the performance of the ML decoder for that code. This is discussed in Chapter 8.

2.5 Bibliographic notes

Low-density parity-check codes, and iterative message-passing decoding algorithms for them, were first introduced by Gallager in his 1962 thesis [10]. In the early 1960s, however, limited computing resources prevented Gallager from demonstrating the capabilities of message-passing decoders for code lengths over 500 bits, and for over 30 years his work was ignored by all but a handful of researchers, most notably Tanner [11]. It was only rediscovered by several researchers, e.g. [12], in the wake of turbo decoding [13], which itself has subsequently been recognized as an instance of the sum–product algorithm. In his work, Gallager used a graphical representation of the bit and parity-check sets of regular LDPC codes to describe the application of iterative decoding. The systematic study of codes on graphs, however, is largely due to Tanner who, in 1981, formalized the use of bipartite graphs for describing families of codes [11].

New generalizations of Gallager's LDPC codes to irregular codes by a number of researchers including Luby, Mitzenmacher, Shokrollahi, Spielman and Stemann [14, 15] produced codes with performances that meet the capacity of the BEC and approach the capacity of the BI-AWGN channel to within a small fraction of a decibel from capacity [16]. To encode them, Richardson and Urbanke soon presented an algorithm for finding a generator matrix for block codes defined by a sparse parity-check matrix [17].

Low-density parity-check codes have been adopted for 10 Gb/s ethernet (IEEE 803.3an), WiFi (IEEE 802.11n) and WiMAX (IEEE 802.16e) wireless LAN standards and are being considered for a range of application areas, from optical networks to digital storage. A very early adoption of LDPC codes by Digital Fountain used LDPC codes designed for erasure correction channels to improve the efficiency and reliability of internet content delivery.

We will consider LDPC codes in more detail in the chapters that follow. In Chapter 3 we discuss the properties of LDPC codes and present some common construction methods. In Chapter 7 we focus on design techniques for finding degree distributions for LDPC codes that approach Shannon's capacity limits, and in Chapter 8 we consider the finite-length analysis of LDPC codes. A good online resource for LDPC codes in general, and C code programs to implement them, is [18].

2.6 Exercises

2.1 For the (7, 4, 3) Hamming code with parity-check matrix

$$H = \begin{bmatrix} 1 & 1 & 1 & 0 & 1 & 0 & 0 \\ 1 & 1 & 0 & 1 & 0 & 1 & 0 \\ 1 & 0 & 1 & 1 & 0 & 0 & 1 \end{bmatrix},$$

(a) give the code rate,
(b) find a generator matrix for this code,
(c) give the codeword produced by the message $\mathbf{u} = [1\ 0\ 1\ 0]$, and
(d) determine whether [0 1 0 1 1 0 1] is a valid codeword for this code.

2.2 For the block code with generator matrix

$$G = \begin{bmatrix} 1 & 0 & 0 & 1 & 0 & 1 & 1 \\ 0 & 1 & 0 & 0 & 1 & 0 & 1 \\ 0 & 0 & 1 & 1 & 1 & 1 & 0 \end{bmatrix},$$

(a) what is the code rate,
(b) how many codewords does this code have?
(c) determine the codeword produced by the message $\mathbf{u} = [1\ 0\ 1]$,
(d) find a parity-check matrix for the code,
(e) determine whether [1 0 1 0 0 0 1] is a valid codeword for this code, and
(f) find the minimum distance of the code.

2.3 Is

$$G = \begin{bmatrix} 1 & 0 & 0 & 0 & 1 & 0 & 1 \\ 0 & 1 & 0 & 0 & 1 & 1 & 0 \\ 0 & 0 & 1 & 0 & 0 & 1 & 1 \\ 0 & 0 & 0 & 1 & 1 & 0 & 1 \end{bmatrix}$$

a valid generator matrix for the (7, 4, 3) Hamming code in Example 2.1?

2.4 For the Hamming code parity-check matrix in Example 2.1,
(a) give the degree distribution of H,
(b) draw the Tanner graph for H and determine its girth.

2.5 Verify that the matrix in Example 2.5 is a valid parity-check matrix for the code in Example 2.3.

2.6 For the LDPC code with parity-check matrix given in Example 3.10:
(a) draw the Tanner graph for H and determine its girth,
(b) find a generator matrix for this code,
(c) give the codeword produced by the all-ones message.

2.7 For the LDPC code with parity-check matrix given in Example 3.11:
(a) draw the Tanner graph for H and determine its girth,
(b) find a generator matrix for this code,
(c) give the codeword produced by the all-ones message.

2.8 For the LDPC code with parity-check matrix given in Example 3.10, determine whether [1 0 1 0 0 0 1 0 1 0 1 1] is a valid codeword for this code.

2.9 For the LDPC code with parity-check matrix given in Example 3.11, determine whether [0 0 0 1 0 1 1 1 1 0 1 0] is a valid codeword for this code.

2.10 Give the sets A_i and B_j for every bit node and check node for the parity-check matrix in Example 2.5.

2.11 Write a Matlab program to implement bit-flipping decoding. Repeat Example 2.21 but now assume that the received vector is
(a) [1 1 0 0 1 0],
(b) [1 1 0 1 1 1].

2.12 Show that the probability that an even number of digits in a length-N codeword are 1s is

$$\frac{1}{2} + \frac{1}{2}\prod_i (1 - 2P_i),$$

where P_i is the probability that the ith bit is a 1. Use this result to derive (2.23).

2.13 Write a program to implement sum–product decoding and repeat Example 2.21 using this type of decoding.

2.14 Which codeword would be returned by ML decoding for Example 2.22?

2.15 Using sum–product decoding, find the decoded codeword for the code in Example 2.22,
(a) assuming a BSC with crossover probability $\epsilon = 0.1$ when

$$\mathbf{y} = [1\ 1\ 1\ 1\ 0\ 1]$$

is received,
(b) for a BI-AWGN channel with signal-to-noise ratio 5 dB when

$$\mathbf{y} = [1.1\ 0.9\ -0.1\ 1.1\ -1.1\ 0.9]$$

is received.

2.16 Repeat Exercise 2.15 using
(a) bit-flipping decoding,
(b) min–sum decoding,
(c) ML decoding.

2.17 Explain the differences in the results for Exercises 2.15 and 2.14.

2.18 Using your program to implement sum–product decoding, determine the bit probabilities for the message corresponding to the received vector

$$\mathbf{y} = [1\ 1\ 1\ 1\ 0\ 1\ 1\ 1\ 1\ 0\ 1\ 0]$$

for the code of Example 3.11. Assume a binary equiprobable source and a BSC with crossover probability $\epsilon = 0.2$.

2.19 Using software, randomly construct a length-100 (3, 6)-regular LDPC code and find a generator matrix for it.
(a) Find the average bit error rate of sum–product decoding on a BI-AWGN channel with signal-to-noise ratios of 1 dB and 3 dB using this code with a maximum of five iterations.
(b) Repeat your simulation using a maximum of 10 and 20 iterations of sum–product decoding.
(c) Repeat your simulation using a length-1000 code and a maximum of 20 iterations of sum–product decoding.

2.20 You are to choose a regular rate-1/2 LDPC code with $w_c \le 5$ to transmit a length-100 message over a BSC. Use BER plots to determine which value of w_c is best for each range of crossover probabilities.

3
Low-density parity-check codes: properties and constructions

3.1 Introduction

The construction of binary low-density parity-check (LDPC) codes simply involves replacing a small number of the values in an all-zeros matrix by 1s in such a way that the rows and columns have the required degree distribution. In many cases, randomly allocating the entries in H will produce a reasonable LDPC code. However, the construction of H can affect the performance of the sum–product decoder, significantly so for some codes, and also the implementation complexity of the code.

While there is no one recipe for a "good" LDPC code, there are a number of principles that inform the code designer. The first obvious decisions are which degree distribution to choose and how to construct the matrix with the chosen degrees, i.e. pseudo-randomly or with some sort of structure. Whichever construction is chosen, the features to consider include the girth of the Tanner graph and the minimum distance of the code.

In this chapter we will discuss those properties of an LDPC code that affect its iterative decoding performance and then present the common construction methods used to produce codes with the preferred properties. Following common practice in the field we will call the selection of the degree distributions for an LDPC code *code design* and the methods to assign the locations in the parity-check matrix for the 1 entries *code construction*.

3.2 LDPC properties

It is important to keep in mind that, when considering the properties of an LDPC code, often we are actually considering the properties of a particular choice of parity-check matrix for that code; a different choice of parity-check matrix for the same code might behave differently with sum–product decoding. This is in contrast with ML decoding, where the probability of incorrect decoding is determined solely by the weight distribution of the codewords, which in turn is independent of which generator or parity-check matrix is chosen to represent the code.

3.2.1 Choosing the degree distribution

Firstly, we seek to determine the role of the column weight in determining the decoding performance of regular LDPC codes and then we will compare regular and irregular codes.

Column weight

Let us consider regular codes and choose the column weight w_c. The row weight w_r is then fixed by the choice of code rate r, since

$$r \approx 1 - w_c/w_r.$$

The expected minimum distance of a randomly chosen block code can increase with the column weight of a sparse H. Thus an LDPC code with a larger column weight can give a better decoding performance as the noise decreases (and the code's performance becomes dominated by the lowest-weight codewords). In higher-noise channels, however, a code's performance can actually suffer if higher-weight columns are chosen, since the sum–product decoder requires a very sparse H matrix in order to perform well.

Figure 3.1 shows the simulated performance of regular LDPC codes, with varying column weight, on a BI-AWGN channel. For these codes there is definitely a performance benefit in low-noise channels for codes with a longer column weight. However, for high noise levels the effect of increasing the column weight is to reduce the error correction performance. The negative effect of larger column weights in high-noise channels can be explained by observing that if any two code bits in a given parity check are in error (or erased) it is more difficult (or impossible) to determine and correct the erroneous bit using that parity-check equation. In codes with a large column weight the row weight must also increase proportionally, so each parity-check equation includes more of the codeword bits, increasing the likelihood that two bits in any given parity check are in error. This is less of a problem in high signal-to-noise-ratio channels, where only very few code bits are affected by the noise. Indeed, in these channels a large column weight can improve the sum–product decoding convergence speed as well as improving the code's minimum distance.

Regular versus irregular codes

By choosing irregular codes the threshold performance can be significantly improved. The idea is that the bits corresponding to high-weight columns are easily decoded, and their good LLRs are then used to decode the low-weight columns.

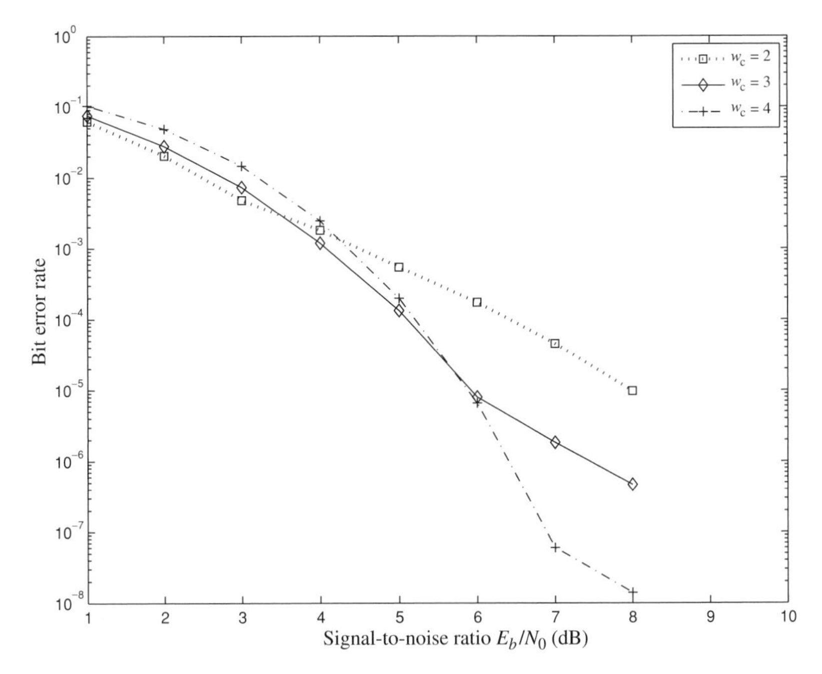

Figure 3.1 The decoding performance on a BI-AWGN channel of length-100 rate-1/2 LDPC codes with varying column weight.

Recall that an irregular parity-check matrix has columns and rows with varying weights (on the Tanner graph, bit nodes and check nodes with varying degrees). We designated the fraction of columns having weight i (bit nodes of degree i) by v_i and the fraction of rows having weight i (check nodes of degree i) by h_i.

To design irregular LDPC codes, an alternative characterization of the degree distribution from the perspective of Tanner graph edges, the *edge-degree distribution*, is used. The fraction of edges that are connected to degree-i bit nodes is denoted λ_i, and the fraction of edges that are connected to degree-i check nodes, is denoted ρ_i. By definition,

$$\sum_i \lambda_i = 1 \tag{3.1}$$

and

$$\sum_i \rho_i = 1. \tag{3.2}$$

The functions

$$\lambda(x) = \sum_{i=2}^{\lambda_{\max}} \lambda_i x^{i-1} = \lambda_2 x + \lambda_3 x^2 + \cdots + \lambda_i x^{i-1} + \cdots, \tag{3.3}$$

$$\rho(x) = \sum_{i=2}^{\rho_{\max}} \rho_i x^{i-1} = \rho_2 x + \rho_3 x^2 + \cdots + \rho_i x^{i-1} + \cdots \tag{3.4}$$

are defined to describe the degree distributions. Translating between node degrees and edge degrees:

$$v_i = \frac{\lambda_i/i}{\sum_j \lambda_j/j},$$

$$h_i = \frac{\rho_i/i}{\sum_j \rho_j/j}.$$

Now substituting into (2.11),

$$r = 1 - \frac{\sum_i v_i i}{\sum_i h_i i} = 1 - \frac{\sum_i \dfrac{\lambda_i}{\sum_j \lambda_j/j}}{\sum_i \dfrac{\rho_i}{\sum_j \rho_j/j}} = 1 - \frac{\sum_j \rho_j/j}{\sum_j \lambda_j/j}.$$

Using functions $\lambda(x)$ and $\rho(x)$ the code rate can be expressed alternatively as

$$r = 1 - \frac{\int_0^1 \rho(x)dx}{\int_0^1 \lambda(x)dx},$$

where $\int_0^1 \rho(x)dx$ replaces $\sum_j \rho_j/j$.

Example 3.1 The parity-check matrix H in Example 2.12,

$$H = \begin{bmatrix} 1 & 1 & 0 & 1 & 1 & 0 & 0 & 1 & 0 & 0 \\ 0 & 1 & 1 & 0 & 1 & 1 & 1 & 0 & 0 & 0 \\ 0 & 0 & 0 & 1 & 0 & 0 & 0 & 1 & 1 & 1 \\ 1 & 1 & 0 & 0 & 0 & 1 & 1 & 0 & 1 & 0 \\ 0 & 0 & 1 & 0 & 0 & 1 & 0 & 1 & 0 & 1 \end{bmatrix},$$

is irregular with node degree distribution $v_2 = 0.7, v_3 = 0.3, h_4 = 0.4, h_5 = 0.6$. For the edge degrees, λ_2 is given by the fraction of edges connected to degree-2 bit nodes, thus $\lambda_2 = 14/23 = 0.6087$; λ_3 is given by the fraction of edges connected to degree-3 bit nodes, thus $\lambda_3 = 9/23 = 0.3913$; ρ_4 is given by the fraction of edges connected to degree-4 check nodes, thus $\rho_4 = 8/23 = 0.3478$; and ρ_5 is given by the fraction of edges connected to degree-5 check nodes, thus $\rho_5 = 15/23 = 0.6522$.

In Chapter 7 we will see how to optimize the degree distribution of irregular LDPC codes so as to improve the threshold of the code (increasing the noise level at which the waterfall region occurs). Here we will simply present the performance of codes that have been optimized in this way, in order to compare their performance with that of regular codes.

Example 3.2 Figures 3.2 and 3.3 show the error correction performance of LDPC codes on the BI-AWGN channel for rate-1/2 codes with code lengths of 1000 and 10 000. These irregular codes have degree distribution

$$\lambda(x) = 0.3282x + 0.1581x^2 + 0.2275x^3 + 0.2862x^8,$$
$$\rho(x) = 0.02448x^4 + 0.35640x^5 + 0.61911x^6,$$

from [19]. As expected, irregular LDPC codes can have greatly improved thresholds; this is particularly evident when long codes are considered. However, for shorter codes the threshold performance improvement is reduced. The trade-off for an improved threshold is an error floor at a much higher bit error rate than for a regular code with the same rate and length.

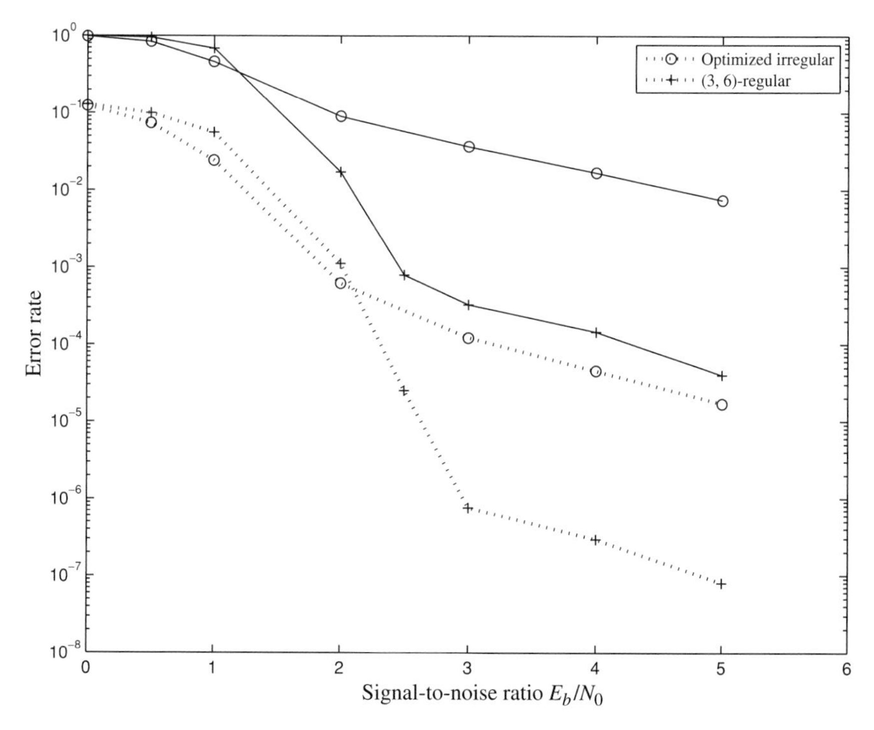

Figure 3.2 The bit error rate (dotted curves) and word error rate (solid curves) for a BI-AWGN channel using sum–product decoding for length-1000 rate-1/2 regular and irregular (optimized for threshold) LDPC codes.

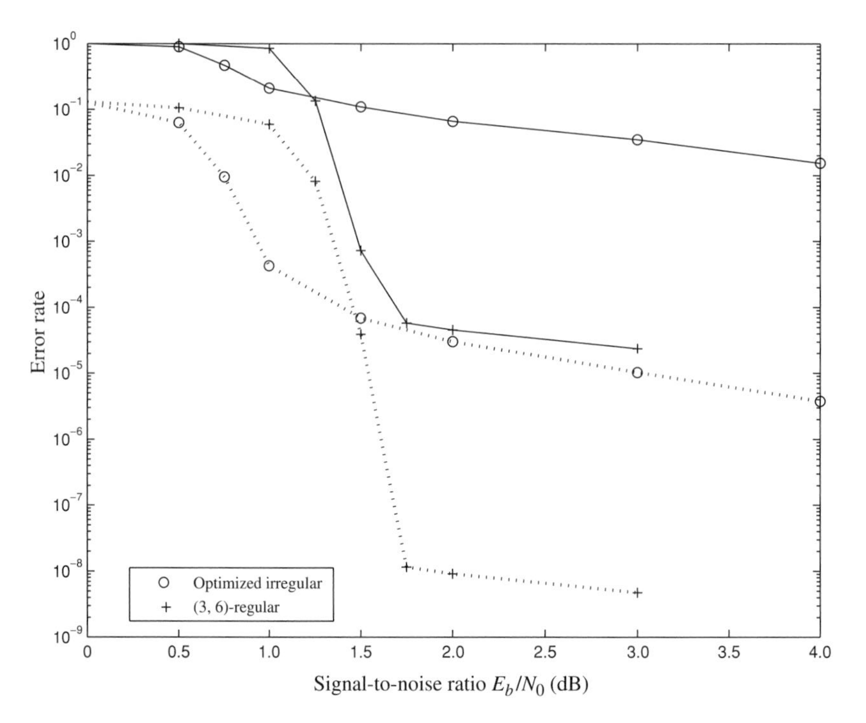

Figure 3.3 The bit error rate (dotted curves) and word error rate (solid curves) for a BI-AWGN channel using sum–product decoding for length-10 000 rate-1/2 regular and irregular (optimized for threshold) LDPC codes. After [19], © IEEE 2006.

In general the threshold gain due to using irregular codes is greater for lower-rate codes; an average column weight of about 3.5 produces the best thresholds for most code rates with a slightly larger average column weight needed for higher-rate codes. While a large range of column weights can improve the code threshold, the row weights do not need to vary much; one or two column weights (to meet the required average) work just as well as a large range of column weights with the same average.

Low-rank parity-check matrices

Recall that the rate r of a code is given by

$$r = \frac{K}{N} = \frac{N - \text{rank}_2\, H}{N}.$$

Only when the code is full rank, i.e.

$$\text{rank}_2\, H = m,$$

can we write

$$r = \frac{N - m}{N} = \frac{N - Nw_c/w_r}{N} = 1 - \frac{w_c}{w_r}$$

or

$$r = 1 - \frac{\sum_i v_i i}{\sum_i h_i i}.$$

Thus, one way in which the column weight can be increased without increasing the row weight or reducing the code rate is to choose parity-check matrices with

$$\text{rank}_2 H < m.$$

Example 3.3 The BER performance on a BI-AWGN channel for two length-255 rate-175/255 LDPC codes is shown in Figure 3.4. Although both codes have the same length and rate, the EG LDPC code (see Section 3.3.2) has significantly more rows in its parity-check matrix than the pseudo-random LDPC code (see Section 3.3.1), allowing a much greater column weight, 16. This, combined with the absence of 4-cycles, gives a much greater minimum distance, 17, and thus an improved performance.

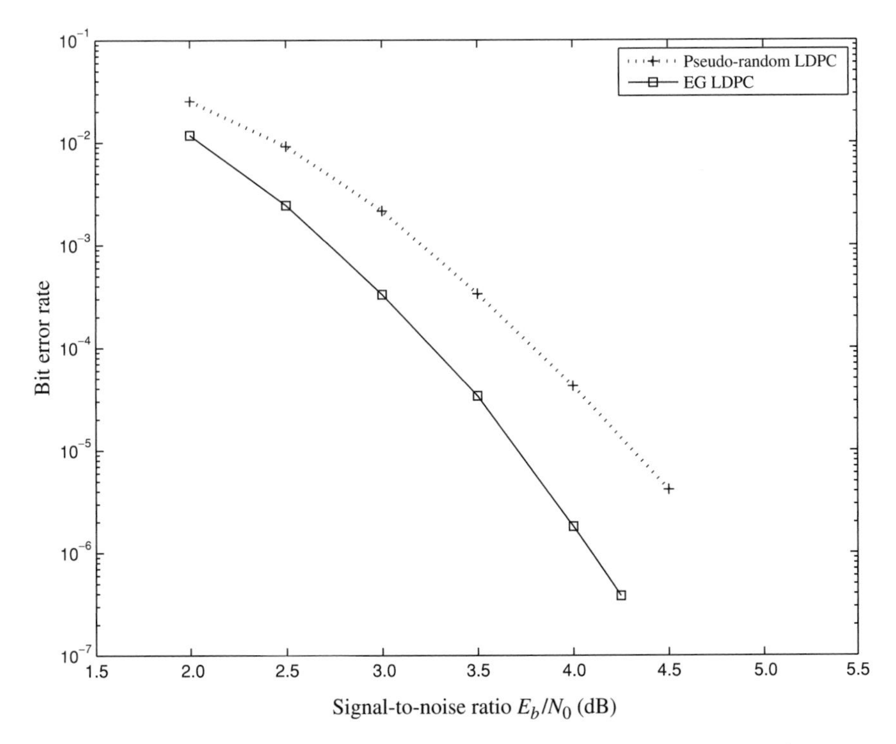

Figure 3.4 The decoding performance of two length-255 rate-175/255 LDPC codes on a BI-AWGN channel using sum–product decoding with a maximum of 200 iterations. After [31], © IEEE 2004.

Unfortunately, however, constructing a rank-deficient parity-check matrix without adding in unwanted configurations (such as small cycles) is not a straightforward task. Nevertheless existing combinatorial constructions for matrices with just such a property have proved ideal for LDPC codes (we will consider these in more detail in Section 3.3.2)

3.2.2 Girth and expansion

The relationship between LDPC codes and their decoding is closely associated with the graph-based representation of the codes. The most obvious example of this is the link between the existence of cycles in the Tanner graph of a code and the performance of sum–product decoding of the code.

Example 2.22 showed the detrimental effect of a 4-cycle on the bit-flipping message-passing decoder. For sum–product decoding, cycles in the Tanner graph lead to correlations in the marginal probabilities passed on by the sum–product decoder: the smaller the cycles the fewer the number of iterations that are correlation free. Thus cycles in the Tanner graph affect the decoding convergence of sum–product decoding similarly, and the smaller the girth the larger the effect. Definite performance improvements can be obtained by avoiding 4-cycles and 6-cycles in LDPC Tanner graphs, but the returns tend to diminish as the girth is increased further.[1]

Indeed, for a small code an improvement in decoding performance can be seen even with the removal of a single cycle.

Example 3.4 The parity-check matrices

$$H_1 = \begin{bmatrix} 1 & 1 & 0 & 1 & 0 & 0 \\ 0 & 1 & 1 & 0 & 1 & 0 \\ 1 & 0 & 0 & 0 & 1 & 1 \\ 1 & 0 & 1 & 1 & 1 & 0 \end{bmatrix}$$

and

$$H_2 = \begin{bmatrix} 1 & 1 & 0 & 1 & 0 & 0 \\ 0 & 1 & 1 & 0 & 1 & 0 \\ 1 & 0 & 0 & 0 & 1 & 1 \\ 0 & 0 & 1 & 1 & 0 & 1 \end{bmatrix}$$

both describe the same code but have Tanner graphs with different girths. Figure 3.5 shows the performance of sum–product decoding using each of these

[1] A small subset of LDPC codes that include 4-cycles have been shown to perform well with sum–product decoding; however, this effect is due to the large number of extra linearly dependent rows in these parity-check matrices, which helps to overcome the negative impact of the cycles.

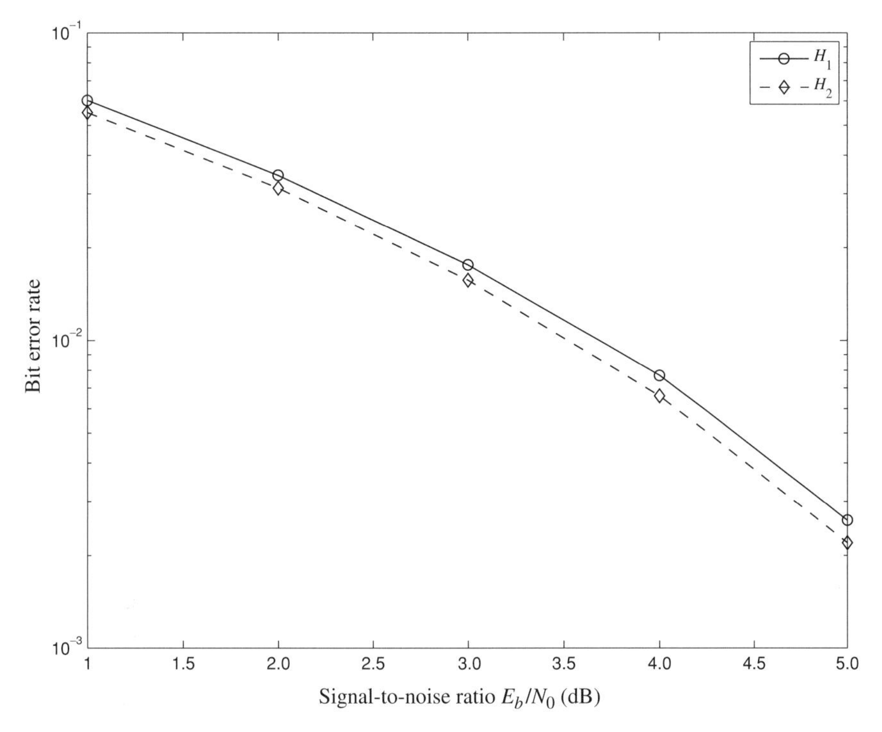

Figure 3.5 The bit error rate performance of sum–product decoding on a BI-AWGN channel using the parity-check matrices from Example 3.4.

parity-check matrices on the same channel. Removing a single 4-cycle from the parity-check matrix H_1 by adding (mod 2) the third row to the fourth to give H_2 noticeably improves the performance of the sum–product decoder even though exactly the same code is being decoded.

In a regular LDPC code with bit node degree w_c and check node degree w_r the expected number of cycles of size g in a randomly constructed LDPC code can be approximated by

$$\frac{(w_c - 1)^{g/2}(w_r - 1)^{g/2}}{g}, \tag{3.5}$$

and so increasing the column and row weights will increase the expected number of cycles. For this reason LDPC codes with large column weights are usually constructed algebraically to ensure the best possible girth for a given column weight (see e.g. Section 3.3.2).

Simulation results have shown that while removing 4-cycles can significantly improve decoding performance and removing 6-cycles can also have a positive effect, removing cycles of larger size tends not to produce much benefit. In many cases the algorithms that remove the larger cycles tend to concentrate the number cycles at the next largest size – which can actually decrease the code's performance slightly. For short or high-rate codes, this can even be true when 4-cycles are removed. In these cases, though, combinatorial constructions can provide the required 4-cycle-free Tanner graphs for short or high-rate codes.

It has been shown that, for a cycle-free graph, the sum–product decoding algorithm is optimal (i.e it returns exact MAP probabilities). However, the avoidance of all cycles is undesirable since a linear code of rate ≥ 0.5 without any cycles has a minimum distance of at most 2. So, while removing all cycles allows the sum–product decoder to decode optimally, the code itself is then so poor that even an optimal decoder does not perform well; better performances can be achieved with a better code and less optimal decoding. The best approach is to avoid small cycles in the Tanner graph, certainly 4-cycles and 6-cycles if this is possible given the code length and rate.

A related concept to the graph's girth is whether it is a *good expander*. In a good expander every subset of vertices has a large number of neighbors that are not in the subset. More precisely, any subset S of bit vertices of size n or less is connected to at least $\epsilon|S|$ constraint vertices for some defined n and ϵ.

If a Tanner graph is a good expander then the bit nodes of a small set of erroneous codeword bits will be connected to a large number of check nodes, all of which will be receiving correct information from an even larger number of correct codeword bits. Generally, randomly constructed graphs tend to be good expanders; adding structure can reduce the expansion of the code.

3.2.3 Codeword and pseudo-codeword weights

We saw in Section 1.3 that the ML decoding performance of a given error correction code on a given channel depends entirely on the weight distribution of the codewords. Furthermore, in channels with low noise the decoding performance can be predicted just by knowing the minimum distance of the code.

Although LDPC codes are not decoded with ML decoding, the aim of an iterative decoder is to approach ML or MAP decoding performance and so the codeword weight distribution can be used to determine bounds on the performance of the iterative decoder for that code. We will consider the ML performance of LDPC codes in Chapter 8. Here we will discuss how the structure of a chosen LDPC code can affect its minimum distance.

Regarding the performance of a code with sum–product decoding, errors can arise not just when the decoder chooses the wrong codeword; the decoder may fail to converge to any codeword. In such cases the decoder can be thought of as having converged not to a codeword but to a code structure called a pseudo-codeword. When using iterative erasure decoding, pseudo-codewords have a straightforward definition and are called *stopping sets*.

In this section we will discuss the role of both codewords and stopping sets in the performance of the sum–product decoder.

Minimum distance

In the absence of any obvious algebraic structure in pseudo-randomly constructed LDPC codes, finding the minimum distance of the code can require an exhaustive search of all 2^K codewords to find the one with smallest Hamming weight. While algorithms exist to approximate the minimum distance of a given code (see e.g. [20]), the exact calculation of the minimum distance is infeasible for long codes. Two approaches to work around this exist. Firstly, the minimum distance of a given code can be lower bounded if the girth of the code is known and, secondly, the average distance function can be calculated for an ensemble of LDPC codes. The latter approach will be used to derive coding theorems in Chapter 8 but the former approach is more useful when it comes to code construction.

Recall that a codeword is any binary vector that satisfies every one of the m parity-check equations in H. Thus, if a set S of codeword bits contains only the non-zero bit locations of a particular codeword for the code, every parity-check equation connected to a bit in S must be connected to an even number of bits in S. The size of the smallest such set is then the minimum distance of the code.

If a codeword bit c_i is checked by w_c parity-check equations, those w_c equations must include at least one other non-zero codeword bit for parity to be satisfied. If H contains no 4-cycles then the second bit in each of those parity-check equations must be a distinct bit. Thus the set S containing c_i must contain at least w_c other bits, and so an LDPC code that can be represented by a 4-cycle-free parity-check matrix with minimum column weight w_c will have minimum distance

$$d_{\min} \geq w_c + 1. \tag{3.6}$$

More generally, for (w_c, w_r)-regular LDPC codes with girth g a lower bound on the minimum distance is given by

$$d_{\min} \geq \begin{cases} 1 + w_c + w_c(w_c - 1) + w_c(w_c - 1)^2 \\ \quad + \cdots + w_c(w_c - 1)^{(g-6)/4} & \text{if } g/2 \text{ is odd,} \\ 1 + w_c + w_c(w_c - 1) + w_c(w_c - 1)^2 \\ \quad + \cdots + w_c(w_c - 1)^{(g-8)/4} & \text{otherwise.} \end{cases} \tag{3.7}$$

(See [21] for a derivation.) Thus we see that the minimum distance of a code can be increased by increasing the code girth or the column weight.

However, note in (3.7) that w_r, and the code rate r, are absent from the bound. Consequently, though a code with the required girth g and column weight w_c must have a minimum distance at least that in (3.7), there is no restriction on the code rate; this can be made very small to ensure the required minimum distance.

Stopping sets

As we saw in Chapter 2, the message-passing decoding of LDPC codes on erasure channels is particularly straightforward since a transmitted bit is either received correctly or completely erased. Message-passing decoding is a process of finding parity-check equations that will check on only one erased bit. In a decode iteration all such parity-check equations are found and the erased bits corrected. After these bits have been corrected, any new parity-check equations checking on only one erased bit are then corrected in the subsequent iteration. The process is repeated until all the erasures are corrected or all the remaining uncorrected parity-check equations check on two or more erased bits. The question for coding theorists is when will this occur.

For the binary erasure channel at least, the answer is known. The message-passing decoder will fail to converge if the erased bits include a set of code bits $\mathcal{S}$ that is a *stopping set*. A stopping set $\mathcal{S}$ is a set of code bits with the property that every parity-check equation that checks on a bit in $\mathcal{S}$ in fact checks on at least two bits in $\mathcal{S}$. The size of a stopping set is the number of bits it includes, and the minimum stopping set size of a parity-check matrix is denoted by $S_{\min}$.

Recall that a set S containing the non-zero bits for a codeword of the code will have the property that every parity-check equation connected to a bit in S is also connected to an even number of bits in S. So any set of non-zero codeword bits is also a stopping set. However, not all stopping sets are codewords since the stopping sets may have parity-check nodes that connect to an odd number of bits in S.

Example 3.5 Figure 3.6 (left) shows the Tanner graph of a length-5 parity-check matrix with three parity-check equations. The filled bit nodes represent a stopping set of size 3.

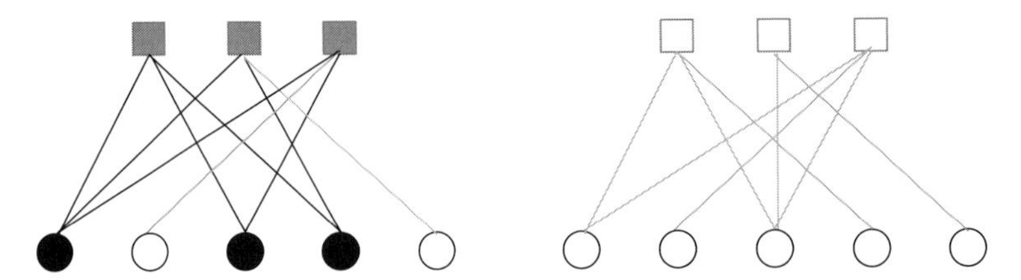

Figure 3.6 The Tanner graphs of two different parity-check matrices for the same code. Circles, bit nodes; squares, check nodes.

Unlike the codeword weight distribution, the stopping set distribution depends on the choice of parity-check matrix for a given code. Figure 3.6 shows as an example the Tanner graphs of two parity-check matrices for the same code. The Tanner graph on the left contains a size-3 stopping set that contains the first, third and fourth bits (shown highlighted) while the Tanner graph on the right does not contain this stopping set. The size-3 stopping set that includes the first, second and fourth bits is a codeword for the code, however, so every possible parity-check matrix for this code will also contain a stopping set in these bit locations.

Example 3.6 The LDPC parity-check matrix used in Example 2.5,

$$H = \begin{bmatrix} 1 & 1 & 0 & 1 & 0 & 0 \\ 0 & 1 & \mathbf{1} & 0 & \mathbf{1} & 0 \\ 1 & 0 & 0 & 0 & \mathbf{1} & \mathbf{1} \\ 0 & 0 & \mathbf{1} & 1 & 0 & \mathbf{1} \end{bmatrix},$$

has a size-3 stopping set on the third, fifth and sixth codeword bits; these are shown in bold.[2]

A generator matrix for the matrix H is used to encode the codeword

$$\mathbf{c} = [0\ 0\ 1\ 0\ 1\ 1],$$

which is sent through an erasure channel; the third, fifth and sixth bits are erased and so the vector

$$\mathbf{y} = [0\ 0\ e\ 0\ e\ e]$$

is received. Message-passing decoding is used to recover the erased bits; see Algorithm 2.1.

At initialization $M_i = y_i$, which gives

$$\mathbf{M} = [0\ 0\ e\ 0\ e\ e].$$

In *Step 1* of the algorithm the check node messages are calculated. The first check node is joined to the first, second and fourth bit nodes and so has no incoming e messages. The second check includes the second, third and fifth bits, and so receives two e messages (M_3 and M_5) and thus cannot be used to correct any codeword bits. The third check includes the first, fifth and sixth bits and so also receives two e messages (M_5 and M_6) and again cannot be used to correct any codeword bits. Finally, the fourth check includes the third, fourth and sixth bits and so receives two e messages (M_3 and M_6) and, once again, cannot be used to correct any codeword bits.

In *Step 2* of the algorithm there are no new messages coming into any of the erased bits and no corrections can be made; regardless of how many iterations are run, an erased bit will never be corrected. By examination of H, it is clear why this is the case: the set of erased codeword bits is a stopping set in H.

---

[2] Note that for parity-check matrices with $w_c = 2$ a cycle of size g will form a stopping set of size $g/2$; however, this is not true in general.

Since every parity-check node connected to a stopping set $\mathcal{S}$ includes at least two erased bits, if all the bits in $\mathcal{S}$ are erased then there will never be a parity-check equation available to correct a bit in $\mathcal{S}$, regardless of the number of iterations employed. In a sense we can say that the decoder has converged to the stopping set.

The stopping set distribution of an LDPC parity-check matrix determines the erasure patterns for which the message-passing decoding algorithm will fail, in the same way that the codeword distribution of a code determines the error patterns for which an ML decoder will fail.

Since the minimum stopping set size determines the minimum number of erased bits that can cause a decoding failure, an error correction code with an excellent d_{min} will nevertheless perform poorly using message-passing decoding on the BEC if it is represented by a parity-check matrix with many small stopping sets. However, a code with a poor minimum distance can never be represented by a parity-check matrix with a good minimum stopping set size because the non-zero bits in every codeword form a stopping set. This observation highlights a common feature of LDPC code design: many properties required for classical error correction codes are necessary, but not sufficient, properties for LDPC codes.

Knowing exactly where the decoding algorithm will fail allows us to predict its performance. Indeed, if the stopping set distribution of an LDPC parity-check matrix were known then the performance of the message-passing decoder on the BEC could be determined exactly by counting the stopping sets. Unfortunately, finding the stopping set distribution of a given LDPC code is no easier than finding its weight distribution. However, an average stopping set distribution can be calculated for the set of all LDPC codes with a particular degree distribution. This process, called finite-length analysis, is presented in Chapter 8.

The bound on the minimum distance (3.6) is based on the requirement that every parity-check equation should contain at least two codeword bits, and so it is general enough to apply to stopping sets as well. Thus, for a 4-cycle-free code with minimum column weight w_c, the minimum stopping set size S_{min} is bounded by

$$S_{\text{min}} \geq w_c + 1. \tag{3.8}$$

More generally, for (w_c, w_r)-regular LDPC codes with girth g the bound in (3.7) is actually a lower bound on the minimum stopping set size (and on the minimum distance since every codeword is also a stopping set).

The role of stopping sets in predicting the performance of message-passing decoding on the BEC tells us that for message-passing decoding, unlike for ML decoding, properties other than the codeword set influence decoder performance. The same is true of message-passing decoding on more general channels;

however, defining the pseudo-codewords that lead to decoder failure in the general case is less straightforward.

Nevertheless, on the erasure channel at least, the message-passing decoding of the code can be improved by choosing a parity-check matrix with as large a value of $S_{\min}$ as possible. Furthermore, the relationship between $d_{\min}$ and $S_{\min}$ makes designing a code with good minimum stopping set size generally desirable.

Example 3.7 Figure 3.7 shows the erasure correction performance of three different length-57 rate-1/3 LDPC codes using column-weight-3 parity-check matrices and varying minimum stopping set sizes. As expected, improving the minimum stopping set size of a code can improve its erasure correction performance.

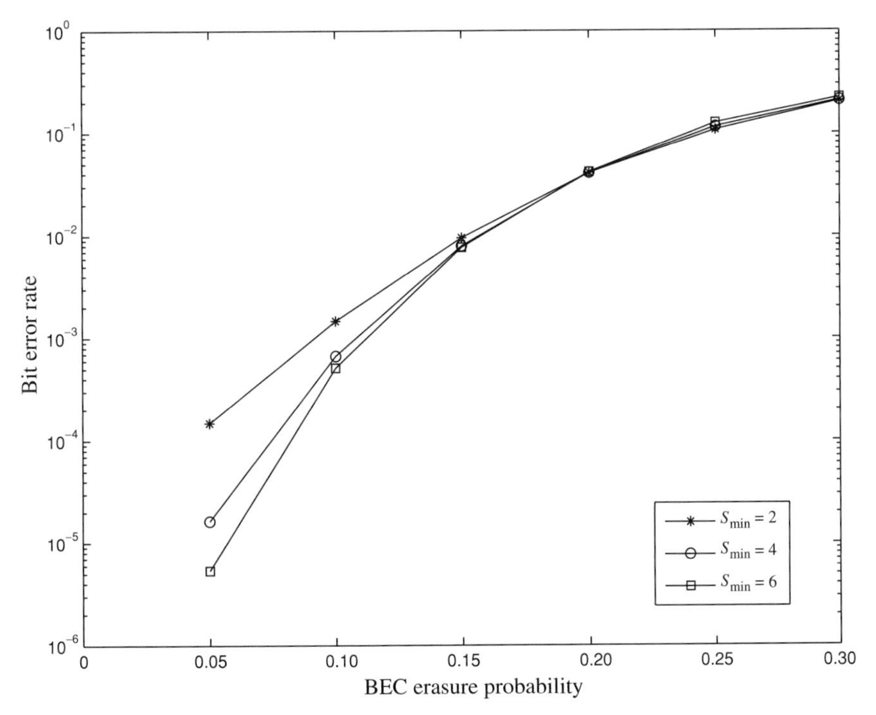

Figure 3.7 The bit erasure rate of three different length-57 rate-1/3 LDPC codes using column-weight-3 parity-check matrices with various minimum stopping set sizes.

3.2.4 Cyclic and quasi-cyclic codes

Rather than trying to convert a parity-check matrix into an encodable form after it has been produced, encodability can be incorporated into the design of the parity-check matrix. For classical codes this has been successfully achieved using cyclic and quasi-cyclic codes and the same ideas can be applied to LDPC codes.

Row-circulant quasi-cyclic codes

A code is *quasi-cyclic* if for any cyclic shift of a codeword by c places the resulting word is also a codeword, and so a cyclic code is a quasi-cyclic code with $c = 1$. The simplest quasi-cyclic codes are row-circulant codes, which are described by a parity-check matrix

$$H = [A_1 \; A_2 \; \cdots \; A_l], \tag{3.9}$$

where $A_1, \ldots, A_l$ are binary $m \times m$ circulant matrices: a circulant matrix A is a square matrix for which each row is a cyclic shift of the one above. For example, the matrix

$$A = \begin{bmatrix} 1 & 1 & 0 & 0 & 0 \\ 0 & 1 & 1 & 0 & 0 \\ 0 & 0 & 1 & 1 & 0 \\ 0 & 0 & 0 & 1 & 1 \\ 1 & 0 & 0 & 0 & 1 \end{bmatrix}$$

is circulant.

Provided that one of the circulant matrices in (3.9), say A_l, is full rank and therefore invertible, the parity-check matrix for the code can be constructed in systematic form as

$$H_{\text{sys}} = [A_l^{-1} A_1 \quad A_l^{-1} A_2 \quad \cdots \quad I_m], \tag{3.10}$$

where I_m is the $m \times m$ identity matrix. We then have for the generator matrix

$$G = \left[\begin{array}{cc} & (A_l^{-1} A_1)^T \\ I_{m(l-1)} & (A_l^{-1} A_2)^T \\ & \vdots \\ & (A_l^{-1} A_{l-1})^T \end{array} \right], \tag{3.11}$$

resulting in a quasi-cyclic code of length $N = ml$ and dimension $K = m(l-1)$. As we have assumed that one of the circulant matrices is invertible, the construction of the generator matrix in this way necessarily requires a full rank parity-check matrix H.

The algebra of $m \times m$ binary circulant matrices is isomorphic to the algebra of polynomials modulo $x^m - 1$ over $GF(2)$. A circulant matrix A is completely characterized by the polynomial $a(x) = a_0 + a_1 x + \cdots + a_{m-1} x^{m-1}$ with coefficients from its first row, and a code C of the form (3.9) is completely characterized by the polynomials $a_1(x), \ldots, a_l(x)$ corresponding to the l matrices in

(3.9). Polynomial transpose is defined as

$$a(x)^T \triangleq \sum_{i=0}^{m-1} a_i x^{m-i} \qquad (x^m = 1).$$

For a binary code of length $N = ml$ and dimension $K = m(l-1)$, the K-bit message $[i_0\ i_1\ \cdots\ i_{K-1}]$ is described by the polynomial $i(x) = i_0 + i_1 x + \cdots + i_{K-1}x^{K-1}$. The codeword for this message is $c(x) = [i(x)\ p(x)]$, where $p(x)$ is given by

$$p(x) = \sum_{j=1}^{l-1} i_j(x) * (a_l^{-1}(x) * a_j(x))^T; \qquad (3.12)$$

here $i_j(x)$ is the polynomial representation of the information bits $i_{m(j-1)}$ to i_{mj-1},

$$i_j(x) = i_{m(j-1)} + i_{m(j-1)+1}x + \cdots + i_{mj-1}x^{m-1}$$

and polynomial multiplication $*$ is multiplication modulo $x^m - 1$.

Example 3.8 The rate-1/2 quasi-cyclic code with $m = 5$ from [22] is made up of a first circulant described by $a_1(x) = 1 + x$ and a second circulant described by $a_2(x) = 1 + x^2 + x^4$. Thus

$$H = \left[\begin{array}{ccccc|ccccc} 1 & 1 & 0 & 0 & 0 & 1 & 0 & 1 & 0 & 1 \\ 0 & 1 & 1 & 0 & 0 & 1 & 1 & 0 & 1 & 0 \\ 0 & 0 & 1 & 1 & 0 & 0 & 1 & 1 & 0 & 1 \\ 0 & 0 & 0 & 1 & 1 & 1 & 0 & 1 & 1 & 0 \\ 1 & 0 & 0 & 0 & 1 & 0 & 1 & 0 & 1 & 1 \end{array}\right].$$

The second circulant is invertible:

$$a_2^{-1}(x) = x^2 + x^3 + x^4,$$

and so the generator matrix G contains a 5×5 identity matrix and the 5×5 matrix described by the polynomial

$$(a_2^{-1}(x) * a_1(x))^T = (1 + x^2)^T = 1 + x^3,$$

that is,

$$G = \left[\begin{array}{ccccc|ccccc} 1 & 0 & 0 & 0 & 0 & 1 & 0 & 0 & 1 & 0 \\ 0 & 1 & 0 & 0 & 0 & 0 & 1 & 0 & 0 & 1 \\ 0 & 0 & 1 & 0 & 0 & 1 & 0 & 1 & 0 & 0 \\ 0 & 0 & 0 & 1 & 0 & 0 & 1 & 0 & 1 & 0 \\ 0 & 0 & 0 & 0 & 1 & 0 & 0 & 1 & 0 & 1 \end{array}\right].$$

Figure 3.8 shows the Tanner graph for H.

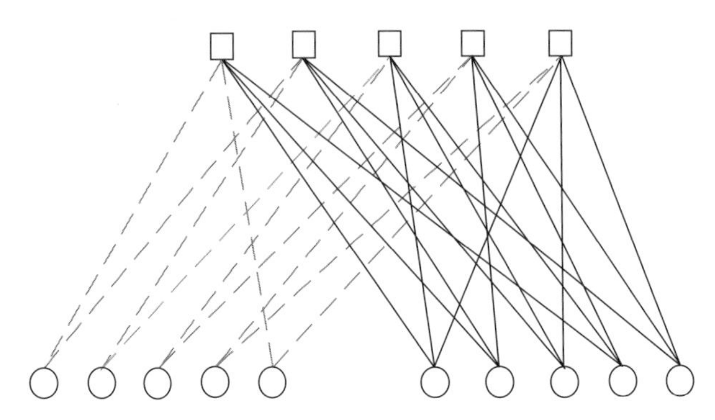

Figure 3.8 A Tanner graph for the quasi-cyclic LDPC code in Example 3.8.

Linear-time encoding can be achieved using $l-1$ m-stage shift registers with separate length-m shift registers for each circulant in G.

Example 3.9 A quasi-cyclic length-108 rate-3/4 LDPC code has the parity-check matrix

$$H = [A_1\ A_2\ A_3\ A_4].$$

The matrix H consists of four circulants, defined by

$$\begin{aligned} a_1(x) &= 1 + x^3 + x^{16}, \\ a_2(x) &= x^2 + x^6 + x^8, \\ a_3(x) &= x + x^8 + x^9, \\ a_4(x) &= x + x^{13} + x^{23}. \end{aligned}$$

The polynomial $a_4(x)$ is invertible, with inverse given by

$$\begin{aligned} a_4^{-1}(x) = x &+ x^4 + x^5 + x^6 + x^7 + x^9 + x^{12} + x^{13} \\ &+ x^{15} + x^{17} + x^{20} + x^{21} + x^{23} + x^{24} + x^{25}, \end{aligned}$$

and so the parity-check matrix can be put into the systematic form

$$H_s = [A_4^{-1}A_1 \quad A_4^{-1}A_2 \quad A_4^{-1}A_3 \quad I_{27}].$$

We thus have

$$\begin{aligned} a_4^{-1}(x)a_1(x) &= 1 + x + x^4 + x^5 + x^7 + x^9 + x^{11} + x^{13} \\ &\quad + x^{15} + x^{18} + x^{19} + x^{22} + x^{23} + x^{25} + x^{26}, \\ a_4^{-1}(x)a_2(x) &= x^4 + x^5 + x^7 + x^9 + x^{12} + x^{17} + x^{19} + x^{22} + x^{26}, \\ a_4^{-1}(x)a_3(x) &= x^4 + x^6 + x^7 + x^{10} + x^{13} + x^{14} + x^{15} + x^{18} \\ &\quad + x^{19} + x^{21} + x^{22} + x^{23} + x^{24} + x^{25} + x^{26}, \end{aligned}$$

and the generator matrix for this code is

$$G = \begin{bmatrix} & (A_4^{-1}A_1)^T \\ I_{81} & (A_4^{-1}A_2)^T \\ & (A_4^{-1}A_3)^T \end{bmatrix}.$$

Using G in this form, the code can be encoded using shift registers. Figure 3.9 shows an encoding circuit for this code.

Note that, although we have used H_s to construct G, we will use the original matrix H for our decoding. Both H and H_s are valid parity-check matrices for the code; however, H has the properties required for better sum–product decoding performance.

Block-circulant quasi-cyclic codes

More general quasi-cyclic codes are the *block-circulant* codes. The parity-check matrix of a block-circulant quasi-cyclic LDPC code is

$$H = \begin{bmatrix} I_p & I_p & I_p & \cdots & I_p \\ I_p & I_p(p_{1,1}) & I_p(p_{1,2}) & \cdots & I_p(p_{1,w_r}) \\ \vdots & & \ddots & & \vdots \\ I_p & I_p(p_{w_c-1,1}) & I_p(p_{w_c-1,2}) & \cdots & I_p(p_{w_c-1,w_r-1}) \end{bmatrix},$$

where I_p represents the $p \times p$ identity matrix and $I_p(p_{i,j})$ represents a circulant shift of the identity matrix by $p_{i,j}$ (mod p) columns to the right which gives the matrix whose rth row has a 1 in the $(r + p_{i,j} \pmod p)$th column.

Generally, block-circulant LDPC codes are preferable to row-circulant codes as they do not have the same limitations of minimum distance and girth as row-circulant codes.

Finding a generator matrix for a block-circulant code requires that we find a square matrix of circulants that is invertible. That is, if H is made up of m/p rows of circulants, we need to find m/p columns of circulants to form a full-rank $m \times m$ square matrix S. Multiplying H by S^{-1} gives the parity-check matrix, and hence the generator matrix, in systematic form.

Of course, G will not be sparse but the encoding of block circulant codes can be done one row of circulants at a time (with each row encoded in exactly the same manner as for row-circulant codes) and the results of each row added modulo 2.

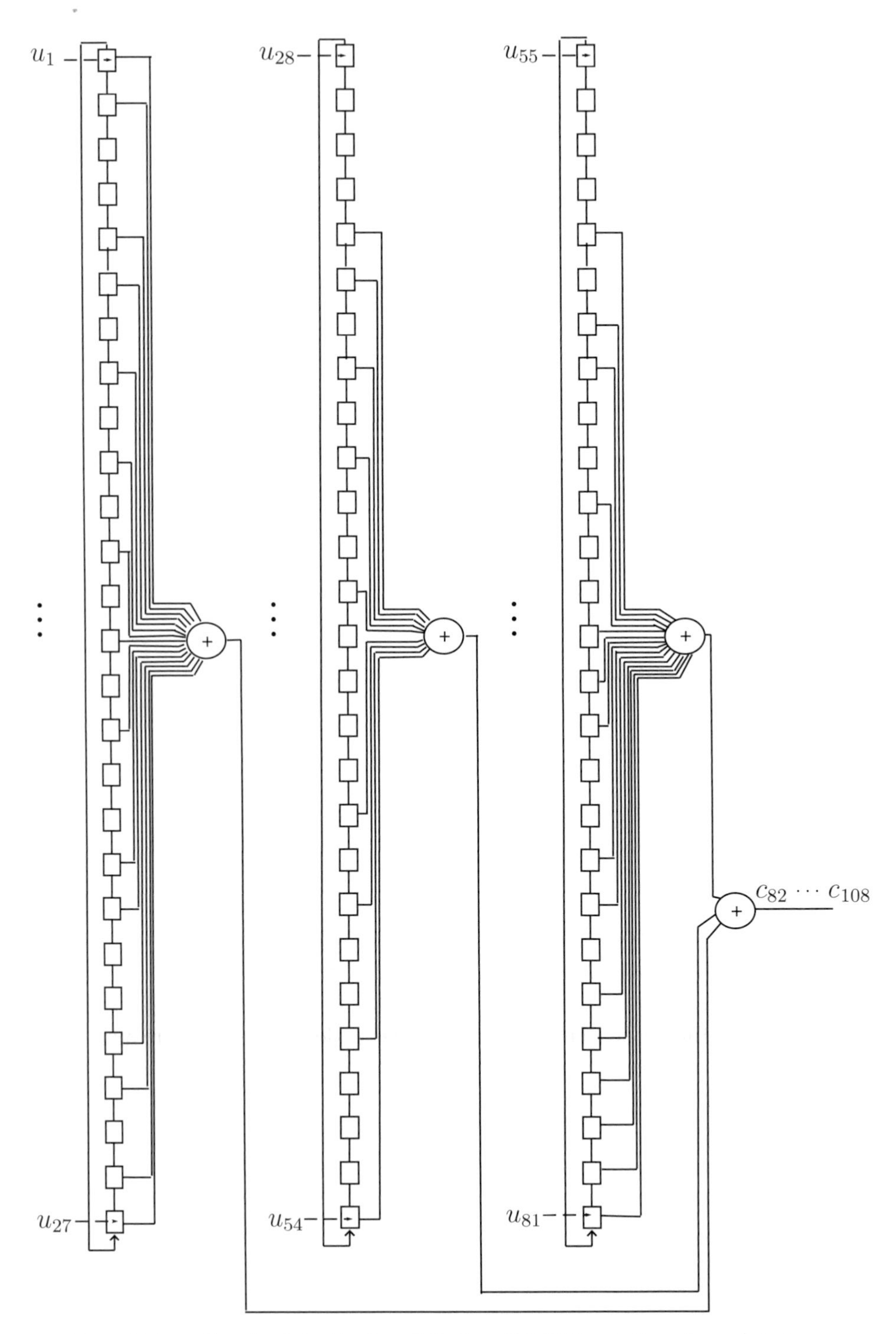

Figure 3.9 Encoding circuit for the $N = 108$, $K = 81$ LDPC code from Example 3.9.

3.2.5 Implementation and complexity

The complexity of the LDPC decoder implementation has three parts: the complexity of the Tanner graph nodes; the complexity of routing for the edges; and the number of iterations required for the decoder.

Node complexity

The node complexity for sum–product decoding depends on the variant of the message-passing decoding algorithm employed. Bit-flipping decoding, which requires just XOR operations at the nodes, will have significantly less computational complexity than sum–product decoding.

For sum–product decoding, the use of log likelihood ratios reduces the evaluation at the bit node to a sum, as in (2.29), but the check node still requires the calculation of exponential and logarithmic functions. In practice, the check node of a sum–product decoder is usually implemented to calculate (2.33) using a lookup table for ϕ. As a compromise between the bit-flipping and sum–product algorithms, min–sum decoding, as in (2.34), simplifies the check node calculation of the sum–product algorithm even further, requiring just minimum and sum operations at the check nodes.

The choice between these three decoding algorithms does not affect the code design (a code good for one is good for all three), so LDPC standardization efforts can specify the codes and allow the receiver designer the flexibility to trade off complexity and performance in their selection of the decoding algorithm.

Routing complexity

Typically, decoder implementations use between three and eight bits for fixed point representation of the messages; the larger the number of edges that need to be routed, the fewer the bits used. Generally, using just five bits per message (four for the magnitude and one for the sign) is sufficient to produce reasonable decoding performances. However, given large code lengths there are many wires to be routed for a standard LDPC application, even with only 3–5 wires per edge (6–10 if the wires are not bidirectional).

A high-throughput decoder can be implemented by directly mapping the corresponding Tanner graph edges in hardware. However, the long random-like structures that work best with LDPC codes are particularly difficult to rout. An LDPC Tanner graph will have Nw_c edges where N is the number of bit nodes and w_c their degree (or average degree if the code is irregular). An LDPC code with Nw_c Tanner graph edges will require $n_b Nw_c$ wires, where n_b is the number of bits used to represent each message, if the wires are bidirectional and $2n_b Nw_c$ otherwise. Further, hard-wiring the Tanner graph in this way precludes any flexibility in implementation. Instead, however, the Tanner graph can be implemented via a memory array with the messages written into and read out of the memory according to the particular edge permutation. The edge permutation can be stored in the memory or, for an algebraic construction (see Section 3.3.2), calculated as required.

Number of iterations

The maximum number of iterations required by a sum–product decoder can vary between 10 and 1000 depending on the design of the code and the level of noise in the channel. (As we will see in Chapter 7, operating close to a code's threshold requires a very large number of iterations for convergence.) However, if the noise is reasonably low then successful decoding usually takes only a couple of iterations of the decoder. The throughput can thus be increased by using the zero-syndrome stopping criterion to check for a valid codeword. This of course assumes that the receiver can handle the variable amount of time taken to decode each codeword with some sort of codeword pipelining. However, even without this, stopping the decoder earlier has the advantage of reducing its energy consumption, which can be very useful in some devices.

Serial versus parallel implementation

The LDPC decoder can be implemented in parallel with a dedicated computational circuit for every Tanner graph node. In this way all the bit node processing is done in one step and all the check node processing in another. This implementation gives an excellent decoding speed but requires a large circuit size to implement. A number of fully parallel LDPC decoders have been implemented (see e.g. [23–26]), all using custom silicon to handle the large number of wires required for the routing.

In a fully serial implementation of message-passing decoding, the individual bit and check nodes are computed one at a time. Such a serial implementation does not require much circuit space to implement the nodes; however, a significant amount of memory will be required to store all the bit-to-check and check-to-bit messages. For example, an LDPC code with Nw_c Tanner graph edges and n_b bits per message will need storage for $2n_bNw_c$ bits.

If we say that a bit node computation takes b clock cycles per input edge and that a check node computation takes c clock cycles per input edge, a single iteration of a serial decoder will take roughly $Nw_c(b + c)$ clock cycles while a fully parallel decoder will take just $bv_{max} + ch_{max}$ cycles. Thus the delay for a serial decoder increases proportionally with the number of Tanner graph edges, Nw_c, and therefore with the code length, while the delay for a parallel decoder increases only if the maximum node weight is increased; it does not change for longer codes. Of course the circuit space needed to implement the parallel decoder increases proportionally with the code length (since new bit and check nodes are required as the code length increases), while the circuit space required by the serial decoder does not change. The fully parallel decoder also requires all $2n_b N w_c$ message wires to be routed at once, so fully parallel decoders usually employ small graphs and use the minimum number of bits per message.

A trade-off between the two approaches is provided by a partially parallel design, where a number of bit and check node circuits are realized in hardware and are shared between groups of bit and check nodes. As for the serial implementation, storage will be required for the bit-to-check and check-to-bit messages and memory access collisions can be a significant issue if the routing is not designed carefully. (For implementations of partially parallel LDPC decoders see e.g. [27–29].)

Overall, we can say that the complexity of message-passing decoding scales roughly with the number of edges in the Tanner graph, regardless of whether we are using a serial implementation (where the memory demands and delay scale with the number of Tanner graph edges) or a parallel implementation (where the required circuit space scales with the number of Tanner graph edges).

3.3 LDPC constructions

In the previous section a number of properties that make a good LDPC code were discussed. In this section some methods used to construct LDPC codes that achieve these properties are outlined. For long codes, randomly constructing a parity-check matrix almost always produces a good code. Nevertheless, in practical applications the codes may be required to be much shorter, and a random code will be much more difficult to implement than a structured one. Most codes are constructed at least pseudo-randomly, where the construction is random but certain undesirable configurations such as 4-cycles are either avoided during construction or removed afterwards. Some of these techniques are considered in the following.

3.3.1 Pseudo-random constructions

The original LDPC codes presented by Gallager were regular and defined by a banded structure in H. The rows of Gallager's parity-check matrices are divided into w_c sets, with m/w_c rows in each set. Each row in the first set of rows contains w_r consecutive 1s ordered from left to right across the columns (i.e. for $i \leq m/w_c$ the ith row has entries equal to 1 in the $((i-1)w_r+1)$th to iw_rth columns). Every other set of rows is a randomly chosen column permutation of this first set. Consequently every column of H has a single entry equal to 1 in each w_c set.

Example 3.10 An example of a length-12 (3, 4)-regular Gallager parity-check matrix is

$$H = \left[\begin{array}{cccccccccccc} 1 & 1 & 1 & 1 & 0 & 0 & 0 & 0 & 0 & 0 & 0 & 0 \\ 0 & 0 & 0 & 0 & 1 & 1 & 1 & 1 & 0 & 0 & 0 & 0 \\ 0 & 0 & 0 & 0 & 0 & 0 & 0 & 0 & 1 & 1 & 1 & 1 \\ \hline 1 & 0 & 1 & 0 & 0 & 1 & 0 & 0 & 0 & 1 & 0 & 0 \\ 0 & 1 & 0 & 0 & 0 & 0 & 1 & 1 & 0 & 0 & 0 & 1 \\ 0 & 0 & 0 & 1 & 1 & 0 & 0 & 0 & 1 & 0 & 1 & 0 \\ \hline 1 & 0 & 0 & 1 & 0 & 0 & 1 & 0 & 0 & 1 & 0 & 0 \\ 0 & 1 & 0 & 0 & 0 & 1 & 0 & 1 & 0 & 0 & 1 & 0 \\ 0 & 0 & 1 & 0 & 1 & 0 & 0 & 0 & 1 & 0 & 0 & 1 \end{array}\right].$$

Another common construction for LDPC codes was proposed by MacKay and Neal. In this method, the columns of H are added one column at a time from left to right. The weight of each column is chosen so as to obtain the correct bit degree distribution and the locations of the non-zero entries in each column are chosen randomly from those rows that are not yet full. If at any point there are rows with more positions unfilled than there are columns remaining to be added, the row degree distributions for H will not be completely correct. Then the process can be started again or backtracked by a few columns until the correct row degrees are obtained.

Example 3.11 An example of a length-12 (3, 4)-regular MacKay–Neal parity-check matrix is

$$H = \left[\begin{array}{cccccccccccc} 1 & 0 & 0 & 0 & 0 & 1 & 0 & 1 & 0 & 1 & 0 & 0 \\ 1 & 0 & 0 & 1 & 1 & 0 & 0 & 0 & 0 & 0 & \mathbf{1} & 0 \\ 0 & 1 & 0 & 0 & 1 & 0 & 1 & 0 & 1 & 0 & 0 & 0 \\ 0 & 0 & 1 & 0 & 0 & 1 & 0 & 0 & 0 & 0 & \mathbf{1} & 1 \\ 0 & 0 & 1 & 0 & 0 & 0 & 1 & 1 & 0 & 0 & 0 & 1 \\ 0 & 1 & 0 & 0 & 1 & 0 & 0 & 0 & 1 & 0 & \mathbf{1} & 0 \\ 1 & 0 & 0 & 1 & 0 & 0 & 1 & 0 & 0 & 1 & 0 & 0 \\ 0 & 1 & 0 & 0 & 0 & 1 & 0 & 1 & 0 & 1 & 0 & 0 \\ 0 & 0 & 1 & 1 & 0 & 0 & 0 & 0 & 1 & 0 & 0 & 1 \end{array}\right].$$

When adding the 11th column, shown in bold, the unfilled rows were the second, fourth, fifth, sixth and ninth, from which the second, fourth and sixth were chosen.

The Mackay–Neal construction method for LDPC codes can be adapted to avoid 4-cycles by checking each pair of columns in H to see whether they

Algorithm 3.1 MacKay–Neal LDPC codes

1: **procedure** MN CONSTRUCTION($N, r, \mathbf{v}, \mathbf{h}$) ▷ Required length, rate and degree distributions
2: $H =$ all zero $N(1-r) \times N$ matrix ▷ Initialization
3: $\boldsymbol{\alpha} = []$;
4: **for** $i = 1$: length($\mathbf{v}$) **do**
5: **for** $j = 1 : v_i \times N$ **do**
6: $\boldsymbol{\alpha} = [\boldsymbol{\alpha}, i]$
7: **end for**
8: **end for**
9: $\boldsymbol{\beta} = []$
10: **for** $i = 1$: length($\mathbf{h}$) **do**
11: **for** $j = 1 : h_i \times N(1-r)$ **do**
12: $\boldsymbol{\beta} = [\boldsymbol{\beta}, i]$
13: **end for**
14: **end for**
15:
16: **for** $i = 1 : N$ **do** ▷ Construction
17: $\mathbf{c} =$ random subset of the set of unfilled rows, of size α_i
18: **for** $j = 1 : \alpha_i$ **do**
19: $H(c_j, i) = 1$
20: **end for**
21: remove the entries in $\mathbf{c}$ from the set of unfilled rows
22: **end for**
23:
24: **repeat**
25: **for** $i = 1 : N - 1$ **do** ▷ Remove 4-cycles
26: **for** $j = i + 1 : N$ **do**
27: **if** the ith and jth columns of H contain a 1 entry in more than one row in common **then**
28: permute the entries in the jth column
29: **end if**
30: **end for**
31: **end for**
32: **until** cycles removed
33: **end procedure**

overlap in two places. The construction of 4-cycle-free codes using this method is given in Algorithm 3.1. The inputs are the code length N, the design rate r and the column and row degree distributions $\mathbf{v}$ and $\mathbf{h}$. The length-N vector $\boldsymbol{\alpha}$ contains an entry i for each column in H of weight i and the length-m vector $\boldsymbol{\beta}$ contains an entry i for each row in H of weight i.

Alternatively, the MacKay–Neal construction method for LDPC codes can be adapted to avoid 4-cycles, without disturbing the row-degree distribution, by checking each column before it is added to see whether it will cause a cycle with any of the columns already chosen and rejecting it if it will.

Example 3.12 If a 4-cycle-free code had been required in Example 3.11 then the fourth column would have had to have been discarded, and a new one chosen, because it causes a 4-cycle with the first column in H.

In principle this process can be extended to include larger cycles and even column configurations that lead to small stopping sets or codewords. However, as more complex configurations are avoided the algorithm becomes significantly less likely to converge to a final parity-check matrix.

Even for 4-cycle-free codes there is no guarantee that the algorithm will converge, while removing the 4-cycles after construction can have the effect of disturbing the row-degree distribution. For long codes H will be very sparse and so 4-cycles will be very uncommon, and the effect on the row degrees of removing 4-cycles will be negligible. However, for short or high-rate codes, 4-cycle-free parity-check matrices can be constructed much more effectively by using algebraic methods, as we will see in Section 3.3.2.

Column or row splitting

In this construction technique, cycles, or indeed any unwanted configurations, in the parity-check matrix are removed by splitting a column or row in half. In column splitting a single column in H is replaced by two columns that share the entries of the original column between them. Since an extra column has been added, the parity-check matrix is made slightly more sparse (the number of 1 entries is the same but more 0 entries have been added) and a new code is produced whose length is greater by one than that of the previous code.

Example 3.13 Figure 3.10 shows column splitting applied to remove a 4-cycle.

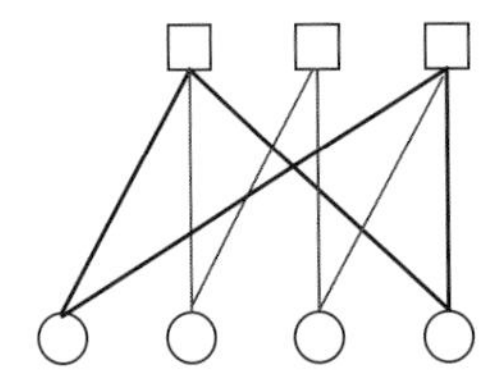

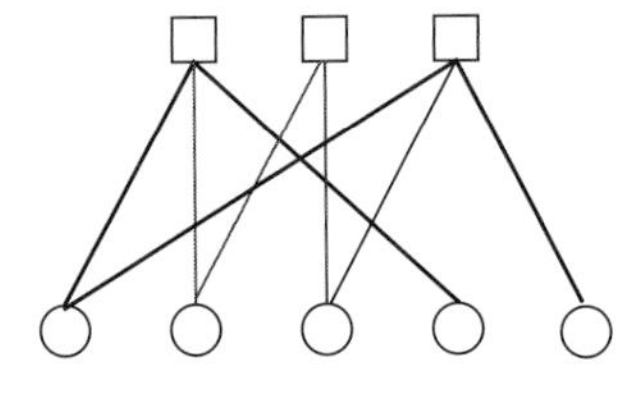

Figure 3.10 Column splitting to remove a 4-cycle.

Alternatively, a single row in H can be replaced by two rows that share the entries of the original row between them. Since an extra row has been added, a new code is produced having one more parity-check equation in H, thus reducing the code rate and producing a parity-check matrix that is slightly more sparse (the number of 1 entries is the same but more 0 entries have been added).

Example 3.14 Figure 3.11 shows row splitting applied to remove a 4-cycle.

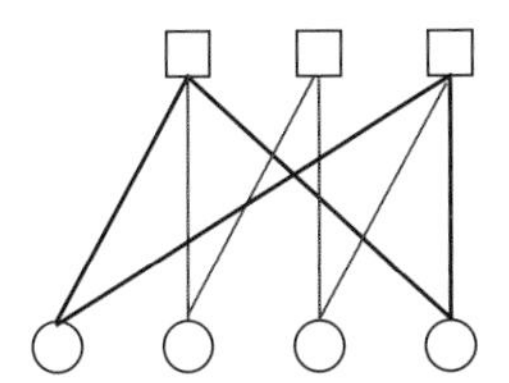
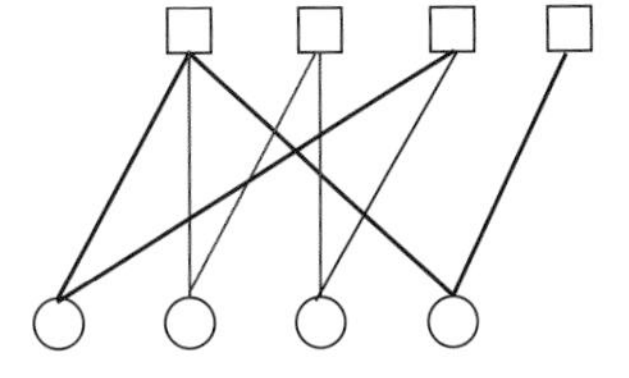

Figure 3.11 Row splitting to remove a 4-cycle.

Bit-filling and progressive edge growth (PEG) Tanner graphs

In *bit filling*, bit nodes are added to the Tanner graph one at a time and edges connecting the new bit nodes to the graph are chosen to avoid cycles of size g. For each new bit node b_i, w_c check nodes are selected to be joined by an edge to b_i. The set of feasible check nodes comprises the nodes that are a distance $g/2$ or more edges away from all the check nodes already connected to b_i. From each set of feasible check nodes, the check node chosen is the one that has been least used so far (i.e. with the lowest degree).

Example 3.15 Figure 3.12 shows how bit filling is used to construct a Tanner graph with 4-cycles avoided. The first bit node is connected to three randomly selected check nodes. Next, the second bit node is connected to three randomly chosen check nodes with degree 0. Then the third bit node also adds three edges. Although the first edge can be connected to any of the check nodes the seventh is selected as it has the lowest degree. The remaining (first–sixth) check nodes (shown shaded in the top left graph) are all valid check nodes for the second edge and they all have the same degree, so any one can be chosen, say the second. When the third edge is added we note that the first and fourth check nodes are not so far connected to the third bit node; however, they are not valid check nodes for the third edge since connecting to them would form a 4-cycle with the first bit node through the second check node. Each valid check node (shown shaded in the top right graph) has the same degree (1) so again one is randomly chosen, the fifth.

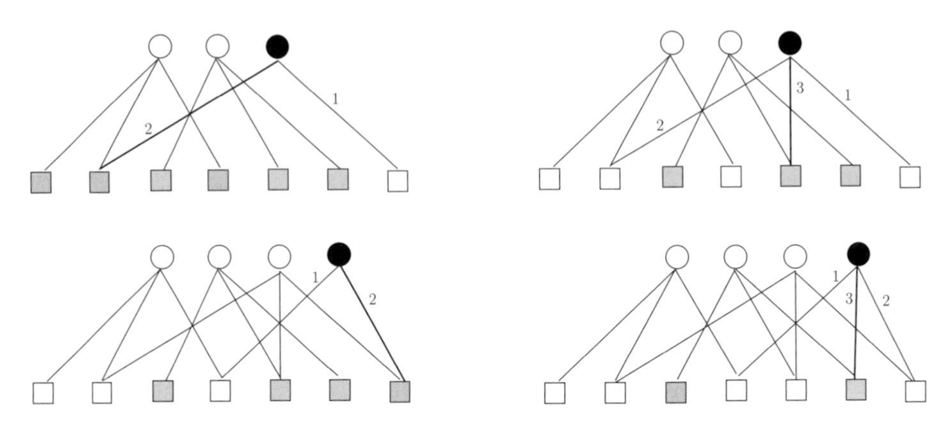

Figure 3.12 Bit filling to avoid cycles of size 4.

Lastly, the fourth bit node also adds three edges. Again the first edge can be connected to any check node, and the fourth check node is randomly chosen. The shaded check nodes in the bottom left graph show valid locations for the second edge. The first and second check nodes are not valid check nodes for this edge as they are both a distance 2 from the fourth check node. Of these the third, sixth and seventh have the lowest degree and the seventh is randomly chosen. The check nodes that are a distance 2 from the seventh check node (the second and fifth) are now removed from the set of valid check nodes for the third edge. The set of valid check nodes for this edge comprises the shaded check nodes in the bottom right graph. This leaves the third and sixth check nodes, and the sixth is randomly chosen.

In *progressive edge growth* Tanner graphs, edges are similarly added to the graph one at a time but, whereas previously all edges that met the minimum girth requirement were allowed, now only the edge that maximizes the local girth at the current bit node is chosen.

In Example 3.15 the bit-filling algorithm connected the second edge from the fourth bit node to the seventh check node (the bottom left graph in Figure 3.12) because it was one of the lowest-degree check nodes which did not add a 4-cycle. However, connecting to this check node did add a 6-cycle. Using a progressive edge growth construction, however, the third and sixth check nodes would be selected over the seventh check node as they add an 8-cycle rather than a 6-cycle.

Superposition and protographs

With the superposition technique, we start with an $m_b \times N_b$ *base matrix* H_b. The entries in H_b are replaced with $v \times v$ binary matrices, called the *superposition matrices*, to create an $m \times N$ LDPC code parity-check matrix H for which $m = m_b v$ and $N = N_b v$. Each zero entry in H_b is replaced by the $v \times v$

all-zeros matrix and each non-zero entry in H_b is replaced by a $v \times v$ binary matrix (which may involve using a different superposition matrix for each non-zero entry). Choosing circulant superposition matrices will produce quasi-cyclic codes.

Protograph LDPC codes are constructed by repeating a small graph multiple times and randomly permuting edges between the graph copies.

Example 3.16 Figure 3.13 shows an example of a protograph construction. A Tanner graph with two check nodes, three bit nodes (labeled A to C) and five edges (labeled 1 to 5) is repeated three times, or "lifted" by a factor 3, to produce a length-9 code. Next the edges are permuted, one edge at a time, between the graph copies. The key point is that one end of an edge is fixed (in our case the end connected to the check node) while the other end is randomly connected to the same bit node in one of the other graph copies. The second row of Figure 3.13

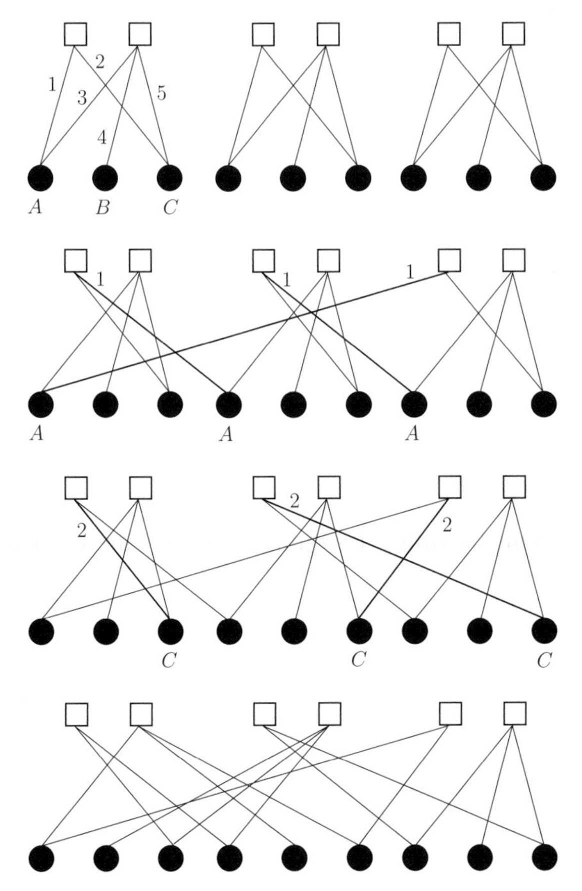

Figure 3.13 Protograph construction of LDPC codes. The bit nodes are solid circles, indicating that the corresponding bits are transmitted.

shows a permutation of the edges labeled 1 between the first check node and the bit nodes A. Next the edges labeled 2 between the first check node and the bit nodes C are similarly permuted. This process is repeated for every edge, giving the final Tanner graph at the bottom of Figure 3.13.

For each edge in the protograph a permutation is required to define the final expanded Tanner graph. Pseudo-random permutations that avoid small cycles can be found using algorithms such as the progressive edge growth algorithm above. Alternatively, cyclic shift permutations can be used to produce block-circulant quasi-cyclic codes with the resulting encoding benefits (see Section 3.2.4).

An advantage of the protograph construction is that the degree distribution of the original Tanner graph is maintained in the final graph (e.g. two-thirds of the bit nodes are degree-2 in the original graph and likewise in the final graph). Protograph constructions are thus particularly useful for constructing long irregular LDPC codes. Just as important, the protograph structure can significantly reduce the routing complexity associated with parallel or partially parallel implementations of the decoder.

If the protograph has no repeated edges, it can be defined by a binary parity-check matrix H_{p}. Then it is easy to see that the protograph type of construction is equivalent to superposition into H_{p} as base matrix, using permutation matrices for the superposition matrices. A *permutation matrix* is simply a square binary matrix where the weight of every row and column is 1. Choosing circulant permutation matrices for the superposition matrices will again return a quasi-cyclic code.

However, the superposition technique is not limited to permutation matrices; circulant matrices with row and column weights greater than 1 can be substituted into the base matrix. Thus, for protographs that have repeated edges, a non-binary matrix can be defined as the base matrix and the superposition matrices chosen to have row and column weights equal to the base matrix entry.

Example 3.17 The protograph for the A4RA code in [30], shown in Figure 3.14, has the base matrix

$$H_{\mathrm{b}} = \begin{bmatrix} 0 & 0 & 2 & 1 & 0 \\ 2 & 3 & 1 & 0 & 0 \\ 0 & 1 & 3 & 0 & 2 \end{bmatrix}.$$

A quasi-cyclic LDPC code can be found by substituting a weight-1 circulant for each 1 entry, a weight-2 circulant for each 2 entry, a weight-3 circulant for each 3 entry and the all-zeros matrix for each zero entry in H_{b}.

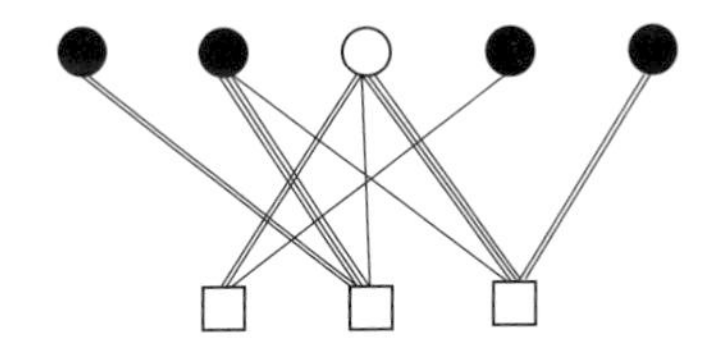

Figure 3.14 The A4RA protograph in Example 3.17. In a protograph the solid bit nodes correspond to transmitted bits while the open bit nodes correspond to bits that are not transmitted.

More successful constructions use a two-step process where each entry is replaced by small superposition matrices of the required weight (but not necessarily circulant) to produce an intermediate matrix and then a second superposition step into the intermediate matrix is performed using circulants for the superposition matrices.

3.3.2 Structured LDPC codes

Designing LDPC codes deterministically will not produce codes with better threshold SNRs (see Section 2.4) but can certainly improve their error floor performance and reduce their implementation complexity. A very large number of constructions have been proposed to produce structured LDPC codes and we will only consider a few of them here.

Codes from combinatorial designs

A combinatorial design is an assignment of a set of objects into subsets subject to some defined condition on the size, structure or composition of the subsets.

Example 3.18 A simple combinatorial problem is to arrange a set of seven academics into seven committees, with three academics in each committee, every academic serving on the same number of committees and each pair of academics serving together in exactly one committee. A solution is found by forming the set of academics ("points")

$$\mathcal{P} = \{1, 2, 3, 4, 5, 6, 7\}$$

into a *design* consisting of a set of committees ("blocks")

$$\mathcal{B} = \{[1, 3, 5], [2, 3, 7], [4, 5, 7], [1, 6, 7], [1, 2, 4], [3, 4, 6], [2, 5, 6]\}. \quad (3.13)$$

The solution to the problem in Example 3.18 of fairly assigning academics into committees is an example of a combinatorial design. Formally, a combinatorial design consists of a finite non-empty set $\mathcal{P}$ of points (academics) and a finite non-empty set $\mathcal{B}$ of subsets of those points, called blocks (committees), with no repeated blocks. A design is *regular* if the number of points in each block, and the number of blocks that contain each point, designated γ and ρ respectively, are the same for every point and block in the design.

For a *t-design* every set of t points occurs in a constant number λ of blocks, and thus for 2-designs every *pair* of points occurs in λ blocks; 2-designs are also called balanced incomplete block designs, and are denoted 2-$(v, b, \rho, \gamma, \lambda)$ or 2-(v, γ, λ) designs (see below for the definitions of v and b). The designs of interest for constructing LDPC codes are 2-designs with λ equal to 1, called *Steiner 2-designs*, in which (as in Example 3.18) every pair of points occurs in exactly one block and so any two blocks intersect in at most one position.

Every design can be represented by a $v \times b$ binary matrix $\mathcal{H}$ called an *incidence matrix*, where $v = |\mathcal{P}|$, $b = |\mathcal{B}|$, each column in $\mathcal{H}$ represents a block B_j of the design and each row a point P_i. The (i, j)th entry of $\mathcal{H}$ is 1 if the ith point is contained in the jth block, otherwise it is 0:

$$\mathcal{H}_{i,j} = \begin{cases} 1 & \text{if } P_i \in B_j, \\ 0 & \text{otherwise.} \end{cases} \tag{3.14}$$

Thus each element 1 of $\mathcal{H}$ corresponds to an occurrence, or *incidence*, of a point. The *incidence graph* of $\mathcal{H}$ has vertex set $\mathcal{P} \bigcup \mathcal{B}$, two vertices x and y being connected if and only if either $x \in \mathcal{P}$, $y \in \mathcal{B}$ and $P_x \in B_y$ or $x \in \mathcal{B}$, $y \in \mathcal{P}$ and $P_y \in B_x$; it is thus a bipartite graph.

The parameters of a 2-$(v, b, \rho, \gamma, \lambda)$ design are constrained by

$$v\rho = b\gamma, \tag{3.15}$$

as $v\rho$ and $b\gamma$ are each equal to the number of non-zero entries in the incidence matrix. Further, any given point P is in a pair with $\gamma - 1$ points in each of the ρ blocks containing it, and so there are $\rho(\gamma - 1)$ pairs involving P. However, P must be paired with each of the $v - 1$ other points exactly λ times and so

$$\lambda(v - 1) = \rho(\gamma - 1). \tag{3.16}$$

Thus the choice of any three of the parameters completely specifies the design, the remaining two parameters being determined by (3.16) and (3.15).

Example 3.19 The design in Example 3.18 can easily be seen to satisfy the regularity constraint, since there are $\gamma = 3$ points in every block (three academics in every committee), each point is in exactly $\rho = 3$ blocks (each academic is on exactly three committees) and each pair of points (academics) occurs in exactly one block (committee) together. Thus the blocks in $\mathcal{B}$ form a 2-design with $v = b = 7$, $\gamma = \rho = 3$ and $\lambda = 1$.

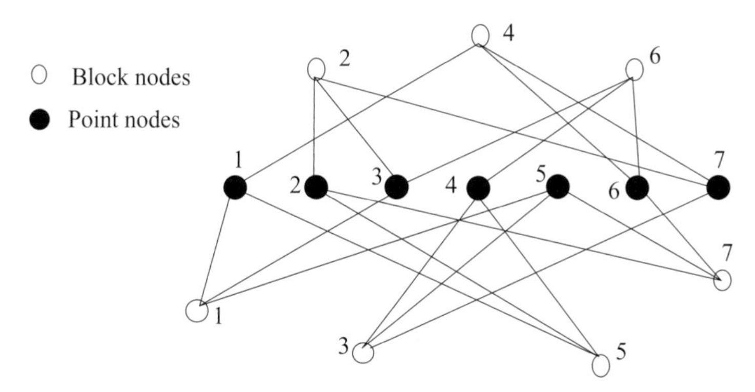

Figure 3.15 An incidence graph for the 2-(7, 3, 1) design in Example 3.18.

The incidence matrix for this design is

$$\mathcal{H} = \begin{bmatrix} 1 & 0 & 0 & 1 & 1 & 0 & 0 \\ 0 & 1 & 0 & 0 & 1 & 0 & 1 \\ 1 & 1 & 0 & 0 & 0 & 1 & 0 \\ 0 & 0 & 1 & 0 & 1 & 1 & 0 \\ 1 & 0 & 1 & 0 & 0 & 0 & 1 \\ 0 & 0 & 0 & 1 & 0 & 1 & 1 \\ 0 & 1 & 1 & 1 & 0 & 0 & 0 \end{bmatrix},$$

and an incidence graph for the design is shown in Figure 3.15, using an ordering of blocks 1–7 from left to right respectively.

An LDPC code is defined by setting the incidence matrix of the design as the parity-check matrix of the code, i.e. $H = \mathcal{H}$. The LDPC code parity-check matrix will thus have length $N = b$, with $m = v$ parity-check equations, and the code rate will be

$$r = \frac{K}{N} = \frac{b - \text{rank}_2 \mathcal{H}}{b}.$$

The incidence matrix in Example 3.19 has rank 4 and so will produce a length-7 rate-3/7 LDPC code.

Designs that are regular give (γ, ρ)-regular LDPC codes, and sparse codes are defined by choosing designs with γ and ρ small relative to v and b. In particular 4-cycle-free LDPC codes are guaranteed by choosing Steiner 2-designs, since then each pair of points cannot occur in more than one block (column of H), and so a 4-cycle cannot be formed. In fact, Steiner 2-designs put each pair of points in exactly one column, so they give the largest possible N value for a 4-cycle-free code with a given m and w_c. Thus, with the possible exception of low-rank parity-check matrices (if they exist), for a given code length Steiner 2-designs provide the largest possible rate codes that are 4-cycle free.

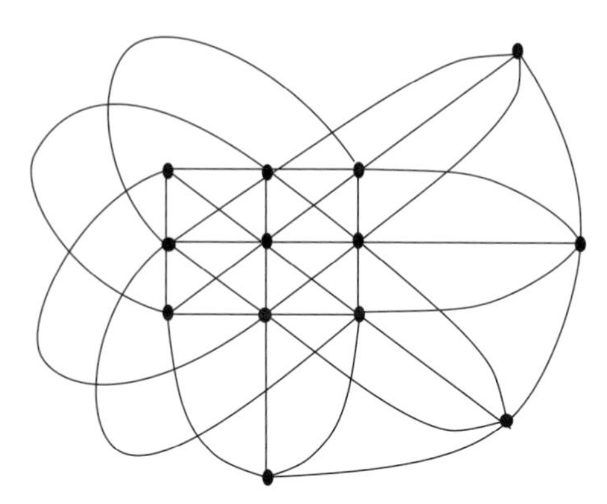

Figure 3.16 The finite projective plane of order 3 consists of 13 points on 13 lines.

Codes from finite geometries

A large class of designs can be constructed using the relationships between the points and lines, planes and hyperplanes of finite geometries to define the assignments of objects into sets. The points and lines of the two-dimensional projective geometry over the field $GF(q = 2^s)$ give projective geometry (PG) designs. The PG designs correspond to the set of $q^2 + q + 1$ lines and $q^2 + q + 1$ points such that every line passes through exactly $q + 1$ points and every point is incident on exactly $q + 1$ lines. Since any pair of points in the plane must define exactly one line, the points and lines of a projective plane are the points and blocks of a 2-$(q^2 + q + 1, q + 1, 1)$ design, the composition of the design being given by the composition of the plane. Figure 3.16 shows a typical representation of the finite projective plane of order 3.

Similarly, the points and lines of the 2-dimensional Euclidean geometry over the field $GF(q = 2^s)$ give Euclidean geometry (EG) designs while the points and lines of the higher-dimension projective geometries over the field GF$(q = 2)$ give 2-$(v, 3, 1)$ designs, commonly called Steiner triple systems (STSs).

Example 3.20 Figure 3.16 shows the finite projective plane of order 3 that consists of 13 points on 13 lines.

Example 3.21 An EG design is formed from the $2^{2s} - 1$ points of the Euclidean geometry on $GF(2^s)$ not including the origin; the blocks of the design are the $2^{2s} - 1$ lines of the geometry that do not pass through the origin. The incidence matrix of the EG design for $s = 4$ is thus a square 255×255 matrix with column and row weights both equal to 16. Although this incidence matrix is square it has a large number of linearly dependent rows, and rank 80. The LDPC code with parity-check matrix $\mathcal{H}$ produces a length-255 rate-175/255 LDPC code with a (16, 16)-regular parity-check matrix.

Figure 3.4 showed the bit error rate performance on a BI-AWGN channel of this EG LDPC code together with that for an LDPC code constructed pseudo-randomly using Algorithm 3.1. Although both codes have the same length and rate the EG code has significantly more rows in its parity-check matrix, and a much greater minimum distance, 17, that gives it its improved performance.

An important outcome of the work on algebraic codes was the demonstration that highly redundant parity-check matrices can lead to very good iterative decoding performances without the need for very long block lengths. While the probability of a random graph having a highly redundant parity-check matrix is vanishingly small, the field of combinatorial design offers a source of algebraic constructions for matrices that are both sparse and redundant.

However, the requirement that every pair of points must occur in exactly one block results in a large number of columns in H for a small number of rows and, for the most part, produces high-rate LDPC codes. Figure 3.17 shows the rates and lengths available for the LDPC codes from Steiner 2-designs and finite geometries.

Lower-rate codes can be produced by relaxing the constraint that every pair of points must occur in exactly one column to the lesser constraint that every pair

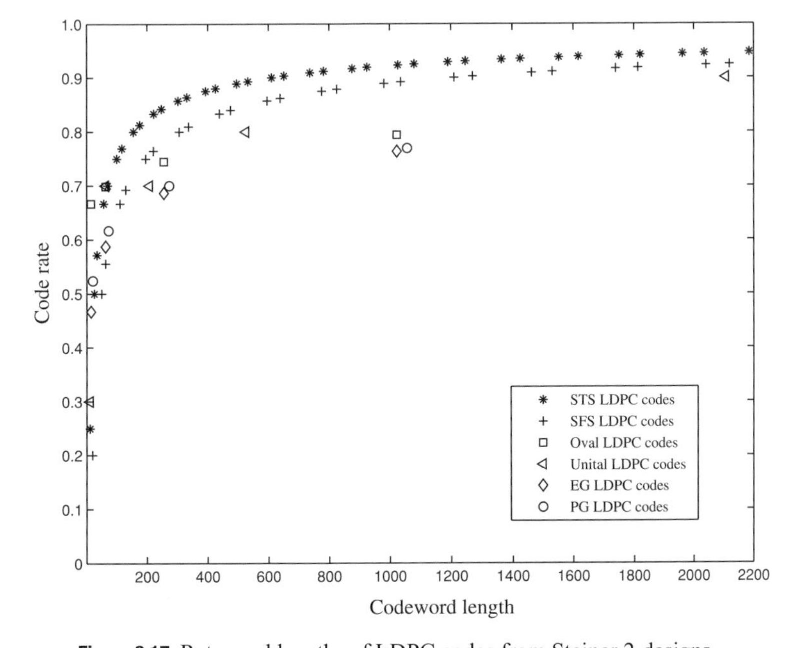

Figure 3.17 Rates and lengths of LDPC codes from Steiner 2-designs.

of points must occur in at most one column. One way of achieving this is by using a class of designs called *partial geometries*.

Partial geometries

A partial geometry, denoted pg(s, t, α), comprises a set of points and subsets of those points called blocks or lines, which satisfy the following properties:

Pi. Each point P occurs in $t + 1$ blocks and each block B has $s + 1$ points.
Pii. Any two blocks have at most one point in common.
Piii. For any non-incident point–block pair (P, B) (i.e. P is not in B) the number of blocks containing P and also intersecting B equals some constant α.

Example 3.22 The incidence matrix of the partial geometry pg(1, 2, 1) is:

$$N = \begin{bmatrix} 1 & 1 & 1 & 0 & 0 & 0 & 0 & 0 & 0 \\ 1 & 0 & 0 & 1 & 0 & 0 & 1 & 0 & 0 \\ 0 & 0 & 0 & 1 & 1 & 1 & 0 & 0 & 0 \\ 0 & 1 & 0 & 0 & 1 & 0 & 0 & 1 & 0 \\ 0 & 0 & 1 & 0 & 0 & 1 & 0 & 0 & 1 \\ 0 & 0 & 0 & 0 & 0 & 0 & 1 & 1 & 1 \end{bmatrix}.$$

Figure 3.18 gives its incidence graph.

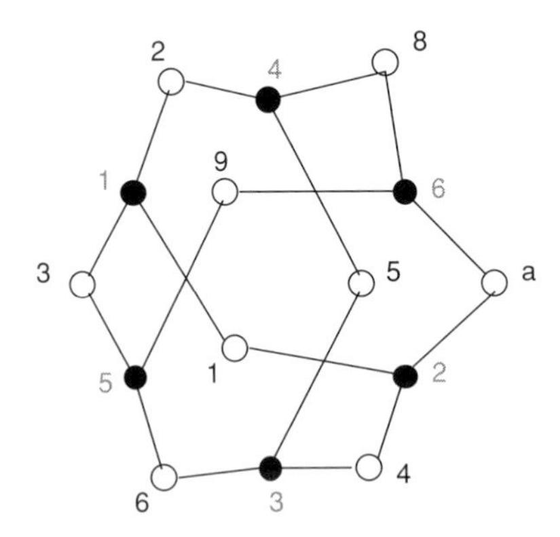

Figure 3.18 The incidence graph for the partial geometry pg(1, 2, 1) in Example 3.22.

The four main classes of partial geometries are as follows.

- A partial geometry with $\alpha = s + 1$ is a Steiner 2-design, since if a point P does not have an occurence in a block B then every one of the $s + 1$ blocks in which P does occur must intersect B and thus every pair of points must occur in a block together.

- A partial geometry with $\alpha = t$ is called a *net* or, dually with $\alpha = s$, a *transversal design*.
- A partial geometry with $\alpha = 1$ is called a *generalized quadrangle*.
- If $1 < \alpha < \min\{s, t\}$ then the partial geometry is *proper*.

Transversal designs, generalized quadrangles, and partial geometries make good LDPC codes. Generalized quadrangles in particular can define LDPC codes with girth 8.

It can be shown that the minimum distance of an LDPC code defined from a partial geometry pg(s, t, α) satisfies

$$d \geq \max\left\{\frac{(t+1)(s+1-t+\alpha)}{\alpha}, \quad \frac{2(s+\alpha)}{\alpha}\right\}, \tag{3.17}$$

(see e.g. [31, 32]). Thus for some codes, most notably those from generalized quadrangles, the lower bound on the minimum distance of the code is up to twice the column weight of H.

Example 3.23 The incidence matrix of the pg(2, 15, 2) transversal design produces the parity-check matrix for a length-256 rate-214/256 (3, 16)-regular

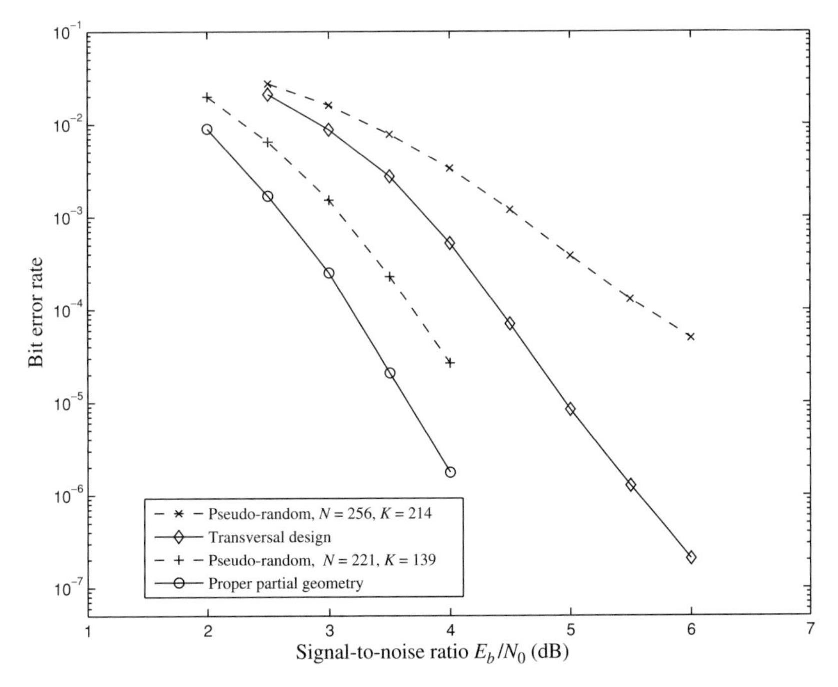

Figure 3.19 The decoding performance of length-255 and length-221 LDPC codes from partial geometries on a BI-AWGN channel using sum–product decoding with a maximum of 200 iterations. After [13], © IEEE 2004.

LDPC code. The incidence matrix of a pg(12, 12, 9) proper partial geometry design produces the parity-check matrix for a length-221 rate-139/221 (13, 13)-regular LDPC code with $d_{\min} \geq 14$.

Figure 3.19 shows the bit error rate performance on a BI-AWGN channel of these codes with, for comparison, LDPC codes with equivalent length and rate constructed pseudo-randomly using Algorithm 3.1.

The LDPC code produced from the pg(12, 12, 9) design significantly outperforms the random code of equivalent length and rate owing to its good minimum distance, 14, and large number (139) of linearly dependent rows. The LDPC code from the pg(2, 15, 2) design significantly outperforms the random code of equivalent length and rate owing to its minimum distance, 6; this performance cannot be achieved, in practice, using random constructions for such a high-rate code.

Resolvable designs

A simplistic approach to obtaining even lower-rate codes using the incidence matrix of a design is to choose a design with more blocks than the required code length and then to remove some columns of H. Alternatively, the rate of a design with the required length can be reduced using row splitting. For each pair of rows in the omitted column, or pair of columns across the split row, the corresponding incidence is now zero and the matrix thus formed is no longer the incidence matrix of a 2-design. However, 4-cycles (or 6-cycles using a partial geometry) will still be avoided in the Tanner graph of the resulting code since removing columns or splitting rows cannot add cycles.

Unfortunately, randomly removing columns from H results in a parity-check matrix with variable row weights and can lead to rows with all entries zero. To retain regular codes we would like to be able to remove a group of columns of H in such a way that we reduce by one the weight of every row in the matrix. This can be achieved with the use of designs that have the property of *resolvability*.

The concept of resolvability was introduced by the Revd T. P. Kirkman when he posed and solved the following problem:

Fifteen young ladies in a school walk out three abreast for seven days in succession: it is required to arrange them daily, so that no two shall walk twice abreast. (T. P. Kirkman, *Lady's and Gentleman's Diary*, 1847.)

If we think of girls as points and each column of three girls as a block, the solution to Kirkman's problem is a 2-(15, 3, 1) design. There are $v = 15$ girls, $\gamma = 3$ girls in each column, each pair of girls must appear together in a column once ($\lambda = 1$), each girl appears in $\rho = 7$ columns, one for each day of the

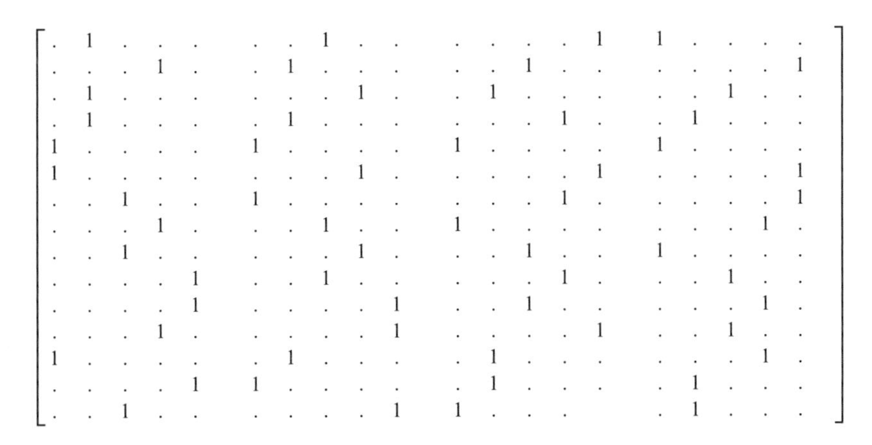

Figure 3.20 Part of the incident matrix of a Kirkman triple system on 15 points. Four of the seven resolution classes are shown. The dots represent zero entries.

week, and there are altogether $b = 35$ columns of girls, five for each day of the week.

The extra property required of the solution to Kirkman's problem that is not guaranteed by the structure of a Steiner triple system is that every set of five blocks that constitutes a single day must contain each point exactly once, since each girl must appear in precisely one of the columns each day. A design with this property is *resolvable*. Figure 3.20 shows the arrangement of Kirkman's schoolgirls for four of the days of the week.

More formally, a design is resolvable if the blocks of the design can be arranged into ρ groups, called resolution classes, such that the v/γ blocks of each resolution class are disjoint and each class contains every point precisely once. Steiner triple systems that are resolvable are called *Kirkman triple systems (KTSs)*.

Kirkman's problem for 15 schoolgirls can be applied to other numbers v of girls, where v is the number of points in a Steiner triple system. However, for the design to be resolvable there must be ρ groups of $v/3$ blocks, since there are $v/3$ blocks in each of r resolution classes. It was not until 1967 that Ray-Chaudhuri and Wilson showed that this necessary condition is also sufficient: Kirkman triple systems exist for any number of points $v \equiv 3 \pmod 6$. More generally, constructions exist for resolvable 2-(v, 4, 1) designs, ovals, Euclidean geometry (EG) and transversal designs.

Resolvable designs provide the solution to our problem of lower-rate LDPC codes. As previously we define the parity-check matrix of an LDPC code by the incidence matrix of a design. The number of blocks in the incidence matrix of a resolvable design, and hence the code length, can be reduced by v/γ blocks at a time while maintaining code regularity by simply removing all the blocks in a resolution class at the same time.

Difference families

In addition to a deterministic construction and guaranteed lower bounds on girth and minimum distance, the LDPC codes from combinatorial designs can also offer straightforward encoders. Many STS, Euclidean and projective geometry designs produce cyclic or quasi-cyclic codes. For example, the quasi-cyclic code in Example 3.9 was derived from a cyclically resolvable STS design.

Further quasi-cyclic LDPC codes can be constructed explicitly using combinatorial structures called *difference families*. A difference family is an arrangement of a group of v elements into not necessarily disjoint subsets of equal size that meet certain difference requirements. More precisely, the t γ-element subsets, called base blocks, of an Abelian group $\mathcal{G}$, $D_1, \ldots, D_t$ with $D_i = \{d_{i,1}, d_{i,2}, \ldots, d_{i,\gamma}\}$ form a (v, γ, λ) difference family if the differences $d_{i,x} - d_{i,y}$ $(i = 1, \ldots t;\ x, y = 1, \ldots, \gamma, x \neq y)$ give each non-zero element of $\mathcal{G}$ exactly λ times. If the Abelian group is Z_v then each translate is a cyclic shift and the difference family is a cyclic difference family.[3]

Example 3.24 The subsets $D_1 = \{1, 2, 5\}$, $D_2 = \{1, 3, 9\}$ of Z_{13} form a (13, 3, 1) difference family having the following differences:

$$\begin{array}{llll} \text{from } D_1, & 2-1=1, & 1-2=12, & 5-1=4, \\ & 1-5=9, & 5-2=3, & 2-5=10; \\ \text{from } D_2, & 3-1=2, & 1-3=11, & 9-1=8, \\ & 1-9=5, & 9-3=6, & 3-9=7. \end{array}$$

Difference families with $\lambda = 1$ allow the design of quasi-cyclic codes free of 4-cycles. To construct a length-vl rate-$(l-1)/l$ regular quasi-cyclic code $H = [a_1(x)\ a_2(x)\ \cdots\ a_l(x)]$ with column weight γ, take l of the base blocks of a $(v, \gamma, 1)$ difference family, and define the jth circulant of H as the transpose of the circulant formed from the jth base block in the difference family, i.e.

$$a_j(x) = x^{d_{j,1}} + x^{d_{j,2}} + \cdots + x^{d_{j,\gamma}}.$$

3.4 LDPC design rules

While there is no one recipe for a "good" LDPC code, there are a number of principles that inform the code designer.

Firstly, a good LDPC code is also a good classical block code. The power of sum–product decoding is that it can decode very long codes, but it is nevertheless

[3] The Abelian group Z_v is the set of residue classes modulo v, e.g. the set of positive integers $1, 2, \ldots, v$, where v, equivalent to zero, is the additive identity and addition is defined modulo v.

a sub-optimal decoding procedure whose performance is upper bounded by that of an optimal (ML and/or MAP) decoder (in many cases, however, it is impractical to implement such an optimal decoder). If an LDPC code has a poor minimum distance then the sum–product decoder will produce an error floor in exactly the same manner as the ML decoder. The reason why LDPC codes often do not show a high error floor is that for very long and very sparse codes it is relatively easy to construct a code pseudo-randomly with a good minimum distance.

A good classical code is, however, not necessarily a good LDPC code. Most critically, the sparsity of the parity-check matrix H is essential to keep the decoding complexity low and the sum–product decoder performance reasonably close to that of the ML decoder. A good LDPC code also has a Tanner graph with a large girth and good expansion. This increases the number of correlation-free iterations and improves the convergence of the decoder.

Other desirable properties of LDPC codes depend on how the codes are to be applied. For capacity-approaching performance on very-low-noise channels, long code lengths and random or pseudo-randomly constructed irregular parity-check matrices produce the best results. However, capacity-approaching performance in the bit error rate equates to poor word error rates and low error floors, making capacity-approaching codes completely unsuitable for some applications. Further, for higher-rate codes irregular constructions are generally not much better than regular ones, which have thresholds much closer to the Shannon capacity limit than lower-rate regular codes. (See Figure 7.9.)

We will see in Chapter 7 how the threshold of an ensemble of codes with a given degree distribution can be found using density evolution. Generally, the more irregular the bit-degree distribution the better. Capacity-approaching LDPC codes are both very long and very irregular. The famous LDPC ensemble with threshold SNR 0.0045 dB from the Shannon limit is 10^7 bits long with nodes having degrees varying from 2 to 8000. Since the overall density of H needs to be low, a large proportion of degree-2 bit nodes is also required, in order to reduce the average node degree. Thus a degree distribution with a good threshold will contain a few very-high-degree bit nodes, many degree-2 nodes and some nodes with degrees in between these.

When it comes to code construction for long codes, particularly for lower-rate codes, a randomly chosen parity-check matrix is almost always good and structured matrices are often much worse. The method of MacKay and Neal (see Example 3.11) almost always produces a fairly good LDPC code. Improved methods include progressive edge growth Tanner graphs, which are still constructed pseudo-randomly but can produce a code with a required minimum girth, and protograph constructions, which are also pseudo-random but can be used to reduce the implementation complexity of irregular codes (both methods were discussed in Section 3.3). Of course, with a 5×10^6 by 10^7 parity-check matrix with no structure, issues of encoding and storage become a problem. Protograph constructions can reduce the decoding complexity, and if cyclic permutations are used then a much more straightforward encoder implementation is also possible (see Section 3.2.4).

For short- and medium-length LDPC codes, particularly for higher rates, algebraic constructions can outperform pseudo-random ones. In addition, the use of structured parity-check matrices can lead to much simpler implementations, particularly for encoding, and can guarantee girth and minimum distance properties difficult to achieve pseudo-randomly for shorter codes.

Overall, in designing new LDPC codes the following code properties are generally taken into account.

- **Girth** Cycles in the Tanner graph affect decoding convergence: the smaller the code girth, the larger the effect on decoding performance. However, the avoidance of all cycles is neither practical nor desirable. The best approach is to avoid small cycles in the Tanner graphs of LDPC codes.
- **Minimum distance** A poor minimum distance will affect the performance of LDPC codes at high signal-to-noise ratios, causing an error floor, so codes with minimum distance as large as possible are preferred.
- **Stopping set distribution** The presence of small stopping sets reduces the effectiveness of erasure message-passing decoders. The minimum stopping set size, $S_{\min}$, is an indication of a code's performance with message-passing decoding just as the minimum distance is an estimate of a code's performance with maximum likelihood (ML) decoding.
- **Density** A sparse parity-check matrix is essential for the low-complexity operation of a sum–product decoder. However, there is a trade-off between the density of H and the code's minimum distance and stopping set size.
- **Encoding** Perhaps the biggest deterrent to the implementation of unstructured LDPC codes is the lack of an efficient encoding algorithm for them. While applying Gaussian elimination to transform the parity-check matrix into lower triangular form and using back substitution is effective, the new matrix is no longer sparse and the computational complexity of encoding is of order N^2. The requirement, therefore, is to produce codes with the structure to enable linear encoding complexity, by enabling a sparse generator matrix to be constructed using only row and column swaps, as in Section 2.3.1, by employing the shift invariance of cyclic codes as in Section 3.2.4 or by using repeat–accumulate codes (see Chapter 6).
- **Implementation** Where randomly constructed LDPC codes are employed the entire matrix must be stored, and this has repercussions for the system implementation requirements. For software implementations, deterministic codes that can be completely described by only a small number of parameters

may prove very useful while hardware implementations may benefit from some regularity.

3.5 Bibliographic notes

By proving the convergence of the sum–product algorithm for codes whose graphs are free of cycles, Tanner was the first to formally recognize the importance of cycle-free graphs in the context of iterative decoding [11]. The effect of cycles on the practical performance of LDPC codes was demonstrated by simulation experiments when LDPC codes were rediscovered by MacKay and Neal [33] (among others) in the mid 1990s, and the beneficial effects of using graphs free of short cycles were shown [12]. Since then the effect of cycles in the Tanner graph on decoding performance have been studied by a number of researchers (see e.g. [34–37]).

Sipser and Spielman [38] showed that the expansion of the graph is a significant factor in the application of iterative decoding. Using only a simple hard decision message-passing algorithm they proved that a fixed fraction of errors can be corrected in linear time provided that the Tanner graph is a sufficiently good expander. Stopping sets were introduced in [39] and used to develop analysis tools for finite-length LDPC ensembles. Stopping sets and finite-length analysis will be considered further in Chapter 8.

Since their introduction by Gallager, LDPC codes have for the most part been constructed pseudo-randomly, at least those that are simulated rather than implemented, and modifications have been made to Gallager's original construction to improve code properties such as girth or expansion [12, 33, 38, 40–44]. However, randomly constructed codes suffer from difficulties associated with implementing both the decoders and encoders, and the codes used in practice are rarely completely random. Indeed, the LDPC codes adopted for 10 Gb/s ethernet (IEEE 803.3an), WiFi (IEEE 802.11n) and WiMAX (IEEE 802.16e) wireless LAN standards are all block-circulant quasi-cyclic codes. Quasi-cyclic block codes were first presented in [45] and [46]. For a good introduction to quasi-cyclic codes in general see [47] or [5] and for more on quasi-cyclic LDPC codes see [30, 48–51].

Superposition and protograph-like constructions were introduced in [11, 52] and [53, 54] respectively and have been extensively used for LDPC codes. Superposition or protograph constructions can produce good codes for channels with memory [55, 56]. The LDPC codes currently proposed for deep space applications (see e.g. [30, 57]) use protograph constructions.

Tanner founded the topic of algebraic methods for constructing graphs suitable for sum–product decoding in [11]. The length-73 finite-geometry code was first implemented on an integrated circuit using iterative decoding by Karplus and Krit [58]. Many subsequent authors have considered the construction of LDPC codes using combinatorial designs [59–64], partial geometries [31], generalized quadrangles [32] and resolvability [62, 64–66]. Combinatorial methods for constructing quasi-cyclic LDPC codes can be found in [22, 61, 67]. Graph-based constructions for codes with good girth were presented by Margulis [68] and extended by Rosenthal and Vontobel [69] and Lafferty and Rockmore [70]. Vontobel and Tanner also used finite generalized polygons to construct LDPC codes that give the maximum possible girth for a graph with given diameter [32]. Other constructions for LDPC codes have been presented that have a mixture of algebraic and randomly constructed portions [71].

An important outcome of the work with algebraic codes has been the demonstration that highly redundant parity-check matrices can lead to very good iterative decoding performances without the need for very long block lengths [32,60]. While the probability that a random construction will produce a highly redundant parity-check matrix is vanishingly small, the field of combinatorial designs offers a rich source of algebraic constructions for matrices that are both sparse and redundant.

The monograph by Assmus and Key [72] gives an excellent treatment of the connection between classical error correction codes and designs. For more on designs see [73]; a good source for constructions is [74].

3.6 Exercises

3.1 For each of the parity-check matrices given in (2.7), (2.5) and (2.3) calculate their
(a) girth,
(b) minimum distance,
(c) minimum stopping set size.

3.2 For the parity-check matrix in Example 3.11 calculate its
(a) girth,
(b) minimum distance,
(c) minimum stopping set size.

3.3 Name an LDPC code property that is important in determining the error floor performance of such a code.

3.4 Name an LDPC code property that is important in determining the threshold performance of such a code.

3.5 Is it more important to ensure that a code is free of 4-cycles or free of 6-cycles? Why?

3.6 Give a potential advantage and a potential disadvantage of

(a) increasing the column weight of a regular LDPC code,
(b) choosing an optimized irregular LDPC code, and
(c) choosing an LDPC code free of all cycles.

3.7 The parity-check matrices given in (2.7), (2.5) and (2.3) are all parity-check matrices for the same code. Which matrix would you use for the iterative decoding algorithm and why?

3.8 What is the design rate of an LDPC code with degree distribution

$$\lambda(x) = \tfrac{1}{3}x + \tfrac{2}{3}x^3, \quad \rho(x) = x^5?$$

3.9 Give the node degree distribution of the irregular LDPC code considered in Example 3.2.

3.10 Construct length-100 rate-1/2 and length-1000 rate-1/2 Gallager LDPC codes. Generate bit error rate plots for bit-flipping and sum–product decoding of this code on a BI-AWGN channel using the pseudo-code from Algorithms 2.2 and 2.3.

3.11 Repeat Exercise 3.10 but use a construction that avoids 4-cycles. Is there a performance difference?

3.12 Construct a length-100 rate-1/2 LDPC code using progressive edge growth Tanner graphs. What is the best girth that you can achieve?

3.13 Design a protograph for a (3, 6)-regular LDPC code. Construct a length-100 rate-1/2 LDPC code using the protograph,
(a) using randomly chosen edge permutations,
(b) using cyclic edge permutations.
Generate bit error rate plots for bit-flipping decoding of this code on a BSC using the pseudo-code from Algorithms 2.2 and 2.3.

3.14 Design a protograph for an LDPC code with the degree distribution in Exercise 3.8. Construct a length-10 code with this degree distribution.

3.15 The incidence matrix of the partial geometry pg(1, 2, 1) from Example 3.22 is used as the parity-check matrix for an LDPC code. Calculate the
(a) rate,
(b) girth,
(c) minimum distance,
(d) minimum stopping set size
of the resulting code.

3.16 What are the lower bounds for the
(a) girth,
(b) minimum distance,
(c) minimum stopping set size

of any LDPC code formed from the resolution classes of any Kirkman triple system? Suppose that a rate-$1/4$ LDPC code is constructed from the four KTS resolution classes shown in Figure 3.20. Calculate its

(a) girth,
(b) minimum distance,
(c) minimum stopping set size.

4
Convolutional codes

4.1 Introduction

In this chapter we introduce convolutional codes, the building blocks of turbo codes. Our starting point is to introduce convolutional encoders and their trellis representation. Then we consider the decoding of convolutional codes using the BCJR algorithm for the computation of maximum a posteriori message probabilities and the Viterbi algorithm for finding the maximum likelihood (ML) codeword. Our aim is to enable the presentation of turbo codes in the following chapter, so this chapter is by no means a thorough consideration of convolutional codes – we shall only present material directly relevant to turbo codes.

4.2 Convolutional encoders

Unlike a block code, which acts on the message in finite-length blocks, a convolutional code acts like a finite-state machine, taking in a continuous stream of message bits and producing a continuous stream of output bits. The convolutional encoder has a memory of the past inputs, which is held in the encoder *state*. The output depends on the value of this state, as well as on the present message bits at the input, but is completely unaffected by any subsequent message bits. Thus the encoder can begin encoding and transmission before it has the entire message. This differs from block codes, where the encoder must wait for the entire message before encoding.

When discussing convolutional codes it is convenient to use time to mark the progression of input bits through the encoder. For example, we say that the input bit at time $t-1$ influences the output bit at time t but the output bit at time t does not depend on the input bits after time t.

A convolutional encoder can be represented by a linear finite-state shift register circuit where each shift register element represents a time delay of one unit. The bit at the output of the shift register element at time t is the bit that was present at the input of the shift register element at time $t-1$. The set of shift register elements of all the shift registers together holds the encoder state.

An encoder can have one or more shift registers, one or more inputs and one or more outputs. The convolutional code is said to have rate $r = k/n$ if, at each time instance t, the convolutional encoder receives k input bits and produces n output bits.

Example 4.1 Figure 4.1 shows a block diagram of a binary convolutional encoder. The blocks represent shift register elements and the circles represent modulo-2 adders. At time t the input to the encoder is one message bit, $u_t^{(1)}$, and the output is two bits, $c_t^{(1)}$ and $c_t^{(2)}$; thus the code rate is $1/2$. The convolutional codeword $\mathbf{c}$ is formed by interleaving the streams of bits $\mathbf{c}^{(1)}$ and $\mathbf{c}^{(2)}$:

$$\mathbf{c} = \left[c_1^{(1)}\ c_1^{(2)};\ c_2^{(1)}\ c_2^{(2)};\ c_3^{(1)}\ c_3^{(2)};\ \cdots;\ c_t^{(1)}\ c_t^{(2)}\right].$$

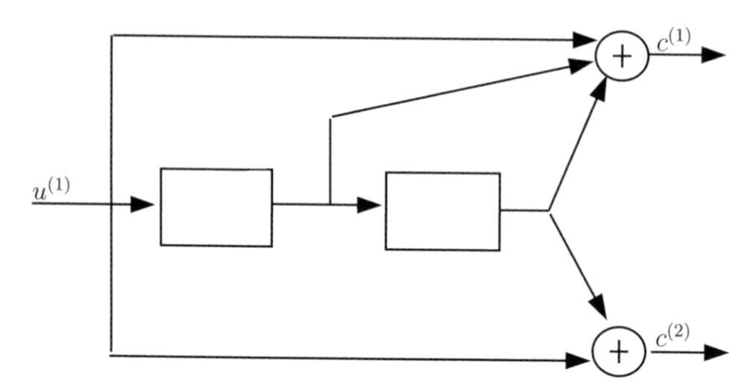

Figure 4.1 A rate-$1/2$ non-systematic non-recursive binary convolutional encoder; $u^{(1)}$ represents an input bit and $c^{(1)}$ and $c^{(2)}$ represent output bits.

For this encoder there is one shift register, which contains two shift register elements. The state of this encoder is given by $S = (s^{(1)}, s^{(2)})$, where $s^{(1)} \in \{1, 0\}$ is the content of the left-hand register element and $s^{(2)} \in \{1, 0\}$ is the content of the right-hand register element. Thus the encoder can be in one of four possible states, $S_0 = (0, 0)$, $S_1 = (0, 1)$, $S_2 = (1, 0)$ and $S_3 = (1, 1)$.

As there is only one input, the message $\mathbf{u}$ is simply given by $[u_1^{(1)}\ u_2^{(1)} \cdots u_t^{(1)}]$, and so here we will drop the superscript (1).

We can see from Figure 4.1 that for this encoder the output bit $c_t^{(1)}$ is the modulo-2 sum of the input bit u_t and the values of both shift register elements:

$$c_t^{(1)} = u_t \oplus s_t^{(1)} \oplus s_t^{(2)}, \tag{4.1}$$

where $\oplus$ represents modulo-2 addition. The output bit $c^{(2)}$ at time t, i.e. $c_t^{(2)}$, is the modulo-2 sum of the input bit u_t and the value of the second shift register element:

$$c_t^{(2)} = u_t \oplus s_t^{(2)}. \tag{4.2}$$

Example 4.2 Suppose that we have as input to the convolutional encoder in Figure 4.1 the 3-bit message $\mathbf{u} = [1\ 1\ 0]$. We start with the content of the shift registers as zero and so the output bits at $t = 1$ are

$$c_1^{(1)} = u_1 \oplus s_1^{(1)} \oplus s_1^{(2)} = 1 \oplus 0 \oplus 0 = 1, \qquad c_1^{(2)} = u_1 \oplus s_1^{(2)} = 1 \oplus 0 = 1.$$

At $t = 2$ the register values are shifted right by one register element and we now have a 1 in the left-hand register element, giving

$$c_2^{(1)} = u_2 \oplus s_2^{(1)} \oplus s_2^{(2)} = 1 \oplus 1 \oplus 0 = 0, \qquad c_2^{(2)} = u_2 \oplus s_2^{(2)} = 1 \oplus 0 = 1.$$

At $t = 3$ the register values are shifted across by one again and we now have a 1 in both shift register elements, giving

$$c_3^{(1)} = u_3 \oplus s_3^{(1)} \oplus s_3^{(2)} = 0 \oplus 1 \oplus 1 = 0, \qquad c_3^{(2)} = u_3 \oplus s_3^{(2)} = 0 \oplus 1 = 1.$$

Thus our codeword is $\mathbf{c} = [c_1^{(1)} c_1^{(2)}; c_2^{(1)} c_2^{(2)}; c_3^{(1)} c_3^{(2)}] = [1\ 1;\ 0\ 1;\ 0\ 1]$.

Example 4.3 The encoder of a rate $r = 2/3$ convolutional code is shown in Figure 4.2. The two input bits at time t, $u_t^{(1)}$ and $u_t^{(2)}$, are mapped to three output bits, $c_t^{(1)}$, $c_t^{(2)}$ and $c_t^{(3)}$.

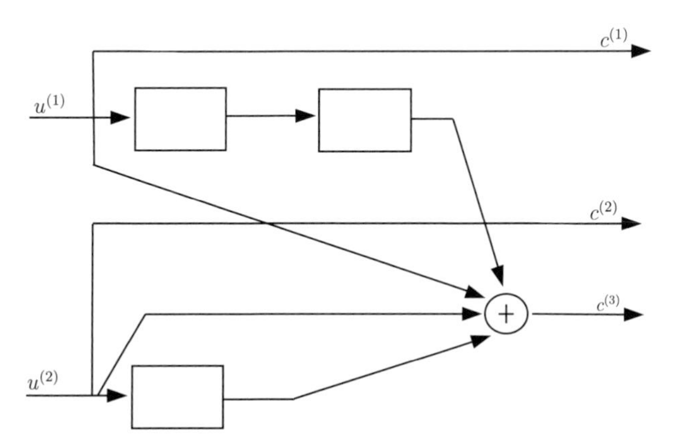

Figure 4.2 A rate-2/3 systematic non-recursive binary convolutional encoder.

For this encoder there are two shift registers, which contain two and one shift register elements respectively. The state of this encoder is given by

$S = (s^{(1)}, s^{(2)}, s^{(3)})$, where $s^{(1)} \in \{1, 0\}$ is the content of the top left-hand register element, $s^{(2)} \in \{1, 0\}$ the content of the right-hand register element in the top register and $s^{(3)} \in \{1, 0\}$ the content of the register element in the bottom register. Thus the encoder can be in one of nine possible states.

In practice, a convolutional code is usually *terminated* once all the message bits have been encoded by returning the encoder state to the zero state. This is done by appending extra *padding* bits to the message. For non-recursive convolutional codes of the form shown in Figure 4.1 and Figure 4.2 the extra bits to terminate the encoder are always zero.

Example 4.4 To terminate the encoder in Example 4.2, two bits, [0 0], are appended to the message to return the encoder to the state S_0.

At $t = 4$ the register values are shifted across by one again and so we now have a 0 in the left-hand register element and a 1 in the right-hand shift register element, giving

$$c_4^{(1)} = u_3 \oplus s_4^{(1)} \oplus s_4^{(2)} = 0 \oplus 0 \oplus 1 = 1, \qquad c_4^{(2)} = u_4 \oplus s_4^{(2)} = 0 \oplus 1 = 1.$$

At $t = 5$ the register values are shifted across by one again and we now have a 0 in both register elements, as required, giving

$$c_5^{(1)} = u_5 \oplus s_5^{(1)} \oplus s_5^{(2)} = 0 \oplus 0 \oplus 0 = 0, \qquad c_5^{(2)} = u_5 \oplus s_5^{(2)} = 0 \oplus 0 = 0.$$

The resulting terminated codeword is

$$\begin{aligned} \mathbf{c} &= [c_1^{(1)}\, c_1^{(2)};\, c_2^{(1)}\, c_2^{(2)};\, c_3^{(1)}\, c_3^{(2)};\, c_4^{(1)}\, c_4^{(2)};\, c_5^{(1)}\, c_5^{(2)}] \\ &= [1\ 1;\ 0\ 1;\ 0\ 1;\ 1\ 1;\ 0\ 0]. \end{aligned}$$

An alternative to adding padding bits is to use a tail-biting convolutional code, where the message is encoded in such a way that the initial and final states are the same. Using this approach there is no need for padding bits but the message does need to be encoded twice.

The number of message bits encoded at a time, before the encoder is terminated, is denoted by K and the total number of output bits generated by the encoder for these message bits is denoted by N. The terminated convolutional code can thus be thought of as a block code with length N and rate K/N. In Example 4.4 a length $K = 3$ message has been encoded into a length $N = 10$ codeword, so the overall code rate is $r = 3/10$. Terminating the encoder can thus reduce the code rate. However, for longer messages this rate reduction will be negligible.

In the encoder in Figure 4.2, the input bits $\mathbf{u}^{(1)}$ are mapped directly to the output bits $\mathbf{c}^{(1)}$ and the input bits $\mathbf{u}^{(2)}$ are mapped directly to the output bits $\mathbf{c}^{(2)}$. A convolutional code with all its input bits mapped directly to output bits is called *systematic*. For systematic convolutional codes the extra outputs that are not message bits are *parity bits*, using the same terminology as for block codes. The codewords from the encoder in Example 4.3 have the message bits $\mathbf{u}^{(1)}$ and $\mathbf{u}^{(2)}$ in positions $\mathbf{c}^{(1)}$ and $\mathbf{c}^{(2)}$ and the parity bits $\mathbf{p}^{(1)}$ in position $\mathbf{c}^{(3)}$.

4.2.1 Memory order and impulse response

The number of register elements in a shift register is called the *memory order* of the register. The memory order of the ith register is denoted by $m^{(i)}$. The *maximum memory order* of an encoder is defined as the maximum memory order of all the shift registers in the encoder. For example, the encoder in Figure 4.2 has memory orders $m^{(1)} = 2$, $m^{(2)} = 1$, and so the maximum memory order $b = 2$.

The total number of shift register elements in the encoder in all registers is called the *total memory* of the encoder. The encoder in Figure 4.2 thus has total memory 3. The total memory indicates the complexity of implementing the encoder, as it specifies how many memory elements will be required. A binary encoder with total memory v will have 2^v possible states; this number is called the *state complexity* of the encoder. For example, the convolutional encoder in Figure 4.1 has a lower state complexity, 4, than the encoder in Figure 4.2, which has state complexity 9. The state complexity will be important in determining the overall decoding complexity.[1]

The encoder memory order also plays an important role in the *impulse response* of the encoder. The impulse response determines which future output bits will be influenced by the current input bit. For the convolutional encoder in Figure 4.1, we saw that at time $t = 1$ the first input bit, $u_1^{(1)}$, affected both output bits $c_1^{(1)}$ and $c_1^{(2)}$. Then, at time $t = 2$, $u_1^{(1)}$ was shifted into the left-hand shift register element, where it played a role in determining $c_2^{(1)}$. At time $t = 3$, $u_1^{(1)}$ was shifted into the right-hand shift register element, where it affected the output bits $c_3^{(1)}$ and $c_3^{(2)}$. At time $t = 4$, $u_1^{(1)}$ was shifted out of the shift register and played no further part in the codeword.

Since the bit $u_1^{(1)}$ affected the output $\mathbf{c}^{(1)}$ for all three time points we say that $\mathbf{u}^{(1)}$ has an *impulse response* of [1 1 1] for the output $\mathbf{c}^{(1)}$. Similarly, since the bit $u_1^{(1)}$ affected the output at $\mathbf{c}^{(2)}$ for the time points $t = 1$ and $t = 3$, we say

[1] Often the implementation complexity of different convolutional encoders is compared by considering their *constraint length*. For an encoder with a single shift register, $v = b$ and the constraint length is usually defined as $v + 1$. However, definitions for the constraint length vary so we will not use this terminology in the present text.

that $\mathbf{u}^{(1)}$ has an impulse response of [1 0 1] for the output $\mathbf{c}^{(2)}$.[2] In general, the impulse response of an input $\mathbf{u}^{(i)}$ can be found by setting $\mathbf{u}^{(i)} = [1\,0\,0 \cdots, 0]$ and observing the encoder outputs.

Example 4.5 For the encoder in Figure 4.2 the input $\mathbf{u}^{(1)} = [u_1^{(1)}\, u_2^{(1)}\, u_3^{(1)}]$ has the impulse responses [1 0 0], [0 0 0] and [1 0 1] to $\mathbf{c}^{(1)}$, $\mathbf{c}^{(2)}$ and $\mathbf{c}^{(3)}$ respectively. The input $\mathbf{u}^{(2)}$ has the impulse responses [0 0], [1 0] and [1 1] to $\mathbf{c}^{(1)}$, $\mathbf{c}^{(2)}$ and $\mathbf{c}^{(3)}$ respectively.

Generator matrix

For each input–output pair of the encoder we can write down its *generator sequence* $g_{i,j}$, which is the impulse response of the ith input bit at the jth output bit. For the code of Example 4.1 the generator sequences are

$$g_{1,1} = [1\ 1\ 1] \qquad \text{and} \qquad g_{1,2} = [1\ 0\ 1],$$

where the leftmost entry of each sequence corresponds to the current input, at time t, and the rightmost entry corresponds to the input at time $t - b$.

Any sequence $y_0, y_1, y_2, \ldots, y_t$ in time can also be represented as a polynomial in D,

$$y_0 + y_1 D + y_2 D^2 + \cdots + y_t D^t,$$

where D is interpreted as a delay of one unit of time. Using this representation the *generator functions* for the code in Example 4.1 are

$$g_{1,1}(D) = 1 + D + D^2 \qquad \text{and} \qquad g_{1,2}(D) = 1 + D^2.$$

Finally, the *generator matrix* G of a convolutional code is a $k \times n$ matrix which has as its (i, j)th entry the polynomial $g_{i,j}(D)$. So the code of Example 4.1 has the generator matrix

$$G = \begin{bmatrix} 1 + D + D^2 & 1 + D^2 \end{bmatrix}. \tag{4.3}$$

We have already seen that the generator sequences for the encoder in Example 4.3 are

$$\begin{aligned} g_{1,1} &= [1\ 0\ 0], & g_{1,2} &= [0\ 0\ 0], & g_{1,3} &= [1\ 0\ 1], \\ g_{2,1} &= [0\ 0\ 0], & g_{2,2} &= [1\ 0\ 0], & g_{2,3} &= [1\ 1\ 0], \end{aligned}$$

[2] For non-recursive convolutional codes of the form shown in Figure 4.1 and Figure 4.2 the length of the impulse response for $\mathbf{u}^{(i)}$ is equal to $m^{(i)} + 1$. However, this observation does not hold for recursive convolutional encoders, which have an infinite impulse response.

and so this encoder has the generator matrix

$$G = \begin{bmatrix} 1 & 0 & 1+D^2 \\ 0 & 1 & 1+D \end{bmatrix}. \tag{4.4}$$

Given the generator-matrix view of systematic convolutional encoders it is easy to see, as in (4.4), the $k \times k$ identity matrix in the first k columns of G, in the same way that systematic block codes include an identity matrix in their generator matrix.

The generator of a convolutional code can also be specified in octal notation. Firstly, the polynomial $g(D)$ is expressed as a binary vector sequence $\mathbf{g}$ where the ith bit in $\mathbf{g}$ is the coefficient of D^{i-1}. For example, $g(x) = 1 + D^3 + D^4$ becomes [1 0 0 1 1]. Then each set of three bits, starting from the right, is written in octal notation: $0, 1, 1$ is written as 3 and $1, 0$ as 2. Thus instead of $1 + D^3 + D^4$ we write 23. Similarly, 31 specifies the polynomial $1 + D + D^4$.

4.2.2 Recursive convolutional encoders

So far we have considered only *non-recursive* convolutional encoders, i.e. convolutional encoders for which the encoder state is not dependent on the encoder output. In the next example we give a *recursive* (or *feedback*) convolutional encoder, where the encoder output is fed back into the encoder state.

Example 4.6 The encoder of a rate $r = 1/2$ systematic recursive convolutional code is given in Figure 4.3. The code is systematic since the input bits $u^{(1)}$ are directly mapped to the output bits $c^{(1)}$ and it is recursive since the output bits $c^{(2)}$ are mapped back into the shift register.

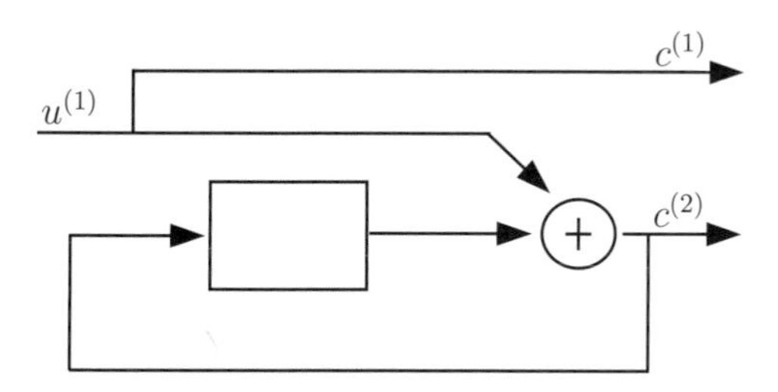

Figure 4.3 A rate-$1/2$ systematic recursive binary convolutional encoder.

The state of the encoder in Figure 4.3 is given by the contents of the single shift register element, $s^{(1)}$. We label the two possible encoder states by S_0, for $s^{(1)} = 0$, and S_1, for $s^{(1)} = 1$. The output bit $c_t^{(1)}$ is $u_t^{(1)}$ and the output bit $c_t^{(2)}$ is the parity bit formed from the modulo-2 sum of the input bit $u_t^{(1)}$ and the value of the shift register element $s_t^{(1)}$, and so we can write

$$c_t^{(1)} = u_t^{(1)} \quad \text{and} \quad c_t^{(2)} = u_t^{(1)} \oplus s_t^{(1)} = u_t^{(1)} \oplus c_{t-1}^{(2)}. \tag{4.5}$$

For this encoder the output at time t is a function of the input at time t and the output at time $t-1$, which is in turn dependent on the input at time $t-1$ and the output at time $t-2$, and so on; it can be seen that $c_t^{(2)}$ is a function of all previous inputs. For this reason recursive convolutional encoders have infinite impulse responses. However, we can still find a generator matrix for the encoder by writing (4.5) as a polynomial in D, which replaces the time index, and solving for $c^{(2)}$:

$$\begin{aligned} c^{(2)} &= u^{(1)} + c^{(2)} D \qquad (\text{mod } 2), \\ c^{(2)}(1+D) &= u^{(1)} \qquad (\text{mod } 2), \\ c^{(2)} &= \frac{u^{(1)}}{1+D}, \end{aligned} \tag{4.6}$$

which gives us the polynomial generator for $c^{(2)}$:

$$g_{1,2}(D) = \frac{1}{1+D}. \tag{4.7}$$

The generator for $c^{(1)}$ is simply $g_{1,1}(D) = 1$ and so we have the generator matrix

$$G = \begin{bmatrix} 1 & \frac{1}{1+D} \end{bmatrix}. \tag{4.8}$$

As with block codes, a single convolutional code can be represented by many different generator matrices; these can be produced by performing elementary row operations on G and so can be encoded by many different encoder circuits. Any non-recursive non-systematic convolutional encoder can be converted to a systematic recursive convolutional encoder by performing appropriate row operations on G.

Example 4.7 The non-systematic non-recursive convolutional encoder in Figure 4.1 has the generator matrix

$$G = [\,1 + D + D^2 \quad 1 + D^2\,]. \tag{4.9}$$

If we multiply G by $1/(1 + D + D^2)$ we obtain the generator matrix of a systematic recursive convolutional encoder for this code,

$$G = \begin{bmatrix} 1 & \dfrac{1+D^2}{1+D+D^2} \end{bmatrix}. \tag{4.10}$$

The systematic recursive convolutional encoder is shown in Figure 4.4.

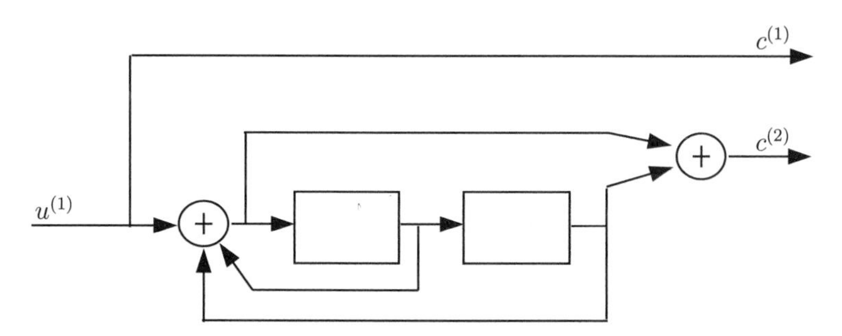

Figure 4.4 A rate-1/2 systematic recursive binary convolutional encoder.

Example 4.8 Suppose that we have as input to the convolutional encoder in Figure 4.4 the 3-bit message $\mathbf{u} = [1\ 1\ 0]$. We start with the content of the shift registers as zero, and so the output bits at $t = 1$ are

$$c_1^{(1)} = u_1 = 1, \qquad c_1^{(2)} = (u_1 \oplus s_1^{(1)} \oplus s_1^{(2)}) \oplus s_1^{(2)} = u_1 \oplus s_1^{(1)} = 1 \oplus 0 = 1.$$

At $t = 2$ the register values are

$$s_2^{(1)} = u_1 \oplus s_1^{(1)} \oplus s_1^{(2)} = 1 \oplus 0 \oplus 0 = 1, \qquad s_2^{(2)} = s_1^{(1)} = 0,$$

and so the output bits are

$$c_2^{(1)} = u_2 = 1, \qquad c_2^{(2)} = u_2 \oplus s_2^{(1)} = 1 \oplus 1 = 0.$$

At $t = 3$ the register values are

$$s_3^{(1)} = u_2 \oplus s_2^{(1)} \oplus s_2^{(2)} = 1 \oplus 1 \oplus 0 = 0, \qquad s_3^{(2)} = s_2^{(1)} = 1;$$

and so the output bits are

$$c_3^{(1)} = u_3 = 0, \qquad c_3^{(2)} = u_3 \oplus s_3^{(1)} = 0 \oplus 0 = 0.$$

Thus our codeword is

$$\mathbf{c} = [c_1^{(1)}\, c_1^{(2)}; c_2^{(1)}\, c_2^{(2)}; c_3^{(1)}\, c_3^{(2)}] = [1\ 1;\ 1\ 0;\ 0\ 0].$$

To determine which bits are required to terminate the code we note that at $t = 4$ the register values are

$$s_4^{(1)} = u_3 \oplus s_3^{(1)} \oplus s_3^{(2)} = 0 \oplus 0 \oplus 1 = 1, \text{ and } s_4^{(2)} = s_3^{(1)} = 0.$$

To terminate the code thus requires firstly an entry 1 to clear the leftmost shift register, producing the output bits

$$c_4^{(1)} = u_4 = 1, \qquad c_4^{(2)} = u_4 \oplus s_4^{(1)} = 1 \oplus 1 = 0.$$

The new state is given by

$$s_5^{(1)} = u_4 \oplus s_4^{(1)} \oplus s_4^{(2)} = 1 \oplus 1 \oplus 0 = 0, \qquad s_5^{(2)} = s_4^{(1)} = 1,$$

and so an entry 1 is required to keep the leftmost register zero while clearing the rightmost shift register, thus producing the output bits

$$c_5^{(1)} = u_5 = 1, \qquad c_5^{(2)} = u_5 \oplus s_5^{(1)} = 1 \oplus 0 = 1.$$

The final state is given by

$$s_6^{(1)} = u_5 \oplus s_5^{(1)} \oplus s_5^{(2)} = 1 \oplus 0 \oplus 1 = 0, \qquad s_6^{(2)} = s_5^{(1)} = 0,$$

as required. Finally, the complete terminated codeword is

$$\mathbf{c} = [1\ 1;\ 1\ 0;\ 0\ 0;\ 1\ 0;\ 1\ 1].$$

As we saw in Example 4.8, the appropriate padding bits to terminate recursive encoders will depend on which state the encoder is in after the final message bits. A simple circuit to terminate recursive codes uses the feedback bits as padding bits once the message has been encoded.

Example 4.9 Figure 4.5 shows the encoder circuit in Figure 4.4 modified to add padding bits. The input to the encoder is switched between point A, where the message is encoded, and point B, where the padding bits are added. By switching the input to point B the feedback bits are canceled and zeros are fed into the shift registers.

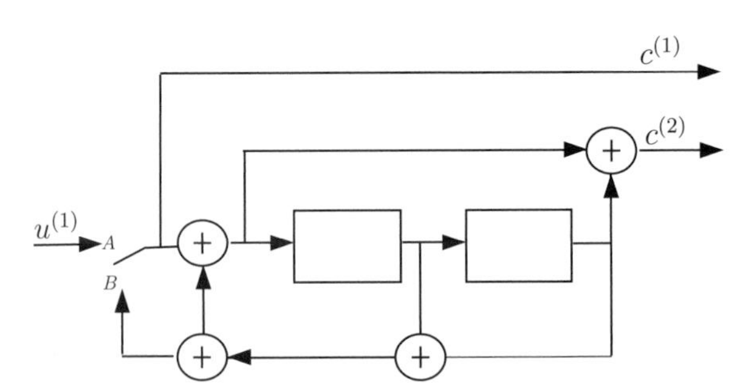

Figure 4.5 A rate-$1/2$ systematic recursive binary convolutional encoder with a circuit to generate padding bits.

Comparing Example 4.8 with Example 4.2 we see that two different encoders for the same convolutional code map the same message to two different codewords. Both encoders produce the same set of codewords; it is just the mapping from messages to codewords that differs. For example, encoding the messages [0 0], [0 1], [1 0] and [1 1] with the encoder in Figure 4.4 and adding padding bits results in the codewords [0 0; 0 0; 0 0; 0 0], [0 0; 1 1; 1 0; 1 1],

[1 1; 0 1; 0 1; 1 1] and [1 1; 1 0; 1 1; 0 0] respectively. Encoding the same four messages with the encoder in Figure 4.1, the codewords are [0 0; 0 0; 0 0; 0 0], [0 0; 1 1; 1 0; 1 1], [1 1; 1 0; 1 1; 0 0] and [1 1; 0 1; 0 1; 1 1] respectively. Both encoders produce the same set of codewords, since they describe the same code; however, the choice of encoder alters the mapping of messages to codewords.

Different encoders can have different state complexities and so the complexity associated with the decoding can vary depending on which encoder is chosen. The state complexity of an encoder with memory order v is 2^v, so increasing the number of registers by one doubles the total number of states and hence the decoding complexity.

4.2.3 Puncturing

The rate of a convolutional code can be reduced using *puncturing*. In a punctured code some subset of the codeword bits is not transmitted over the channel.

Example 4.10 Encoding the message [0 1 0 1 0 0] with the encoder in Figure 4.4 gives the codeword bits

$$\mathbf{c}^{(1)} = [0\ 1\ 0\ 1\ 0\ 0],$$
$$\mathbf{c}^{(2)} = [0\ 1\ 1\ 0\ 1\ 0].$$

Puncturing every third bit in the output $\mathbf{c}^{(2)}$ will produce the sequence

$$\mathbf{c}^{(2)} = [0\ 1\ x\ 0\ 1\ x],$$

where x indicates that the corresponding bit is not transmitted. The punctured encoder now produces 10 code bits for every six message bits and so the punctured code rate is $3/5$.

The puncturing pattern is specified by a *puncturing matrix* P. For an encoder with n output bits the matrix P will have n rows, one for each output stream. The number of columns in P is the number of bits over which the puncturing pattern repeats. For the puncturing pattern in Example 4.10 the puncturing matrix is

$$P = \begin{bmatrix} 1 & 1 & 1 \\ 1 & 1 & 0 \end{bmatrix}.$$

The zero entry in the third column of the second row indicates that every third bit in the output $\mathbf{c}^{(2)}$ is to be punctured.

4.2.4 Convolutional code trellises

One of the main advantages of convolutional codes is that they can be represented by a trellis. It is this trellis representation that enables optimal decoding with reasonable complexity. In this section we show how the evolution of a convolutional encoder in time can be succinctly represented by a trellis diagram.

We have already mentioned that a convolutional encoder can be thought of as a finite-state machine. Like any finite-state machine, a convolutional encoder can be represented by a *state diagram* detailing the relationship between input, state and output. The state diagram of an encoder with memory v (i.e. which contains v shift register elements) shows all 2^v encoder states and all the possible transitions between states.

Consider the systematic recursive convolutional encoder in Figure 4.4. The state of the encoder is given by the pair $S = (s^{(1)}, s^{(2)})$, where $s^{(1)} \in \{1, 0\}$ is the content of the left-hand shift register element and $s^{(2)} \in \{1, 0\}$ is the content of the right-hand shift register element. The encoder output at time t is given by

$$c_t^{(1)} = u_t^{(1)}, \qquad c_t^{(2)} = (u_t^{(1)} \oplus s_t^{(1)} \oplus s_t^{(2)}) \oplus s_t^{(2)} = u_t^{(1)} \oplus s_t^{(1)}, \tag{4.11}$$

and the state at time $t + 1$ is given by

$$s_{t+1}^{(1)} = u_t^{(1)} \oplus s_t^{(1)} \oplus s_t^{(2)}, \qquad s_{t+1}^{(2)} = s_t^{(1)}. \tag{4.12}$$

If the encoder is in the state $S_0 = (0, 0)$ and an input bit 1 is received then the new state can be calculated from (4.12):

$$s_{t+1}^{(1)} = u_t^{(1)} \oplus s_t^{(1)} \oplus s_t^{(2)} = 1 \oplus 0 \oplus 0 = 1, \qquad s_{t+1}^{(2)} = s_t^{(1)} = 0.$$

Similarly, the output bits can be calculated using (4.11):

$$c_t^{(1)} = u_t^{(1)} = 1, \qquad c_t^{(2)} = u_t^{(1)} \oplus s_t^{(1)} = 1 \oplus 0 = 1.$$

Repeating this calculation for all possible states and all possible inputs gives the state transitions shown in Table 4.1. The input bits that cause the state to transition from state S_r to S_s are labeled $\mathbf{u}_{r,s} = [u_{r,s}^{(1)} \cdots u_{r,s}^{(k)}]$ and the output codeword bits produced by the same state transition are labeled $\mathbf{c}_{r,s} = [c_{r,s}^{(1)} \cdots c_{r,s}^{(n)}]$. A graphical representation of the state transition table is given in a *state diagram* in Figure 4.6(a). Each of the four states is represented by a node. The edges between nodes represent the possible state transitions. Each edge is labeled with the input bit that produced the transition and the output bits generated.

The state diagram describes the convolutional encoder state and input–output relationship completely. However, it does not provide a record of how the state has evolved with time. For this we use a *trellis diagram*. Figure 4.6(b) shows the state diagram expanded in time to produce a trellis segment. On the left each state is represented for time t and on the right a copy of each state is represented

Table 4.1 The state transitions for the convolutional encoder in Figure 4.4.

Current state S_r		Input $\mathbf{u}_{r,s}$	Next state S_s		Output $\mathbf{c}_{r,s}$	
S_r	$(s_r^{(1)} s_r^{(2)})$	$u_{r,s}^{(1)}$	S_s	$(s_s^{(1)} s_s^{(2)})$	$c_{r,s}^{(1)}$	$c_{r,s}^{(2)}$
S_0	00	0	S_0	00	0	0
S_0	00	1	S_2	10	1	1
S_1	01	0	S_2	10	0	0
S_1	01	1	S_0	00	1	1
S_2	10	0	S_3	11	0	1
S_2	10	1	S_1	01	1	0
S_3	11	0	S_1	01	0	1
S_3	11	1	S_3	11	1	0

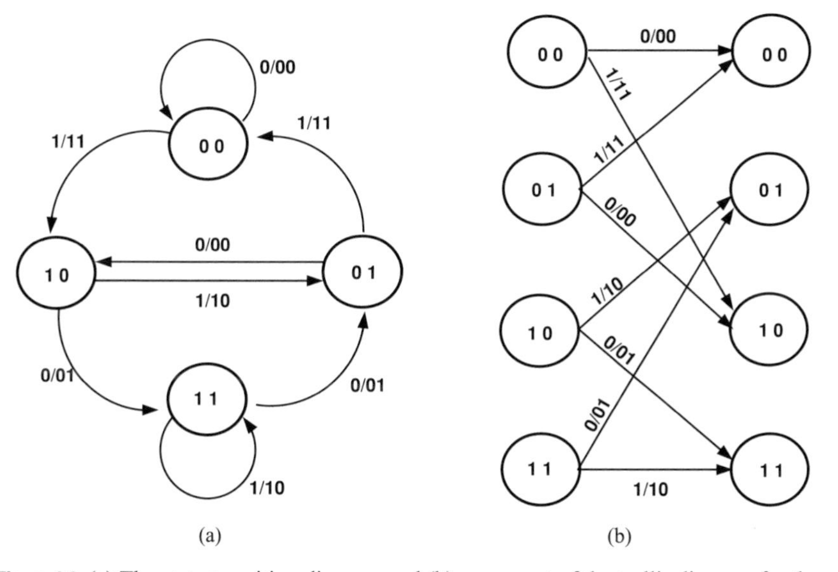

Figure 4.6 (a) The state transition diagram and (b) a segment of the trellis diagram for the rate-$1/2$ systematic recursive binary convolutional encoder in Figure 4.4.

for time $t + 1$. The state transition edges are joined from a state at time t to a state at time $t + 1$ to show the changes of state with time.

A convolutional encoder is assumed to be in the zero state S_0 prior to encoding and so the encoder trellis is drawn to start in state zero. Then, moving from left to right, every possible state transition is added to the trellis for each subsequent time instant. For example, at $t = 1$ an input bit 1 will move the encoder to state S_2 while a 0 input will move the encoder back to state S_0. Thus at $t = 1$ the states S_0 and S_2 are added to the trellis. Figure 4.7 shows the trellis diagram for the encoder in Figure 4.4 extended out to time $t = 5$. Each path through the trellis is an evolution of the convolutional encoder for one of the 2^K possible input

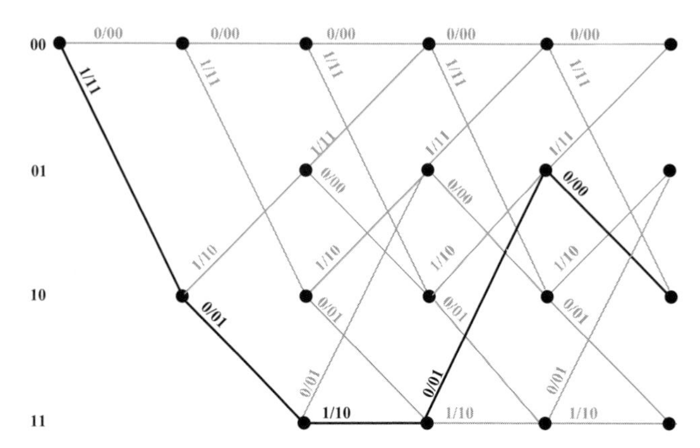

Figure 4.7 A trellis diagram for the rate-$1/2$ systematic recursive binary convolutional encoder in Figure 4.4. The state transitions for the message [1 0 1 0 0] are shown in bold.

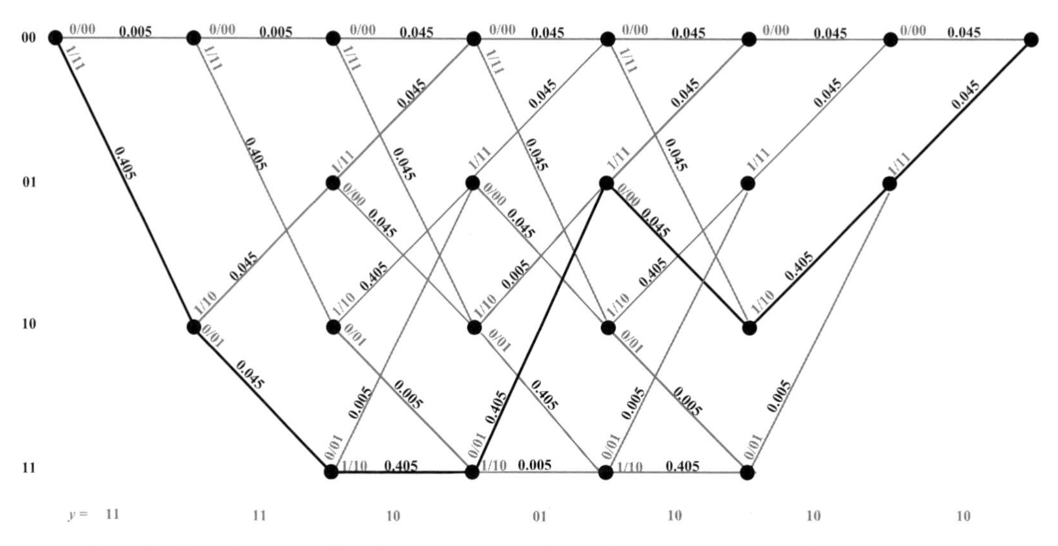

Figure 4.8 A terminated trellis. The γ values (in bold type) are for the decoder in Example 4.11.

streams. Consequently the set of codewords for a convolutional code is the set of all possible paths through its trellis. For example, the trellis path for the message $\mathbf{u}^{(1)} = [1\ 0\ 1\ 0\ 0]$ is shown in bold in Figure 4.7. Reading from the trellis edges we see that the state transitions for this message are S_0, S_2, S_3, S_3, S_1, S_2 and that the codeword generated by this message is $\mathbf{c} = [1\ 1;\ 0\ 1;\ 1\ 0;\ 0\ 1;\ 0\ 0]$.

The trellis of a terminated convolutional code ends in the zero state, and only the edges leading to the zero state are included for the trellis sections corresponding to the padding bits. Figure 4.8 shows an example of a terminated trellis.

4.2.5 Minimum free distance

The *minimum free distance* d_f of a convolutional code is the minimum Hamming distance between all pairs of terminated convolutional codewords. Because convolutional codes are linear, every codeword has the same number of other codewords differing from it in d bit locations, for every possible value of d. Thus the minimum free distance of a convolutional code is the minimum Hamming weight of the non-zero terminated codewords. Since terminated codewords must start and end in S_0, the minimum free distance can be defined as the minimum weight output over all trellis paths that start and end at the zero state. For example, the convolutional code having the trellis shown in Figure 4.7 has $d_f = 5$, which is the weight of the path $S_0 - S_2 - S_1 - S_0$. Importantly, the definition of the minimum free distance is independent of message length. Consider the weight-5 path $S_0 - S_2 - S_1 - S_0$, which corresponds to an input sequence [1 1 1]. We can choose the message length K by adding zeros before and/or after this sequence. These messages will always result in a weight-5 output, as the encoder will remain in the zero state, outputting zeros, before and after this [1 1 1] sequence.

Therefore, unlike for block codes, large minimum distances and, thus, low error probabilities are not obtained for convolutional codes by increasing the code length. Instead d_f can be increased by increasing the memory order v, thereby increasing the total number of encoder states and thus (if the code is designed well) increasing the number of states through which a trellis path must pass before it can return to the zero state. However, recall that increasing the memory order by just one doubles the state complexity, 2^v, of the decoder, so there is a trade-off between decoding performance and complexity.

Since different encoder realizations produce the same set of codewords, d_f is independent of the particular encoder circuit chosen for the code. When considering the performance differences of different encoders for the same code, the mapping of messages to codewords is important.

Non-recursive codes map weight-1 messages to finite-weight outputs (i.e. impulse responses); these are *finite impulse response (FIR)* encoders. Recursive codes map weight-1 messages to infinite-weight outputs; these are *infinite impulse response (IIR)* encoders. This difference is particularly important for turbo codes and IIR codes are employed for this reason.

If a low-weight codeword is mapped to a high-weight message, an error in that codeword will cause a large number of message bit errors. Since, for a given code, infinite impulse response encoders map higher-weight messages to the two lowest-weight codewords than finite impulse response encoders, they are poorer performers in high-SNR channels where low-weight codeword errors dominate. For this reason stand-alone convolutional codes operating at high SNRs generally use finite impulse response encoders. However, for the next few codeword weights the trend is reversed: the next few low-weight codewords are mapped to lower-weight messages by infinite impulse response encoders, making the latter better performers in low-SNR channels, where higher-weight codewords play an increasing role in the overall error rate.

4.3 Decoding convolutional codes

An advantage of convolutional codes is the availability of very efficient decoding algorithms operating on the code trellis. Each path in the trellis corresponds to a codeword, and so the maximum likelihood (ML) decoder (which finds the most likely codeword) searches for the most likely path in the trellis. Alternatively, each edge in the trellis can correspond to a particular input: the bit-wise maximum a posteriori (MAP) decoder, which searches for the maximum-probability input bit, calculates the probability of each trellis edge.

Both decoders receive a noisy version of the transmitted codeword from which to determine the message. We will denote the original message by

$$\mathbf{u} = [u_1 \cdots u_K] = [u_1^{(1)} \cdots u_1^{(k)};\ \cdots;\ u_T^{(1)} \cdots u_T^{(k)}],$$

where the number of bits in the message, K, is equal to the total number of encoder state transitions T multiplied by the number of input bits per state transition k, i.e. $K = kT$. The resulting codeword is denoted by

$$\mathbf{c} = [c_1 \cdots c_N] = [c_1^{(1)} \cdots;\ c_1^{(n)};\ \cdots;\ c_T^{(1)} \cdots c_T^{(n)}],$$

where the number of bits in the codeword, N, is equal to the number of encoder state transitions multiplied by the number of codeword bits per state transition, i.e. $N = nT$. We will denote the noisy values received from the channel by

$$\mathbf{y} = [y_1 \cdots y_N] = [y_1^{(1)} \cdots y_1^{(n)}; \cdots ; y_T^{(1)} \cdots y_T^{(n)}],$$

where we assume a binary input discrete memoryless channel.

4.3.1 Maximum a posteriori (BCJR) decoding

As we saw in Chapter 1, the binary symbol MAP decoder will output the probability $p(u_t^{(i)}|\mathbf{y})$ that the message bit $u_t^{(i)}$ was a 0 or 1 given all the information from the received vector $\mathbf{y}$ and the structure of the code. An efficient algorithm for performing MAP decoding using a trellis was first proposed by Bahl, Cocke, Jelenik and Raviv [75] and is called the *BCJR algorithm* after its authors.

Since the convolutional encoder has memory, the codeword bit output at time t is influenced by the codeword bits sent before it and may itself influence the codeword bits sent after it. Thus all the bits in $\mathbf{y}$ may tell us something about the message bit at time t. To incorporate the information from both the bits transmitted before time t and the bits transmitted after time t the BCJR decoding uses two passes through the trellis: a forward pass that predicts the current message bit on the basis of only the codeword bits that were transmitted before it, and a backwards pass that predicts the current message bit on the basis of only the codeword bits that were transmitted after it.

Since the message bits are the input to a binary convolutional encoder we can determine which message bit was sent by finding out which state transition occurred. We denote by $S+$ the set of state transitions (S_r, S_s) that correspond to a 1 input bit and by $S-$ the set of state transitions that correspond to a 0 input bit. For example, the convolutional encoder in Figure 4.4 (see Table 4.1) has

$$S+ = \{(S_0, S_2), (S_1, S_0), (S_2, S_1), (S_3, S_3)\}$$

and

$$S- = \{(S_0, S_0), (S_1, S_2), (S_2, S_3), (S_3, S_1)\}.$$

If $k > 1$ there will be a different set $S^{(i)}+$ and a different set $S^{(i)}-$ for each input.

The probability that $u_t^{(i)}$ was 1 is the probability that a state transition in the set $S^{(i)}+$ occurred at time t:

$$p(u_t^{(i)} = 1|\mathbf{y}) = \sum_{(S_r,S_s)\in S^{(i)}+} p(X_{t-1} = S_r, X_t = S_s|\mathbf{y}), \tag{4.13}$$

where X_t is the variable for the state at time t and $\{S_0, S_1, \ldots, S_{2^v-1}\}$ are the possible values that the state can take. We represent by S_r and S_s the values of the state at times $t-1$ and time t respectively. Similarly, the probability that $u_t^{(i)}$ was 0 is the probability that a state transition in the set $S^{(i)}-$ occurred at time t:

$$p(u_t^{(i)} = 0|\mathbf{y}) = \sum_{(S_r,S_s)\in S^{(i)}-} p(X_{t-1} = S_r, X_t = S_s|\mathbf{y}). \tag{4.14}$$

For convenience we will denote $p(X_{t-1} = S_r)$ as $p(S_r)$ and $p(X_t = S_s)$ as $p(S_s)$ when the context is clear.

Thus, determining the message bit probabilities $p(u_t^{(i)}|\mathbf{y})$ requires that we determine the probability of each state transition, $p(S_r, S_s|\mathbf{y})$, given that we have only the received vector $\mathbf{y}$. Using Bayes' rule we can rewrite $p(S_r, S_s|\mathbf{y})$ as

$$p(S_r, S_s|\mathbf{y}) = p(S_r, S_s, \mathbf{y})/p(\mathbf{y}).$$

Leaving the $p(\mathbf{y})$ term for now, we will focus on calculating $p(S_r, S_s, \mathbf{y})$, which we rewrite by splitting the received vector into three sets:

$$p(S_r, S_s, \mathbf{y}) = p(S_r, S_s, \mathbf{y}_t^-, \mathbf{y}_t, \mathbf{y}_t^+),$$

where $\mathbf{y}_t^+$ represents the values received for the set of bits sent after time t and $\mathbf{y}_t^-$ the values received for the set of bits sent before time t. The values received for the set of bits sent at time t, $\mathbf{y}_t$, are written as a vector since the convolutional code may output more than one codeword bit at each time point (i.e. n is often greater than 1).

Applying Bayes' rule again;

$$\begin{aligned} p(S_r, S_s, \mathbf{y}) &= p(S_r, S_s, \mathbf{y}_t^-, \mathbf{y}_t, \mathbf{y}_t^+) \\ &= p(S_r, S_s, \mathbf{y}_t^-, \mathbf{y}_t) p(\mathbf{y}_t^+ | S_r, S_s, \mathbf{y}_t^-, \mathbf{y}_t,) \\ &= p(S_r, \mathbf{y}_t^-) p(S_s, \mathbf{y}_t | S_r, \mathbf{y}_t^-) p(\mathbf{y}_t^+ | S_r, S_s, \mathbf{y}_t^-, \mathbf{y}_t). \end{aligned}$$

Since a convolutional encoder has been used to generate the codeword bits, we know that the codeword bits output at time t are completely determined by the state transition, S_r to S_s, at time t. Also, since we are considering a memoryless channel, the channel output $\mathbf{y}_t$ depends only on the transmitted codeword bit $\mathbf{c}_t$ and the channel noise at time t and is not affected by anything previously or subsequently transmitted through the channel. Putting these together, if we know the probability of the state transition S_r to S_s then the probability of $\mathbf{y}_t$ is completely independent of $\mathbf{y}_t^+$ and $\mathbf{y}_t^-$ and so $p(S_s, \mathbf{y}_t | S_r, \mathbf{y}_t^-) = p(S_s, \mathbf{y}_t | S_r)$. Similarly, if we know the probability of the encoder state at time t then the probability of $\mathbf{y}_t^+$ is independent of both the past states and past outputs and so $p(\mathbf{y}_t^+ | S_r, S_s, \mathbf{y}_t^-, \mathbf{y}_t,) = p(\mathbf{y}_t^+ | S_s)$. Thus, finally,

$$p(S_r, S_s, \mathbf{y}) = p(S_r, \mathbf{y}_t^-) p(S_s, \mathbf{y}_t | S_r) p(\mathbf{y}_t^+ | S_s). \tag{4.15}$$

When using BCJR decoding each term in (4.15) is given its own label:

$$\begin{aligned} \alpha_{t-1}(S_r) &= p(S_r, \mathbf{y}_t^-), \\ \beta_t(S_s) &= p(\mathbf{y}_t^+ | S_s), \\ \gamma_t(S_r, S_s) &= p(S_s, \mathbf{y}_t | S_r), \\ m_t(S_r, S_s) &= p(S_r, S_s, \mathbf{y}), \end{aligned}$$

and thus (4.15) is written as

$$m_t(S_r, S_s) = \alpha_{t-1}(S_r) \gamma_t(S_r, S_s) \beta_t(S_s). \tag{4.16}$$

Equation (4.16) tells us that the probability of the state transition from state S_r at time $t-1$ to state S_s at time t is a function of three terms:

(i) $\alpha_{t-1}(S_r)$, which is the probability that the encoder is in state S_r at time $t-1$ based on what we know about $\mathbf{y}_t^-$,
(ii) $\beta_t(S_s)$, which is the probability that the encoder is in state S_s at time t based on what we know about $\mathbf{y}_t^+$, and
(iii) γ_t, which is the probability of a transition between the states S_r and S_s based on what we know about $\mathbf{y}_t$.

The calculation of the α values is called the forward recursion of the BCJR decoder while the β values are calculated in the backward recursion.

Looking more closely at the γ term and using Bayes' rule, $\gamma_t(S_r, S_s)$ can be rewritten as follows:

$$\gamma_t(S_r, S_s) = p(S_s, \mathbf{y}_t | S_r) = p(S_s | S_r) p(\mathbf{y}_t | S_r, S_s).$$

In this form it is easier to see that $p(S_s, \mathbf{y}_t | S_r)$ has two parts. The first part, $p(S_s|S_r)$, is the probability that the state of the encoder moves to S_s at time t if it started in state S_r at time $t-1$. Since the encoder will have moved from S_r to S_s for an input $\mathbf{u}_{r,s}$, we know that $p(S_s|S_r) = p(\mathbf{u}_t = \mathbf{u}_{r,s})$, which is given by the probability distribution of the message source. The second part, $p(\mathbf{y}_t|S_r, S_s)$, is the probability that $\mathbf{y}_t$ was received given the state transition S_r to S_s. Since this state transition produces the codeword bits $\mathbf{c}_{r,s}$, the probability that $\mathbf{y}_t$ is received is equal to the probability that the channel turned $\mathbf{c}_{r,s}$ into $\mathbf{y}_t$, i.e. $p(\mathbf{y}_t|S_r, S_s) = p(\mathbf{y}_t|\mathbf{c}_t = \mathbf{c}_{r,s})$. Thus $\gamma_t(S_r, S_s)$ is a function of the source probability $p(\mathbf{u}_t = \mathbf{u}_{r,s})$ and the channel transition probability $p(\mathbf{y}_t|\mathbf{c}_{r,s})$:

$$\gamma_t(S_r, S_s) = p(\mathbf{u}_t = \mathbf{u}_{r,s}) p(\mathbf{y}_t | \mathbf{c}_{r,s}) \tag{4.17}$$

$$= \prod_i^k p(u_t^{(i)} = u_{r,s}^{(i)}) \prod_i^n p(y_t^{(i)} | c_{r,s}^{(i)}). \tag{4.18}$$

The key to the BCJR decoder, which is summarized in Algorithm 4.1, is that the values of α and β can be calculated recursively. Again using Bayes' rule:

$$\begin{aligned}
\alpha_{t-1}(S_r) &= p(S_r, \mathbf{y}_t^-) \\
&= \sum_{i=0}^{2^v-1} p(X_{t-2} = S_i, S_r, \mathbf{y}_t^-) \\
&= \sum_{i=0}^{2^v-1} p(X_{t-2} = S_i, \mathbf{y}_{t-1}^-) p(S_r | X_{t-2} = S_i, \mathbf{y}_{t-1}^-) \\
&\quad \times p(\mathbf{y}_{t-1} | X_{t-2} = S_i, S_r, \mathbf{y}_{t-1}^-) \\
&= \sum_{i=0}^{2^v-1} p(X_{t-2} = S_i, \mathbf{y}_{t-1}^-) p(S_r | X_{t-2} = S_i) p(\mathbf{y}_{t-1} | X_{t-2} = S_i, S_r) \\
&= \sum_{i=0}^{2^v-1} \alpha_{t-2}(S_i) \gamma_{t-1}(S_i, S_r).
\end{aligned} \tag{4.19}$$

Thus, for the forward recursion, the probability that the encoder is in state S_r at time t is the sum, over all of the states S_i at time $t-1$, of the probability that it is in state S_i times the probability of its moving from S_i to S_r. The encoder starts in the zero state and so at initialization $\alpha_0(S_0) = 1$.

Algorithm 4.1 BCJR decoding

```
1: procedure BCJR(y)
2:
3:     for t = 1 : T do                                              ▷ Initialization
4:         for r = 0 : 2^v − 1 do
5:             for s = 0 : 2^v − 1 do
6:                 if S_r to S_s is a valid trellis edge then
7:                     γ_t(S_r, S_s) = p(u_t = u_{r,s}) p(y_t | c_{r,s})
8:                 else
9:                     γ_t(S_r, S_s) = 0
10:                end if
11:            end for
12:        end for
13:    end for
14:
15:    α_0(S_0) = 1 and α_0(S_{i≠0}) = 0                             ▷ Forward recursion
16:    for t = 1 : T − 1 do
17:        for r = 0 : 2^v − 1 do
18:            for s = 0 : 2^v − 1 do
19:                α_t(S_s) = α_t(S_s) + α_{t−1}(S_r) γ_t(S_r, S_s)
20:            end for
21:        end for
22:    end for
23:
24:    if terminated then                                            ▷ Backward recursion
25:        β_T(S_0) = 1 and β_T(S_{i≠0}) = 0
26:    else
27:        β_T(S_i) = 1/2^v
28:    end if
29:
30:    for t = T − 1 : 1 do
31:        for r = 0 : 2^v − 1 do
32:            for s = 0 : 2^v − 1 do
33:                β_t(S_s) = β_t(S_s) + β_{t+1}(S_r) γ_{t+1}(S_r, S_s)
34:            end for
35:        end for
36:    end for
37:
38:    Continued over the page
```

Algorithm 4.1 BCJR decoding continued

```
39:     for t = 1 : T do                                          ▷ Bit probabilities
40:         for i = 1 : k do
41:             for r = 0 : 2^v − 1 do
42:                 for s = 0 : 2^v − 1 do
43:                     if (S_r, S_s) ∈ S^(i)+ then
44:                         p(u_t^(i) = 1|y) = p(u_t^(i) = 1|y) + m_t(S_r, S_s)
45:                     else p(u_t^(i) = 0|y) = p(u_t^(i) = 0|y) + m_t(S_r, S_s)
46:                     end if
47:                 end for
48:             end for
49:         end for
50:     end for
51: end procedure
```

Similarly, using Bayes' rule for β:

$$\begin{aligned}
\beta_t(S_s) &= p(\mathbf{y}_t^+|S_s) \\
&= \sum_{i=0}^{2^v-1} p(X_{t+1} = S_i, \mathbf{y}_t^+|S_s) \\
&= \sum_{i=0}^{2^v-1} p(\mathbf{y}_{t+1}|S_s, X_{t+1} = S_i)p(X_{t+1} = S_i|S_s) \\
&\quad \times p(\mathbf{y}_{t+1}^+|S_s, \mathbf{y}_{t+1}, X_{t+1} = S_i) \\
&= \sum_{i=0}^{2^v-1} p(\mathbf{y}_{t+1}^+|X_{t+1} = S_i)p(X_{t+1} = S_i|S_s)p(\mathbf{y}_{t+1}|S_s, X_{t+1} = S_i) \\
&= \sum_{i=0}^{2^v-1} \beta_{t+1}(S_i)\gamma_{t+1}(S_s, S_i).
\end{aligned} \tag{4.20}$$

For the backward recursion, the probability that the encoder is in state S_s at time t is the sum, over all the states S_i at time $t+1$, of the probability that it is in state S_i times the probability of its moving from S_s to S_i. If the encoder is terminated then it finishes in the zero state and so $\beta_T(S_0) = 1$. Alternatively, if the encoder has not been terminated, β is initialized in such a way that every state is equally likely, i.e. $\beta_T(S_i) = 1/2^v$ for all i.

The calculation of α and β involves multiplying together small numbers, and numerical stability can become a problem. However, this can be avoided by normalizing α and β at each step so that they sum to unity:

$$\alpha_t'(S_i) = \frac{\alpha_t(S_i)}{\sum_j \alpha_t(S_j)} \quad \text{and} \quad \beta_t'(S_i) = \frac{\beta_t(S_i)}{\sum_j \beta_t(S_j)}.$$

The BCJR algorithm calculates the γ, α and β values and then puts them together to find the state transition probabilities using (4.16). However, through (4.16) we have calculated $p(S_r, S_s, \mathbf{y})$, but we recall that we actually require

$$p(S_r, S_s|\mathbf{y}) = p(S_r, S_s, \mathbf{y})/p(\mathbf{y}).$$

Fortunately, given a probability distribution $p(a, b)$ we can easily calculate $p(b)$ since

$$p(b) = \sum_a p(a, b)$$

and so

$$p(a|b) = \frac{p(a, b)}{p(b)} = \frac{p(a, b)}{\sum_{a'} p(a', b)}.$$

Thus $p(\mathbf{y})$ is a scaling factor equal to

$$\sum_{S_r, S_s} p(S_r, S_s, \mathbf{y})$$

which affects each (S_r, S_s) pair equally, and so a normalizing factor can be applied to $m_t(S_r, S_s)$, just as for $\alpha_t(S_r)$ and $\beta_t(S_s)$:

$$m'_t(S_r, S_s) = \frac{m_t(S_r, S_s)}{\sum_{i,j} m_t(S_i, S_j)}. \tag{4.21}$$

Finally, then, the BCJR algorithm consists of an initialization in (4.17), a forward pass to calculate α in (4.19), a backward pass to calculate β in (4.20), the calculation of each state probability using (4.16) and lastly the calculation of message bit probabilities using (4.13) and (4.14).

The source message bit probabilities $p(u_t^{(i)})$ are called the a priori probabilities for $\mathbf{u}$ because they are known in advance before the BCJR decoder is run. The message bit probabilities $p(u_t^{(i)}|\mathbf{y})$ returned by the BCJR decoder are called the a posteriori probabilities for $\mathbf{u}$.

In Algorithm 4.1 the inputs are the a priori message probabilities $p(\mathbf{u}_t)$, the channel transition probabilities $p(\mathbf{y}_t|\mathbf{c}_{r,s})$ and information about the length-T convolutional trellis with 2^v states. This trellis information comprises the set of allowed state transitions S, the input bits $\mathbf{u}_{r,s} = [u_{r,s}^{(1)} \cdots u_{r,s}^{(k)}]$ for each state transition and the codeword bits $\mathbf{c}_{r,s} = [\mathbf{c}_{r,s}^{(1)} \cdots \mathbf{c}_{r,s}^{(n)}]$ produced by each state transition. The algorithm outputs the a posteriori bit probabilities of the kT message bits.

Example 4.11 The message $\mathbf{u} = [1\ 0\ 1\ 0\ 0]$ was generated by an equiprobable binary source and, using the convolutional encoder from Figure 4.4, was encoded into the codeword

$$\mathbf{c} = [1\ 1;\ 0\ 1;\ 1\ 0;\ 0\ 1;\ 0\ 0;\ 1\ 0;\ 1\ 1]$$

by the addition of two padding bits. This codeword was sent through a binary output binary-symmetric channel with crossover probability $\epsilon = 0.1$, and the vector

$$\mathbf{y} = [1\ 1;\ 1\ 1;\ 1\ 0;\ 0\ 1;\ 1\ 0;\ 1\ 0;\ 1\ 0]$$

was received. The BCJR decoder is used to find the MAP message bit probabilities.

For decoding, the a priori source probabilities $p(\mathbf{u}_t)$ and the channel transition probabilities $p(\mathbf{y}_t|\mathbf{c}_{r,s})$ are required. The source is equiprobable and so, for the message bits, $p(\mathbf{u}_t = \mathbf{u}_{r,s}) = 1/2$ for both $\mathbf{u}_{r,s} = 1$ and $\mathbf{u}_{r,s} = 0$. Since the encoder is IIR the padding bits could be either 1 or 0, so they are also assigned equal probabilities.

Since the channel is binary symmetric, the probability that $\mathbf{c}_{r,s}$ was sent if $\mathbf{y}_t = \mathbf{c}_{r,s}$ is received is the probability that no crossover occurred, which is $(1-\epsilon)^2$. Similarly, the probability that $\mathbf{y}_t$ differs in both digits from $\mathbf{c}_{r,s}$ is the probability that for both bits a crossover occurred, which is ϵ^2. Finally, the probability that $\mathbf{y}_t$ differs in one digit from $\mathbf{c}_{r,s}$ is the probability that only one crossover occurred, which is $\epsilon(1-\epsilon)$. Thus the channel transition probabilities $p(\mathbf{y}_t|\mathbf{c}_{r,s})$ for each $\mathbf{y}_t$ and $\mathbf{c}_{r,s}$ are as in the following table.

	$\mathbf{y}_t$			
$\mathbf{c}_{r,s}$	[0 0]	[0 1]	[1 0]	[1 1]
[0 0]	$(1-\epsilon)^2$	$(1-\epsilon)\epsilon$	$\epsilon(1-\epsilon)$	ϵ^2
[0 1]	$(1-\epsilon)\epsilon$	$(1-\epsilon)^2$	ϵ^2	$\epsilon(1-\epsilon)$
[1 0]	$\epsilon(1-\epsilon)$	ϵ^2	$(1-\epsilon)^2$	$(1-\epsilon)\epsilon$
[1 1]	ϵ^2	$\epsilon(1-\epsilon)$	$(1-\epsilon)\epsilon$	$(1-\epsilon)^2$

Step 1 of BCJR decoding (see Algorithm 4.1) involves calculating the γ's using (4.17). From Table 4.1 (or Figures 4.6(a) and 4.7) we see that the branch between S_0 and S_0 has $\mathbf{c}_{r,s} = [0\ 0]$. Since $\mathbf{y}_1 = [1\ 1]$ the branch metric for the S_0 to S_0 branch at $t = 1$ is

$$\begin{aligned}\gamma_1(S_0, S_0) &= p(\mathbf{u}_1 = \mathbf{u}_{0,0})p(\mathbf{y}_1|\mathbf{c}_{0,0}) \\ &= p(\mathbf{u}_1 = 0)p(\mathbf{y}_1 = [1\ 1]|\mathbf{c}_t = [0\ 0]) \\ &= 0.5 \times (0.1)^2 \\ &= 0.005.\end{aligned}$$

For the branch between S_0 and S_2 we see that $\mathbf{c}_{r,s} = [1, 1]$, giving for the branch metric at time $t = 1$

$$\begin{aligned}\gamma_1(S_0, S_2) &= p(\mathbf{u}_1 = \mathbf{u}_{0,2})p(\mathbf{y}_1|\mathbf{c}_{0,2}) \\ &= p(\mathbf{u}_1 = 1)p(\mathbf{y}_1 = [1\ 1]|\mathbf{c}_t = [1\ 1]) \\ &= 0.5 \times (1 - 0.1)^2 \\ &= 0.405.\end{aligned}$$

Continuing for $t = 2, \ldots, 7$ gives the complete branch metrics shown in Figure 4.8 on p. 134.

Step 2 is the forward pass using (4.19). Since the encoder is always initialized in the zero state, $\alpha_0(S_0) = 1$ and $\alpha_0(S_i) = 0, i \neq 0$. For time $t = 1$ the trellis diagram shows that the encoder can have moved to one of two states, S_0 or S_2. Thus

$$\begin{aligned}\alpha_1(S_0) &= \sum_{i=0}^{2^v-1} \alpha_0(S_i)\gamma_1(S_i, S_0) \\ &= \alpha_0(S_0)\gamma_1(S_0, S_0) + \alpha_0(S_1)\gamma_1(S_1, S_0) + \alpha_0(S_2)\gamma_1(S_2, S_0) \\ &\quad + \alpha_0(S_3)\gamma_1(S_3, S_0) \\ &= 0.005 + 0 + 0 + 0 \\ &= 0.005. \\ \alpha_1(S_2) &= \sum_{i=0}^{2^v-1} \alpha_0(S_i)\gamma_1(S_i, S_2) \\ &= \alpha_0(S_0)\gamma_1(S_0, S_2) + \alpha_0(S_1)\gamma_1(S_1, S_2) + \alpha_0(S_2)\gamma_1(S_2, S_2) \\ &\quad + \alpha_0(S_3)\gamma_1(S_3, S_2) \\ &= 1 \times 0.405 + 0 + 0 + 0 \\ &= 0.405.\end{aligned}$$

It is not possible for the encoder to be in states S_1 or S_3 at $t = 1$, and so $\alpha_1(S_1) = \alpha_1(S_3) = 0$. Finally we scale the α's to sum to unity by dividing through by $\alpha_1(S_0) + \alpha_1(S_1) + \alpha_1(S_2) + \alpha_1(S_3) = 0.41$ and thus obtain $\alpha_1(S_0) = 0.0122$, $\alpha_1(S_2) = 0.9878$ and $\alpha_1(S_1) = \alpha_1(S_3) = 0$. Continuing this process for all T time points gives the α values in the following table:

	$t = 0$	$t = 1$	$t = 2$	$t = 3$	$t = 4$	$t = 5$	$t = 6$	$t = 7$
S_0	1	0.0122	0.0006	0.0826	0.0221	0.3007	0.1198	1
S_1	0	0	0.4734	0.0916	0.8508	0.0726	0.8802	0
S_2	0	0.9878	0.0526	0.0826	0.0221	0.3007	0	0
S_3	0	0	0.4734	0.7432	0.1049	0.3260	0	0

Step 3 is the backward pass using (4.20). Since extra padding bits have been added so that the encoder will finish in the zero state, the default at time $t = 7$ is $\beta_7(S_0) = 1$ and $\beta_7(S_i) = 0, i \neq 0$.

Applying (4.20) gives the β values:

	$t=0$	$t=1$	$t=2$	$t=3$	$t=4$	$t=5$	$t=6$	$t=7$
S_0	1	0.6595	0.0286	0.1365	0.3320	0.0900	0.5000	1
S_1	0	0	0.0286	0.1365	0.3320	0.0900	0.5000	0
S_2	0	0.3405	0.1717	0.0798	0.2992	0.8100	0	0
S_3	0	0	0.7711	0.6472	0.0369	0.0100	0	0

For illustration, we calculate the β values for $t = 2$ explicitly:

$$\begin{aligned}
\beta_2(S_0) &= \sum_{i=0}^{2^v-1} \beta_3(S_i)\gamma_3(S_0, S_i) \\
&= \beta_3(S_0)\gamma_3(S_0, S_0) + \beta_3(S_1)\gamma_3(S_0, S_1) \\
&\quad + \beta_3(S_2)\gamma_3(S_0, S_2) + \beta_3(S_3)\gamma_3(S_0, S_3) \\
&= 0.137 \times 0.045 + 0.137 \times 0 + 0.080 \times 0.045 + 0.647 \times 0 \\
&= 9.765 \times 10^{-3},
\end{aligned}$$

$$\begin{aligned}
\beta_2(S_1) &= \sum_{i=0}^{2^v-1} \beta_3(S_i)\gamma_3(S_1, S_i) \\
&= \beta_3(S_0)\gamma_3(S_1, S_0) + \beta_3(S_1)\gamma_3(S_1, S_1) \\
&\quad + \beta_3(S_2)\gamma_3(S_1, S_2) + \beta_3(S_3)\gamma_3(S_1, S_3) \\
&= 0.137 \times 0.045 + 0.137 \times 0 + 0.080 \times 0.045 + 0.647 \times 0 \\
&= 9.765 \times 10^{-3},
\end{aligned}$$

$$\begin{aligned}
\beta_2(S_2) &= \sum_{i=0}^{2^v-1} \beta_3(S_i)\gamma_3(S_2, S_i) \\
&= \beta_3(S_0)\gamma_3(S_2, S_0) + \beta_3(S_1)\gamma_3(S_2, S_1) \\
&\quad + \beta_3(S_2)\gamma_3(S_2, S_2) + \beta_3(S_3)\gamma_3(S_2, S_3) \\
&= 0.137 \times 0 + 0.137 \times 0.405 + 0.080 \times 0 + 0.647 \times 0.005 \\
&= 5.872 \times 10^{-2},
\end{aligned}$$

$$\begin{aligned}\beta_2(S_3) &= \sum_{i=0}^{2^v-1} \beta_3(S_i)\gamma_3(S_3, S_i) \\ &= \beta_3(S_0)\gamma_3(S_3, S_0) + \beta_3(S_1)\gamma_3(S_3, S_1) \\ &\quad + \beta_3(S_2)\gamma_3(S_3, S_2) + \beta_3(S_3)\gamma_3(S_3, S_3) \\ &= 0.137 \times 0 + 0.137 \times 0.005 + 0.080 \times 0 + 0.647 \times 0.405 \\ &= 0.26272.\end{aligned}$$

The β values are normalized by scaling by their sum, $9.765 \times 10^{-3} + 9.765 \times 10^{-3} + 5.872 \times 10^{-2} + 2.627 \times 10^{-1} = 0.341$, which gives

$$\begin{aligned}\beta_2(S_0) &= 0.029, \\ \beta_2(S_1) &= 0.029, \\ \beta_2(S_2) &= 0.172, \\ \beta_2(S_3) &= 0.771.\end{aligned}$$

Step 4 involves combining the α, β and γ values to find the probability of each state transition using (4.16). For illustration, we calculate the state transition probabilities for $t = 6$:

$$\begin{aligned}m_6(S_0, S_0) &= \alpha_5(S_0)\gamma_6(S_0, S_0)\beta_6(S_0) = 0.3007 \times 0.09 \times 0.5 = 1.353 \times 10^{-2}, \\ m_6(S_1, S_0) &= \alpha_5(S_1)\gamma_6(S_1, S_0)\beta_6(S_0) = 0.0726 \times 0.09 \times 0.5 = 3.268 \times 10^{-3}, \\ m_6(S_2, S_1) &= \alpha_5(S_2)\gamma_6(S_2, S_1)\beta_6(S_1) = 0.3007 \times 0.810 \times 0.5 = 1.218 \times 10^{-1}, \\ m_6(S_3, S_1) &= \alpha_5(S_3)\gamma_6(S_3, S_1)\beta_6(S_1) = 0.3260 \times 0.010 \times 0.5 = 1.63 \times 10^{-3}.\end{aligned}$$

The state transitions $m_6(S_0, S_2)$, $m_6(S_1, S_2)$, $m_6(S_2, S_3)$ and $m_6(S_3, S_3)$ are all zero since these are not valid trellis edges at $t = 6$ and so the m_6 values for these edges are zero. Again, we normalize by the sum of the m_6 values, $1.353 \times 10^{-2} + 3.268 \times 10^{-3} + 1.218 \times 10^{-1} + 1.63 \times 10^{-3} = 0.1402$, obtaining

$$\begin{aligned}m_6(S_0, S_0) &= 0.097, \\ m_6(S_1, S_0) &= 0.023, \\ m_6(S_2, S_1) &= 0.869, \\ m_6(S_3, S_1) &= 0.012.\end{aligned}$$

Continuing for all t produces the state transition probabilities shown in Figure 4.9.

Lastly, the bit probabilities are calculated using (4.13) and (4.14). For example, to find the probability that the fourth input bit was 1 we sum the probabilities

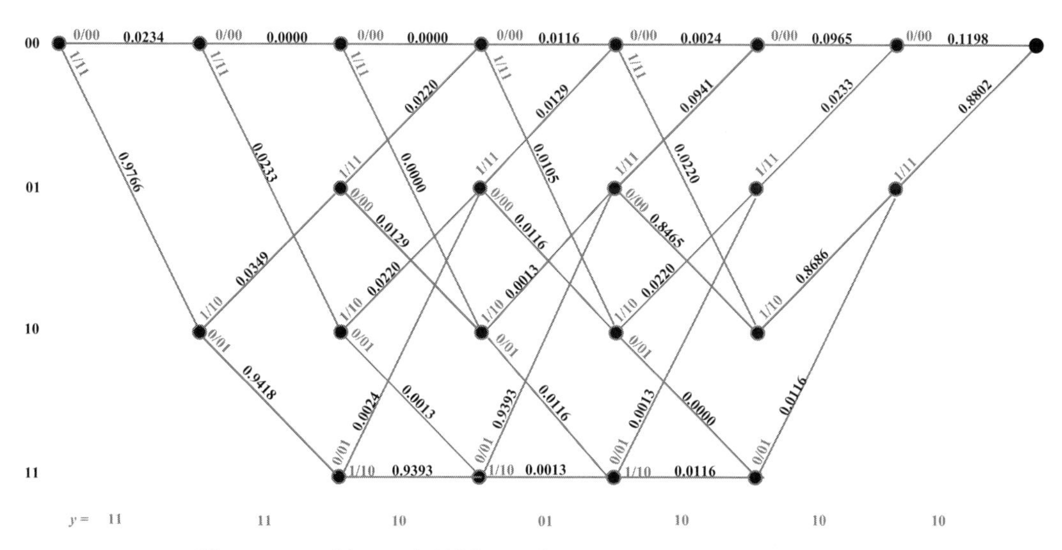

Figure 4.9 The state transition probabilities (to four decimal places) for the BCJR decoder in Example 4.11.

of all the state transitions caused by a 1. It can be seen from Table 4.1 (or Figures 4.6(a) and 4.7) that the state transitions caused by a 1 message bit input are $S+ = \{(S_0, S_2), (S_1, S_0), (S_2, S_1), (S_3, S_3)\}$. Thus

$$p(u_4 = 1|\mathbf{y}) = m_4(S_0, S_2) + m_4(S_1, S_0) + m_4(S_2, S_1) + m_4(S_3, S_3)$$
$$= 0.0105 + 0.0129 + 0.0013 + 0.0013 = 0.026.$$

Similarly, $S- = (S_0, S_0), (S_1, S_2), (S_2, S_3), (S_3, S_1)$ and so

$$p(u_4 = 0|\mathbf{y}) = m_4(S_0, S_0) + m_4(S_1, S_2) + m_4(S_2, S_3) + m_4(S_3, S_1)$$
$$= 0.0116 + 0.0116 + 0.0116 + 0.9393 = 0.974.$$

Repeating for all t gives the message probabilities in the following table:

	$t = 1$	$t = 2$	$t = 3$	$t = 4$	$t = 5$	$t = 6$	$t = 7$
$p(u_t^{(1)} = 0)$	0.023	0.942	0.017	0.974	0.850	0.108	0.120
$p(u_t^{(1)} = 1)$	0.977	0.058	0.983	0.026	0.150	0.892	0.880

The hard decision on the decoded message is the message bit with greater probability. For this example the decoded message is:

$$\hat{\mathbf{u}} = [1\ 0\ 1\ 0\ 0\ 1\ 1].$$

4.3.2 Log MAP decoding

In Example 4.11, at each step of the BCJR algorithm a normalization operation was carried out to obtain the final probability metrics. Furthermore, each step involved many multiplication operations over real numbers. This can make the algorithm very computationally demanding to implement, especially when longer codes are considered. Both problems can be overcome by using log metrics rather than probabilities directly. The benefit of the logarithmic representation of probabilities is that whereas probabilities need to be multiplied, log metrics need only be added:

$$\log\, p(a)p(b) = \log p(a) + \log p(b).$$

Probability values are easily recovered from the log metrics by using the inverse of the log function, the exponential function:

$$p(a) = e^{\log p(a)}.$$

We will use the capital letters $\mathcal{A}$, $\mathcal{B}$ and Γ for the log metrics of α, β and γ and write, for states S_r, S_s (see the discussion before (4.14) and Algorithm 4.1),

$$\begin{aligned}
\Gamma_t(S_r, S_s) &= \log \gamma_t(S_r, S_s) = \log p(\mathbf{u}_t = \mathbf{u}_{r,s}) p(\mathbf{y}_t|\mathbf{c}_{r,s}) \\
&= \log p(\mathbf{u}_t = \mathbf{u}_{r,s}) + \log p(\mathbf{y}_t|\mathbf{c}_{r,s}), \qquad (4.22)\\
\mathcal{A}_t(S_r) &= \log \alpha(S_r) = \log \sum_i \alpha_{t-1}(S_i)\gamma_t(S_i, S_r) \\
&= \log \sum_i e^{\mathcal{A}_{t-1}(S_i) + \Gamma_t(S_i, S_r)}, \qquad (4.23)\\
\mathcal{B}_t(S_s) &= \log \beta(S_s) = \log \sum_i \beta_{t+1}(S_i)\gamma_{t+1}(S_s, S_i) \\
&= \log \sum_i e^{\mathcal{B}_{t+1}(S_i) + \Gamma_{t+1}(S_s, S_i)}, \qquad (4.24)
\end{aligned}$$

and the state transition log metrics are then

$$M_t(S_r, S_s) = \log m_t(S_r, S_s) = \mathcal{A}_{t-1}(S_r) + \Gamma_t(S_r, S_s) + \mathcal{B}_t(S_s). \qquad (4.25)$$

Notice that, in (4.23) and (4.24), whenever probabilities were previously added the log metrics were converted back into probability values to carry out the addition. However, so as to simplify the metric calculations in (4.23) and (4.24), the Jacobian logarithm,

$$\max(a, b) = \log \frac{e^a + e^b}{1 + e^{-|a-b|}}, \qquad (4.26)$$

can be used to compute the log of the sum of exponentials. Thus we set

$$\overset{*}{\max}(a, b) \triangleq \max(a, b) + \log(1 + e^{-|a-b|}) = \log(e^a + e^b). \qquad (4.27)$$

Summing over more than two variables is then a concatenation of $\overset{*}{\max}$ operations:

$$\log \sum_i e^{a_i} = \overset{*}{\max_i}(a_i) = \overset{*}{\max}(\cdots \overset{*}{\max}(\overset{*}{\max}(a_i, a_2), a_3), \cdots a_i). \quad (4.28)$$

So, in the simplified log BCJR algorithm one calculates

$$\mathcal{A}_t(S_r) = \overset{*}{\max_i}(\mathcal{A}_{t-1}(S_i) + \Gamma_t(S_i, S_r)), \quad (4.29)$$

$$\mathcal{B}_t(S_s) = \overset{*}{\max_i}(\mathcal{B}_{t+1}(S_i) + \Gamma_{t+1}(S_s, S_i)). \quad (4.30)$$

Implementation of the $\overset{*}{\max}$ function involves only a max operation and a lookup table for the correction term $\log(1 + e^{|a-b|})$.

Log likelihood ratios

When using log BCJR decoding the probabilities input and output from the decoder are usually expressed as log likelihood ratios (LLRs).

For a binary variable x it is easy to find $p(x = 1)$ given $p(x = 0)$, since $p(x = 1) = 1 - p(x = 0)$, and so we only need to store one value in order to represent the set of probabilities for x. A useful method of representing the metrics for a binary variable by a single value is using LLRs to represent the metrics for a binary variable by a single value (1.2):

$$L(x) = \log \frac{p(x = 0)}{p(x = 1)},$$

where by log we mean $\log_e$. The sign of $L(x)$ provides a hard decision on x and the magnitude $|L(x)|$ gives the reliability of this decision. If $p(x = 1) > p(x = 0)$ then $L(x)$ is negative; furthermore, the greater the difference between $p(x = 1)$ and $p(x = 0)$, i.e. the more sure we are that $p(x = 1)$, the larger is the negative value for $L(x)$. Conversely, if $p(x = 0) > p(x = 1)$ then $L(x)$ is positive, and the greater the difference between $p(x = 1)$ and $p(x = 0)$ the larger the positive value for $L(x)$. Translating from LLRs back to probabilities, (1.3), (1.4), we obtain

$$p(x = 1) = \frac{e^{-L(x)}}{1 + e^{-L(x)}},$$

$$p(x = 0) = \frac{e^{L(x)}}{1 + e^{L(x)}}.$$

The a posteriori probabilities (APPs) output by the log BCJR decoder can be returned as an LLR:

$$\begin{aligned}
L(u_t^{(i)}|\mathbf{y}) &= \log \frac{p(u_t^{(i)} = 0|\mathbf{y})}{p(u_t^{(i)} = 1|\mathbf{y})} \\
&= \log \frac{\sum_{S-} p(S_r, S_s|\mathbf{y})/p(\mathbf{y})}{\sum_{S+} p(S_r, S_s|\mathbf{y})/p(\mathbf{y})} \\
&= \log \frac{\sum_{S-} e^{M_t(S_r,S_s)}}{\sum_{S+} e^{M_t(S_r,S_s)}} \\
&= \log \sum_{S-} e^{M_t(S_r,S_s)} - \log \sum_{S+} e^{M_t(S_r,S_s)} \\
&= \overset{*}{\max_{S-}}(M_t(S_r, S_s)) - \overset{*}{\max_{S+}}(M_t(S_r, S_s)).
\end{aligned} \tag{4.31}$$

As previously, we write $S+$ for the set (S_r, S_s) such that $(S_r, S_s) \in S$ is caused by $u_t^{(i)} = 1$ and $S-$ for the set (S_r, S_s) such that $(S_r, S_s) \in S$ is caused by $u_t^{(i)} = 0$. Using LLRs it is easy to see that $p(\mathbf{y})$ cancels out and so a separate normalization, as in (4.21), is no longer required.

Note that the BCJR algorithm can also output APPs for the codeword bits, and this will be required by serially concatenated turbo codes. The modification required to the log BCJR algorithm to return APP LLRs for the codeword bits is straightforward. Equation (4.31) becomes

$$\begin{aligned}
L(c_t^{(i)}|\mathbf{y}) &= \log \frac{p(c_t^{(i)} = 0|\mathbf{y})}{p(c_t^{(i)} = 1|\mathbf{y})} \\
&= \overset{*}{\max_{S'-}}(M_t(S_r, S_s)) - \overset{*}{\max_{S'+}}(M_t(S_r, S_s)),
\end{aligned} \tag{4.32}$$

where now $S'+$ is the set (S_r, S_s) such that $(S_r, S_s) \in S$ produces $c^{(i)} = 1$ and $S'-$ is the set (S_r, S_s) such that $(S_r, S_s) \in S$ produces $c^{(i)} = 0$.

Example 4.12 The message $\mathbf{u} = [1\ 0\ 1]$ is generated by a biased source that produces a 1 with greater probability $(p(u_t = 1) = 0.6)$ than a 0 $(p(u_t = 0) = 0.4)$. Using the convolutional encoder from Figure 4.3, $\mathbf{u}$ is encoded into the codeword

$$\mathbf{c} = [1\ 1;\ 0\ 1;\ 1\ 0;\ 0\ 0],$$

by the addition of a padding bit 0. The trellis for this encoder is shown in Figure 4.10. The vector

$$\mathbf{x} = [-1\ \ -1;\ +1\ \ -1;\ -1,\ +1;\ +1\ \ +1]$$

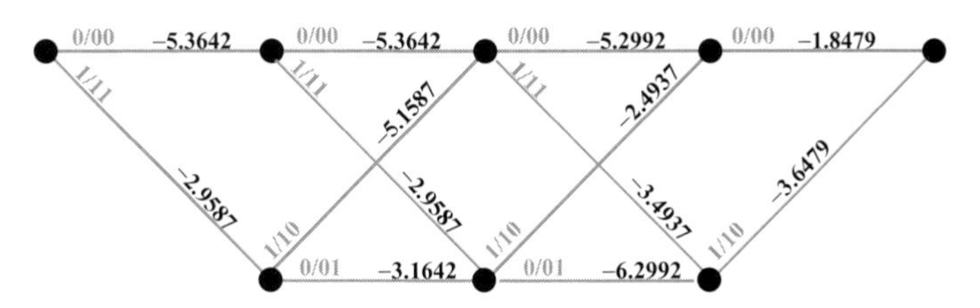

Figure 4.10 The Γ values for the BCJR decoder in Example 4.12.

is sent through a BI-AWGN channel with variance $\sigma^2 = 1$, and the vector

$$\mathbf{y} = [-1.1 \ +0.1; +0.1 \ -1.1; -1.2 \ +0.5; +0.9 \ +1.1]$$

is obtained at the receiver. Decoding is achieved with a log BCJR decoder.

From the source probabilities, the APPs for the message bits at $t = 1, 2, 3$ are

$$\log p(u_t^{(1)} = 1) = \log 0.6 = -0.5108,$$
$$\log p(u_t^{(1)} = 0) = \log 0.4 = -0.9163.$$

Since we have a recursive code, the value of the padded bit depends not on the source but on the encoder state after the last message bit, which is as of yet unknown. Thus the padded bits are assumed to be equiprobable:

$$\log p(u_4^{(1)} = 1) = \log 0.5 = -0.6931,$$
$$\log p(u_4^{(1)} = 0) = \log 0.5 = -0.6931.$$

Since the channel is BI-AWGN we have

$$\begin{aligned}\log p(\mathbf{y}_t|\mathbf{x}_t) &= \log\left(\left(\sqrt{\frac{1}{2\pi\sigma^2}}\right)^n \exp\left(-\frac{1}{2\sigma^2}\|\mathbf{y}_t - \mathbf{x}_t\|^2\right)\right)\\ &= -\frac{1}{2\sigma^2}\|\mathbf{y}_t - \mathbf{x}_t\|^2 + \log\left(\sqrt{\frac{1}{2\pi\sigma^2}}\right)^n. \qquad (4.33)\end{aligned}$$

Using these metrics we can calculate the Γ's from (4.22):

$$\begin{aligned}\Gamma_1(S_0, S_0) &= \log p(\mathbf{u}_1 = \mathbf{u}_{0,0}) + \log p(\mathbf{y}_1|\mathbf{c}_{0,0})\\ &= \log p(u_1 = 0) + \log p(\mathbf{y}_1 = [-1.1 \ 0.1]|\mathbf{x}_1 = [1 \ 1])\end{aligned}$$

$$
\begin{aligned}
&= \log 0.4 - \frac{1}{2}\|[-1.1\ 0.1] - [1\ 1]\|^2 + \log\left(\sqrt{\frac{1}{2\pi}}\right)^2 \\
&= -0.9163 - \frac{1}{2}\|[-2.1\ -0.9]\|^2 + \log\frac{1}{2\pi} \\
&= -0.9163 - \frac{1}{2}5.22 - 1.8378 \\
&= -5.3642,
\end{aligned}
$$

$$
\begin{aligned}
\Gamma_1(S_0, S_1) &= \log p(\mathbf{u}_1 = \mathbf{u}_{0,1}) + \log p(\mathbf{y}_1|\mathbf{c}_{0,1}) \\
&= \log p(u_1 = 1) + \log p(\mathbf{y}_1 = [-1.1\ 0.1]|\mathbf{x}_1 = [-1\ -1]) \\
&= \log 0.6 - \frac{1}{2}\|[-1.1\ 0.1] - [-1\ -1]\|^2 + \log\left(\sqrt{\frac{1}{2\pi}}\right)^2 \\
&= -0.5108 - \frac{1}{2}\|[-0.1\ 1.1]\|^2 + \log\frac{1}{2\pi} \\
&= -0.5108 - \frac{1}{2}1.22 - 1.8378 \\
&= -2.9586.
\end{aligned}
$$

Repeating these calculations for all t gives the Γ metrics in Figure 4.10.

The $\log\alpha$ metrics are calculated using (4.29):

$$
\begin{aligned}
\mathcal{A}_0(S_0) &= 0, \\
\mathcal{A}_1(S_0) &= \overset{*}{\max}\left(\mathcal{A}_0(S_0) + \Gamma_1(S_0, S_0),\ \mathcal{A}_0(S_1) + \Gamma_1(S_1, S_0)\right) \\
&= \overset{*}{\max}(0 + (-5.3642),\ -\infty) \\
&= -5.3641, \\
\mathcal{A}_1(S_1) &= \overset{*}{\max}\left(\mathcal{A}_0(S_0) + \Gamma_1(S_0, S_1),\ \mathcal{A}_0(S_1) + \Gamma_1(S_1, S_1)\right) \\
&= \overset{*}{\max}(0 + (-2.9587),\ -\infty) \\
&= -2.9587, \\
\mathcal{A}_2(S_0) &= \overset{*}{\max}\left(\mathcal{A}_1(S_0) + \Gamma_2(S_0, S_0),\ \mathcal{A}_1(S_1) + \Gamma_2(S_1, S_0)\right) \\
&= \overset{*}{\max}(-5.3642 + (-5.3642),\ -2.9587 + (-5.1587)) \\
&= -8.0465,
\end{aligned}
$$

$$\begin{aligned}
\mathcal{A}_2(S_1) &= \overset{*}{\max}\left(\mathcal{A}_1(S_0)+\Gamma_2(S_0,S_1),\ \mathcal{A}_1(S_1)+\Gamma_2(S_1,S_1)\right)\\
&= \overset{*}{\max}(-5.3642+(-2.9587),\ -2.9587+(-3.1642))\\
&= -6.0178,\\
\mathcal{A}_3(S_0) &= \overset{*}{\max}\left(\mathcal{A}_2(S_0)+\Gamma_3(S_0,S_0),\ \mathcal{A}_2(S_1)+\Gamma_3(S_1,S_0)\right)\\
&= \overset{*}{\max}(-8.0465+(-5.2992),\ -6.0178+(-2.4937))\\
&= -8.5036,\\
\mathcal{A}_3(S_1) &= \overset{*}{\max}\left(\mathcal{A}_2(S_0)+\Gamma_4(S_0,S_1),\ \mathcal{A}_2(S_1)+\Gamma_4(S_1,S_1)\right)\\
&= \overset{*}{\max}(-8.0465+(-3.4937),\ -6.0178+(-6.2992))\\
&= -11.1618.
\end{aligned}$$

The log β metrics are calculated using (4.30):

$$\begin{aligned}
\mathcal{B}_4(S_0) &= 0,\\
\mathcal{B}_3(S_0) &= \overset{*}{\max}\left(\mathcal{B}_4(S_0)+\Gamma_4(S_0,S_0),\ \mathcal{B}_4(S_1)+\Gamma_4(S_0,S_1)\right)\\
&= \overset{*}{\max}(0+(-1.8479),\ -\infty)\\
&= -1.8478,\\
\mathcal{B}_3(S_1) &= \overset{*}{\max}\left(\mathcal{B}_4(S_0)+\Gamma_4(S_1,S_0),\ \mathcal{B}_4(S_1)+\Gamma_4(S_1,S_1)\right)\\
&= \overset{*}{\max}(0+(-3.6479),\ -\infty)\\
&= -3.6479,\\
\mathcal{B}_2(S_0) &= \overset{*}{\max}\left(\mathcal{B}_3(S_0)+\Gamma_3(S_0,S_0),\ \mathcal{B}_3(S_1)+\Gamma_3(S_0,S_1)\right)\\
&= \overset{*}{\max}(-1.8479+(-5.2992),\ -3.6479+(-3.4937))\\
&= -6.4512,\\
\mathcal{B}_2(S_1) &= \overset{*}{\max}\left(\mathcal{B}_3(S_0)+\Gamma_2(S_1,S_0),\ \mathcal{B}_3(S_1)+\Gamma_2(S_1,S_1)\right)\\
&= \overset{*}{\max}(-1.8479+(-2.4937), -3.6479+(-6.2992))\\
&= -4.3379,\\
\mathcal{B}_1(S_0) &= \overset{*}{\max}\left(\mathcal{B}_2(S_0)+\Gamma_2(S_0,S_0),\ \mathcal{B}_2(S_1)+\Gamma_2(S_0,S_1)\right)\\
&= \overset{*}{\max}(-6.4512+(-5.3642),\ -4.3379+(-2.9587))\\
&= -7.2858,
\end{aligned}$$

$$\begin{aligned}\mathcal{B}_1(S_1) &= \overset{*}{\max}\left(\mathcal{B}_2(S_0) + \Gamma_2(S_1, S_0),\ \mathcal{B}_2(S_1) + \Gamma_2(S_1, S_1)\right)\\ &= \overset{*}{\max}(-6.4512 + (-5.1587),\ -4.3379 + (-3.1642))\\ &= -10.3399.\end{aligned}$$

Finally the bit metrics are calculated using (4.31):

$$\begin{aligned}L_t = L(u_t|\mathbf{y}) &= \overset{*}{\max_{S-}}\left(\mathcal{A}_{t-1}(S_r) + \Gamma_t(S_r, S_s) + \mathcal{B}_t(S_s)\right)\\ &\quad - \overset{*}{\max_{S+}}\left(\mathcal{A}_{t-1}(S_r) + \Gamma_t(S_r, S_s) + \mathcal{B}_t(S_s)\right).\end{aligned}$$

For this encoder $S+ = (S_0, S_1), (S_1, S_0)$ and $S- = (S_0, S_0), (S_1, S_1)$, and so

$$\begin{aligned}L_t = &-\overset{*}{\max}\left(\mathcal{A}_{t-1}(S_0) + \Gamma_t(S_0, S_1) + \mathcal{B}_t(S_1),\ \mathcal{A}_{t-1}(S_1) + \Gamma_t(S_1, S_0) + \mathcal{B}_t(S_0)\right)\\ &+ \overset{*}{\max}\left(\mathcal{A}_{t-1}(S_0) + \Gamma_t(S_0, S_0) + \mathcal{B}_t(S_0),\ \mathcal{A}_{t-1}(S_1) + \Gamma_t(S_1, S_1) + \mathcal{B}_t(S_1)\right),\end{aligned}$$

which gives

$$\begin{aligned}L_1 = &-\overset{*}{\max}\left(\mathcal{A}_0(S_0) + \Gamma_1(S_0, S_1) + \mathcal{B}_1(S_1),\ \mathcal{A}_0(S_1) + \Gamma_1(S_1, S_0) + \mathcal{B}_1(S_0)\right)\\ &+ \overset{*}{\max}\left(\mathcal{A}_0(S_0) + \Gamma_1(S_0, S_0) + \mathcal{B}_1(S_0),\ \mathcal{A}_0(S_1) + \Gamma_1(S_1, S_1) + \mathcal{B}_1(S_1)\right)\\ = &-\overset{*}{\max}(0 + (-2.9587) + (-7.4585), -\infty)\\ &+ \overset{*}{\max}(0 + (-5.3642) + (-7.2858), -\infty)\\ = &-2.2055,\end{aligned}$$

$$\begin{aligned}L_2 = &-\overset{*}{\max}\left(\mathcal{A}_1(S_0) + \Gamma_2(S_0, S_1) + \mathcal{B}_2(S_1),\ \mathcal{A}_1(S_1) + \Gamma_2(S_1, S_0) + \mathcal{B}_2(S_0)\right)\\ &+ \overset{*}{\max}\left(\mathcal{A}_1(S_0) + \Gamma_2(S_0, S_0) + \mathcal{B}_2(S_0),\ \mathcal{A}_1(S_1) + \Gamma_2(S_1, S_1) + \mathcal{B}_2(S_1)\right)\\ = &-\overset{*}{\max}(-5.3642 - 2.9587 - 4.3379,\ -2.9587 - 5.1587 - 6.4512)\\ &+ \overset{*}{\max}(-5.3642 - 5.3646 - 64512,\ -2.9587 - 3.1642 - 4.3379)\\ = &\ 2.0628,\end{aligned}$$

$$\begin{aligned}L_3 = &-\overset{*}{\max}\left(\mathcal{A}_2(S_0) + \Gamma_3(S_0, S_1) + \mathcal{B}_3(S_1),\ \mathcal{A}_2(S_1) + \Gamma_3(S_1, S_0) + \mathcal{B}_3(S_0)\right)\\ &+ \overset{*}{\max}\left(\mathcal{A}_2(S_0) + \Gamma_3(S_0, S_0) + \mathcal{B}_3(S_0),\ \mathcal{A}_2(S_1) + \Gamma_3(S_1, S_1) + \mathcal{B}_3(S_1)\right)\\ = &-\overset{*}{\max}(-8.0465 - 3.4937 - 3.6479,\ -6.0178 - 2.4937 - 1.8479)\\ &+ \overset{*}{\max}(-8.0465 - 5.2992 - 1.8279,\ -6.0178 - 6.2992 - 6.4512)\\ = &-4.4621.\end{aligned}$$

A hard decision on the message bits returns the decoded message $\hat{\mathbf{u}} = [1\ 0\ 1]$.

The reader may have noticed that when the metric for Γ in Example 4.12 was calculated the term $\log(1/2\pi)$, appeared in every Γ value. Since this term scales every Γ metric equally we can simplify the calculations by leaving it out without changing the result. Further, the term $\|\mathbf{y}_t - \mathbf{x}_t\|^2$ can be rewritten as $\|\mathbf{y}_t\|^2 - 2\mathbf{x}_t \cdot \mathbf{y}_t + \|\mathbf{x}_t\|^2$, which can then be simplified to $-2\mathbf{x}_t \cdot \mathbf{y}_t$ by noting that $\|\mathbf{y}_t\|^2 + \|\mathbf{x}_t\|^2 = \|\mathbf{y}_t\|^2 + n$ is the same for every Γ value with the same t and can similarly be left out. The new metric for (4.33) is now

$$\log p(\mathbf{y}_t|\mathbf{x}_t) \approx \frac{2E_s}{N_0}\mathbf{y}_t \cdot \mathbf{x}_t. \tag{4.34}$$

The resulting metrics for $\Gamma_t(S_r, S_s)$ are no longer directly the log of a probability and so can be positive; however, since the metrics are scaled equally the decoder result will not change.

The calculation of the Γ values from (4.34) requires that the decoder knows which channel was used to transmit the codeword. To avoid this requirement, the received and a priori probabilities are instead calculated beforehand and passed into the BCJR decoder, usually as the LLRs

$$R_t^{(i)} = \log \frac{p(y_t|c_t^{(i)} = 0, \text{channel})}{p(y_t|c_t^{(i)} = 1, \text{channel})}$$

and

$$A_t^{(i)} = \log \frac{p(u_t^{(i)} = 0|\text{source})}{p(u_t^{(i)} = 1|\text{source})}$$

respectively, making the decoder itself channel independent. The expression for Γ then becomes

$$\Gamma_t(S_r, S_s) = \sum_i^k \log p\left(u_t^{(i)} = u_{r,s}^{(i)}\right) + \sum_i^n \log p\left(c_t^{(i)} = c_{r,s}^{(i)}\right), \tag{4.35}$$

where

$$\log p\left(c_t^{(i)} = c_{r,s}^{(i)}\right) = \begin{cases} \log \dfrac{e^{R_t^{(i)}}}{1+e^{R_t^{(i)}}} & \text{if } c_{r,s}^{(i)} = 0, \\ \log \dfrac{e^{-R_t^{(i)}}}{1+e^{-R_t^{(i)}}} & \text{if } c_{r,s}^{(i)} = 1 \end{cases}$$

and

$$\log p\left(u_t^{(i)} = u_{r,s}^{(i)}\right) = \begin{cases} \log \dfrac{e^{A_t^{(i)}}}{1+e^{A_t^{(i)}}} & \text{if } u_{r,s}^{(i)} = 0, \\ \log \dfrac{e^{-A_t^{(i)}}}{1+e^{-A_t^{(i)}}} & \text{if } u_{r,s}^{(i)} = 1. \end{cases}$$

Using LLRs as input and output, the pseudo code for the log BCJR decoding of a binary convolutional code is given in Algorithm 4.2. The inputs are the a priori message bit LLRs $\mathbf{A}$, the received LLRs $\mathbf{R}$ and the length-T convolutional trellis with 2^v states. The Γ, $\mathcal{A}$ and $\mathcal{B}$ values are calculated using (4.22), (4.29) and (4.30) respectively. The outputs are the APPs of the kT message bits calculated using (4.31).

Max log MAP decoding

The log BCJR decoder can be made even less computationally complex, albeit with some reduction in performance, by replacing the $\overset{*}{\max}$ operation with the simpler max operation. The resulting algorithm is called *max log MAP* decoding. In this case the metric returned by the max operation will be incorrect by the value of the correction term, $\log(1 + e^{-|x-y|})$. However, because the correction term is bounded,

$$0 < \log(1 + e^{-|x-y|}) \leq \log 2 \approx 0.6931,$$

the performance loss will be small if the value of $\max(x, y)$ is large. Using a relatively small lookup table for the correction term can produce performances with a loss of only around 0.5 dB compared with the full log MAP algorithm.

4.3.3 Maximum likelihood (Viterbi) decoding

The maximum likelihood (ML) decoder finds the codeword that is closest to the received vector. For block codes this can mean 2^K length-N vector comparisons. However, for a convolutional code, finding the most likely codeword is equivalent to finding the most likely path in the trellis, which can be done in a much cleverer way. An ML algorithm for convolutional codes was first presented by Viterbi [76] and is called the *Viterbi algorithm* after its author. The benefit of the Viterbi algorithm is that complete path probabilities do not need to be calculated for every possible codeword. At each node in the trellis the best path so far, called the *survivor* path for that node, is stored and the remaining paths discarded. This drastically reduces the complexity of finding the ML codeword for convolutional codes.

The ML decoder will output the codeword $\hat{\mathbf{c}}$ that maximizes $p(\mathbf{y}|\mathbf{c})$ over all possible codewords $\mathbf{c}$. Since we have a convolutional code on a memoryless channel,

$$p(\mathbf{y}|\mathbf{c}) = \prod_{i=1}^{nT} p(y_i|c_i) = \prod_{t} p(\mathbf{y}_t|\mathbf{c}_{r,s}). \tag{4.36}$$

As in BCJR decoding we will use log metrics and define

$$B_t(S_r, S_s) = \log p\left(\mathbf{y}_t|\mathbf{c}_{r,s}\right) \tag{4.37}$$

Algorithm 4.2 Log BCJR decoding

```
1: procedure LOGBCJR(trellis,A,R)
2:                                                        ▷ Initialization
3:     for t = 1 : T do
4:         for r = 0 : 2^v − 1 do
5:             for s = 0 : 2^v − 1 do
6:                 Γ_t(S_r, S_s) = −∞
7:                 if S_r to S_s is a trellis edge then
8:                     Γ_t(S_r, S_s) = 0
9:                     for i = 1 : k do
10:                        if u^(i)_{r,s} = 0 then
11:                            Γ_t(S_r, S_s) = Γ_t(S_r, S_s) + log( e^{A_t^(i)} / (1 + e^{A_t^(i)}) )
12:                        else
13:                            Γ_t(S_r, S_s) = Γ_t(S_r, S_s) + log( e^{−A_t^(i)} / (1 + e^{−A_t^(i)}) )
14:                        end if
15:                    end for
16:
17:                    for i = 1 : n do
18:                        if c^(i)_{r,s} = 0 then
19:                            Γ_t(S_r, S_s) = Γ_t(S_r, S_s) + log( e^{R_t^(i)} / (1 + e^{R_t^(i)}) )
20:                        else
21:                            Γ_t(S_r, S_s) = Γ_t(S_r, S_s) + log( e^{−R_t^(i)} / (1 + e^{−R_t^(i)}) )
22:                        end if
23:                    end for
24:                end if
25:            end for
26:        end for
27:    end for
28:    𝒜_t(S_i) = −∞ ∀i, t except
29:    𝒜_0(S_0) = 0                                        ▷ Forward recursion
30:    for t = 1 : T − 1 do
31:        for r = 0 : 2^v − 1 do
32:            for s = 0 : 2^v − 1 do
33:                𝒜_t(S_s) = max*(𝒜_t(S_s), 𝒜_{t−1}(S_r) + Γ_t(S_r, S_s))
34:            end for
35:        end for
36:    end for
37:
38:    Continued over the page
```

Algorithm 4.2 Log BCJR decoding continued

39: **if** terminated **then** ▷ Backward recursion
40: $\mathcal{B}_T(S_0) = 0$ and $\mathcal{B}_T(S_{i \neq 1}) = -\infty$
41: **else**
42: $\mathcal{B}_T(S_i) = \log(1/2^v)$
43: **end if**
44: $\mathcal{B}_t(S_i) = -\infty \ \forall i, \ t < T$
45: **for** $t = T - 1 : 1$ **do**
46: **for** $r = 0 : 2^v - 1$ **do**
47: **for** $s = 0 : 2^v - 1$ **do**
48: $\mathcal{B}_t(S_r) = \max^*(\mathcal{B}_t(S_r), \ \mathcal{B}_{t+1}(S_s) + \Gamma_{t+1}(S_r, S_s))$
49: **end for**
50: **end for**
51: **end for**
52:
53: **for** $t = 1 : T$ **do** ▷ State transition probabilities
54: **for** $r = 0 : 2^v - 1$ **do**
55: **for** $s = 0 : 2^v - 1$ **do**
56: $M_t(S_r, S_s) = \mathcal{A}_{t-1}(S_r) + \Gamma_t(S_r, S_s) + \mathcal{B}_t(S_s)$
57: **end for**
58: **end for**
59: **end for**
60:
61: **for** $t = 1 : T$ **do** ▷ Bit probabilities
62: **for** $i = 1 : k$ **do**
63: $L_{\text{plus}} = L_{\text{minus}} = -\infty$
64: **for** $r = 0 : 2^v - 1$ **do**
65: **for** $s = 0 : 2^v - 1$ **do**
66: **if** $(S_r, S_s) \in S^{(i)}+$ **then**
67: $L_{\text{plus}} = \max^*(L_{\text{plus}}, \ M_t(S_r, S_s))$
68: **else**
69: $L_{\text{minus}} = \max^*(L_{\text{minus}}, \ M_t(S_r, S_s))$
70: **end if**
71: **end for**
72: **end for**
73: $L(u_t^{(i)}|\mathbf{y}) = L_{\text{plus}} - L_{\text{minus}}$
74: **end for**
75: **end for**
76: **end procedure**

Algorithm 4.3 Viterbi decoding

```
1: procedure VITERBI(trellis,R)
2:     M_0(S_0) = 0                                          ▷ Initialization
3:     Path_0(S_0) is empty
4:     for t, s ≠ 0 do
5:         M_t(S_s) = −∞                 ▷ or some other large metric as a default
6:     end for
7:     for t = 1 : T − 1 do
8:         for r = 0 : 2^v − 1 do
9:             for s = 0 : 2^v − 1 do
10:                if S_r to S_s is a valid trellis edge then
11:                    B(r, s) = − log p(y_t|c_{r,s})
12:                    if M_{t−1}(S_r) + B(r, s) < M_t(S_s) then
13:                        M_t(S_s) = M_{t−1}(S_r) + B(r, s)
14:                        Path_t(S_s) = [Path_{t−1}(S_r), u_{r,s}]
15:                    end if
16:                end if
17:            end for
18:        end for
19:    end for
20:    output = Path_T(S_0)
21: end procedure
```

so that we can compute (4.36) as the sum of logs. The value $B_t(S_r, S_s)$ is called the *branch metric* for the trellis edge between state S_r at time t and state S_s at time $t+1$. The *path metric* for a codeword is the sum of all the branch metrics that make up the trellis path corresponding to that codeword. Thus the codeword that maximizes (4.36) will be the codeword with the largest path metric. Then the survivor path at a node is the path into that node with the best path metric so far. In the case of identical metrics for two paths into a node the survivor path is chosen at random.

The pseudo code for the Viterbi decoding of a convolutional code is given in Algorithm 4.3. The inputs are the channel transition probabilities $p(\mathbf{y}_t|\mathbf{c}_{r,s})$ and the length-T convolutional trellis with 2^v states. At each time t the best path into each trellis node S_s is chosen and stored in $\text{Path}_t(S_s)$ and the path metric for that path is stored in $M_t(S_s)$. Since the trellis is terminated there will be only one final surviving path at time T. This path will correspond to the ML codeword. The output is the length-kT message corresponding to this surviving path. Note that our implementation of Viterbi decoding sacrifices efficiency for clarity. The process can be made more memory efficient by storing the node metrics only for the current and previous time points and not storing the branch metrics at all.

It is easy to see that the Viterbi-algorithm branch metrics (4.37) are the Γ metrics in BCJR decoding without the a priori bit probabilities. Furthermore, the Viterbi algorithm operates from left to right on the trellis using a max operation to select the best path metrics into each node, which is exactly the same as the forward pass in the max log MAP BCJR decoder. Thus the Viterbi algorithm can be considered as the forward recursion in the max log MAP BCJR decoder in the case when the input message bits are equally likely. So, while the Viterbi algorithm will select the best overall path through the trellis, the probabilities it calculates for each branch metric will be an estimation of the actual probabilities for each edge. The Viterbi algorithm can still return these estimated probabilities, however, and is called the *soft output Viterbi algorithm (SOVA)* in this case.

If message bit probabilities are not required, so that only the ML codeword is returned, the Viterbi algorithm can use much simpler metrics than (4.37) in some cases. For example, a Viterbi decoder with a binary received vector uses the Hamming distance between the codeword and the received vector as its path metric and is called a hard decision Viterbi decoder.

4.4 Bibliographic notes

Convolutional codes were introduced by Elias in 1955 [77] not long after Hamming introduced block codes. A sequential decoding algorithm for them was presented soon after by Wozencraft and Reiffen [78], and a rate-$1/2$ convolutional code with sequential decoding was used in the Pioneer 9 solar orbiter, becoming the first code in space. A later, less computational demanding, decoding method presented by Massey [79], called "threshold decoding", enabled the use of convolutional codes in high-speed applications.

In 1967 Viterbi presented a trellis-based decoding algorithm, which was later shown to maximum-likelihood by Forney in 1973 [80]. The Viterbi algorithm became the decoder of choice for deep space and satellite communications, cellular telephony and digital video broadcasting. When combined with trellis-coded modulation the Viterbi algorithm also helped to improve the data rate of digital modems from 9600 bits per second to 33 600 bits per second. In early 2005 Forney [81] wrote that Viterbi decoders were in use in about one billion cellular telephones and that about 10^{15} bits per second were being decoded by a Viterbi decoder in TV sets around the world.

The latest deep space systems and digital video and cellular standards are now adopting iterative decoding algorithms for turbo and LDPC codes, and it may be that the use of the Viterbi decoder has seen its peak and will be replaced, in the case of turbo codes, by the BCJR decoder. The BCJR algorithm was first introduced in 1974 [75] but not widely used owing to its complexity, which is about three times that of the Viterbi decoder. However, with the introduction of turbo codes by Berrou *et al.* [82] the BCJR decoder has come into its own: the symbol-wise MAP values output by the BCJR algorithm are ideal for the operation of the turbo decoder.

Although convolutional codes will not be covered further in this book, except as part of a turbo code, there is a wealth of material available for the interested reader. Three texts devoted solely to convolutional codes are [83], [84] and [85], and most classical error correction texts include chapters on convolutional codes; these include [5] and [7] or, for those interested in a more mathematical treatment, [8] and [9]. A combined treatment of trellis and turbo codes is given in [86] and a very engaging historical treatment of coding applications is presented in [87].

4.5 Exercises

4.1 The convolutional encoder used in the IS-95 code division multiple access (CDMA) system for mobile phones has the generators 753, 561.
(a) Determine the code rate and maximum memory order,
(b) give the generator matrix,
(c) determine the codeword produced by the message $\mathbf{u}^{(1)} = [1\ 0\ 1\ 1\ 0]$,
(d) draw this encoder's block diagram.

4.2 For the convolutional encoder with generator sequences

$$g_{1,1} = [1\ 1\ 1],\quad g_{1,2} = [1\ 1\ 0],\quad g_{1,3} = [1\ 1\ 0],$$
$$g_{2,1} = [1\ 0\ 1],\quad g_{2,2} = [1\ 1\ 0] \text{ and } g_{2,3} = [1\ 0\ 0],$$

(a) draw the convolutional encoder block diagram,
(b) determine the code rate and encoder maximum memory order,
(c) determine the terminated codeword produced by the message $\mathbf{u}^{(1)} = [1\ 0\ 1\ 1]$, $\mathbf{u}^{(2)} = [1\ 0\ 0\ 1]$,
(d) find a puncturing pattern to produce a rate-$4/5$ code.

4.3 The systematic recursive convolutional encoder used in the first turbo code has the generator matrix

$$G = \begin{bmatrix} 1 & \dfrac{1 + D^4}{1 + D + D^2 + D^3 + D^4} \end{bmatrix}.$$

(a) Draw the convolutional encoder block diagram,
(b) determine the code rate and encoder maximum memory order,

(c) determine the first eight bits in the codeword produced by the message $\mathbf{u} = [1\ 0\ 1\ 1\ 1]$,

(d) find a puncturing pattern to produce a rate-3/4 code.

4.4 The 1977 Voyager missions employed a rate-1/2 convolutional code with generators 171, 133. Determine the codeword produced by the Voyager code for the message $\mathbf{u} = [1\ 0\ 1\ 1\ 0\ 1]$.

4.5 The Galileo mission to Jupiter used a rate-1/4 convolutional code with generators 46321, 51271, 63667, 70535 and a rate-1/2 convolutional code with generators 171, 133. What is the state complexity of these encoders and what is the approximate difference in their decoding complexities?

4.6 The convolutional code used in the global system for mobile communications (GSM) standard, and employed in over 80 million cell phones, has the generator matrix

$$G(D) = \left[1 + D^3 + D^4 \quad 1 + D + D^3 + D^4\right].$$

Find a systematic encoder for this code.

4.7 Draw the state transition diagram for the non-recursive convolutional encoder in Example 4.2.

4.8 Find a non-recursive convolutional encoder for the code in Example 4.6 and draw its trellis up to time $T = 4$. What is the minimum free distance of this code?

4.9 For the convolutional encoder with generators

$$g_{1,1} = [1\ 1\ 1\ 1], \quad g_{1,2} = [1\ 0\ 0\ 0] \text{ and } g_{1,3} = [1\ 1\ 0\ 0],$$

draw its trellis and show the trellis path corresponding to the terminated codeword generated by the message $\mathbf{u} = [1\ 0\ 1\ 1]$.

4.10 Why are the input bits required to terminate the trellis of Example 4.6 sometimes non-zero?

4.11 Repeat the decoding procedure in Example 4.11 but now assuming that the source was biased, with input probabilities given by the following table:

	$t=1$	$t=2$	$t=3$	$t=4$	$t=5$
$p(u_t^{(1)} = 0)$	0.9	0.8	0.5	0.8	0.8
$p(u_t^{(1)} = 1)$	0.1	0.2	0.5	0.2	0.2

4.12 A 3-bit message was produced by an equiprobable source and encoded with the convolutional encoder in Figure 4.3 using the puncturing pattern

$$P = \begin{bmatrix} 1 & 1 & 0 \\ 1 & 0 & 1 \end{bmatrix}$$

and no padding bits. The codeword was transmitted on a BI-AWGN channel with signal-to-noise ratio $1/2$, and the vector

$$y = [-1.1 \; 0.9; \; -1 \; x; \; x \; +0.5]$$

was received. Find the transmitted message using log BCJR decoding.

4.13 Find the MAP message using BCJR decoding (with probability metrics) for the encoder in Figure 4.3 assuming an equiprobable binary source:

(a) for an 8-bit terminated codeword sent on the BSC with crossover probability 0.02 when $\mathbf{y} = [1 \; 1; \; 1 \; 1; \; 0 \; 1; \; 1 \; 1]$ is received,

(b) for an 8-bit terminated codeword sent on the BI-AWGN channel with signal-to-noise ratio 5 dB when $\mathbf{y} = [1.1 \; 0.9; \; -0.1 \; 1.1; \; -1.1 \; 0.9; \; 1.1 \; 1.1]$ is received, for a biased source with $p(u_t) = 1) = 0.4$.

4.14 Repeat Exercise 4.13,

(a) using log BCJR decoding,

(b) using max log MAP BCJR decoding,

(c) using soft output Viterbi decoding,

(d) using hard decision Viterbi decoding.

4.15 Explain the differences in the results for Exercises 4.13 and 4.14.

4.16 Using the encoder from Exercise 4.3 a 3-bit message from an equiprobable source was encoded into an unterminated 6-bit codeword and transmitted over a binary symmetric channel with crossover probability $p = 0.2$. If the vector $\mathbf{y} = [1 \; 0; \; 0 \; 1; \; 1 \; 0]$ was received, what is the maximum a posteriori message?

4.17 Write a Matlab program to implement log BCJR decoding.

(a) Using your program determine the MAP bit probabilities for the unterminated message corresponding to the received vector

$$\mathbf{y} = [1 \; 1 \; 1; \; 1 \; 0 \; 1; \; 1 \; 1 \; 1; \; 0 \; 1 \; 0; \; 0 \; 0 \; 1; \; 0 \; 1 \; 1]$$

from the convolutional encoder of Figure 4.2. Assume a binary equiprobable source and a BSC with crossover probability 0.2.

(b) Using your program find the average bit error rate of the IS-95 convolutional code (see Exercise 4.1) operating on a BI-AWGN channel with signal-to-noise ratios of 1 dB and 3 dB for message lengths of 10 bits and 100 bits respectively. Assume an equiprobable binary source.

4.18 Generate bit error rate plots for log BCJR decoding and for Viterbi decoding operating on the convolutional encoder in Exercise 4.3 with codeword lengths of 100 bits sent on the BI-AWGN channel using an equiprobable binary source. Explain the results.

5 Turbo codes

5.1 Introduction

In this chapter we introduce turbo codes, the ground-breaking codes introduced by Berrou, Glavieux and Thitimajshima in 1993 [82], which sparked a complete rethink of how we do error correction. Turbo codes are the parallel concatenation of two convolutional codes which, at the receiver, share information between their respective decoders. Thus most of what we need for a turbo code has already been presented in the previous chapter. The turbo encoder uses two convolutional encoders (see Section 4.2) while the turbo decoder uses two copies of the log BCJR decoder (see Section 4.3).

The exceptional performance of turbo codes is due to the long pseudo-random interleaver, introduced below, which produces codes reminiscent of Shannon's noisy channel coding theorem, and to the low-complexity iterative algorithm that makes their implementation feasible.

In the first part of this chapter we discuss how the component convolutional codes and interleaver combine to form a turbo code. We then consider the properties of a turbo code that affect its iterative decoding performance and describe turbo code design strategies. Our aim here is to convey basic information about the encoding, decoding and design of turbo codes; a deeper understanding of these codes and of the decoding process is left to later chapters.

5.2 Turbo encoders

At the encoder, turbo codes use a parallel concatenation of two convolutional component encoders, as shown in Figure 5.1. The length-K message $\mathbf{u}$ is encoded directly by the first component encoder, which produces the parity bits $\mathbf{p}^{(1)}$; however, it is interleaved (see below) before being encoded by the second convolutional encoder, which produces the parity bits $\mathbf{p}^{(2)}$. Turbo codes are systematic codes (see Section 2.3); thus the bits in $\mathbf{u}$, $\mathbf{p}^{(1)}$ and $\mathbf{p}^{(2)}$ are all transmitted and the turbo codeword is

$$\mathbf{c} = \left[\mathbf{u}_1 \; \mathbf{p}_1^{(1)} \; \mathbf{p}_1^{(2)} \cdots \mathbf{u}_K \; \mathbf{p}_K^{(1)} \; \mathbf{p}_K^{(2)}\right].$$

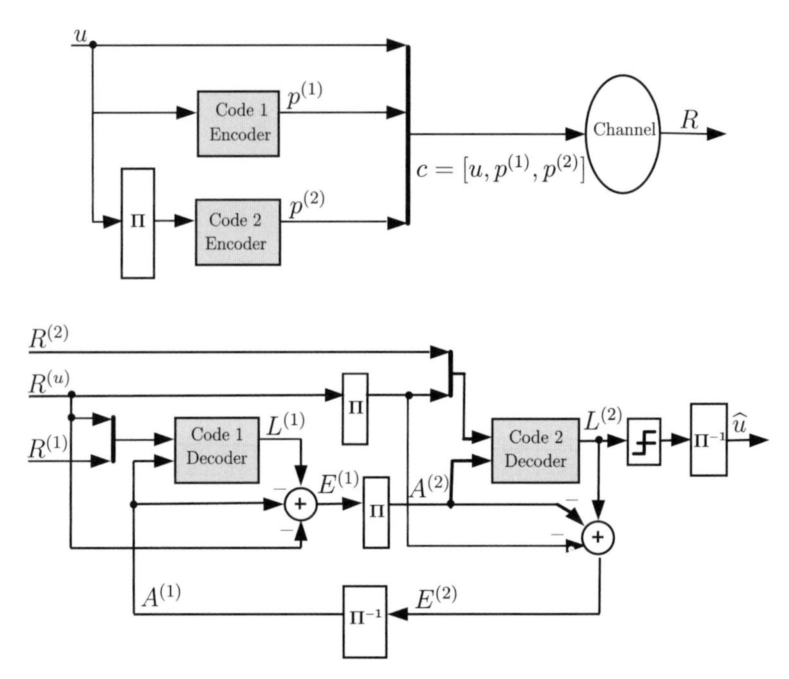

Figure 5.1 Block diagram of a turbo code system. Encoding (top) and turbo decoding (bottom).

The two component convolutional encoders are usually identical, but *asymmetric turbo codes*, which use two different convolutional encoders, are also possible. While many turbo codes employ component convolutional encoders for which the number of input bits per state transition $k = 1$ and puncture the parity bits to increase the code rate, *duobinary turbo codes* use component encoders with $k = 2$.

Since both encoders encode the same message bits, their decoders can cooperate with one another at the receiver to decode the message. Importantly, the set of message bits is permuted, or *interleaved*, before reaching the second convolutional encoder. Using interleaving, the parity bits output by the two encoders will be different even if both encoders are the same.

Interleavers

An interleaver is represented by a permutation sequence

$$\Pi = [\pi_1, \pi_2, \ldots, \pi_n],$$

where the sequence $[\pi_1, \pi_2, \ldots, \pi_n]$ is a permutation of the integers 1 to n.

Example 5.1 The interleaver represented by $\Pi = [4, 2, 5, 3, 1]$ acting on the input vector $\mathbf{u} = [u_1\ u_2\ u_3\ u_4\ u_5]$ will produce the output $\Pi(\mathbf{u}) = [u_4\ u_2\ u_5\ u_3\ u_1]$. Thus if $\mathbf{u}$ is $[1\ 0\ 1\ 1\ 0]$ then $\Pi(\mathbf{u})$ will be the vector $[1\ 0\ 0\ 1\ 1]$.

For a convolutional encoder there is no requirement to specify a message length; the encoder can continue encoding indefinitely. However, for a turbo encoder the existence of the interleaver makes a predefined message length, i.e. a predefined number of bits interleaved at a time, necessary.

Interleaving ensures that the parity bits produced by the second encoder are completely different from the parity bits produced by the first. If a low-weight parity sequence is produced by the first encoder then the different pattern of message bits passing into the second encoder should produce a high-weight output for the second parity sequence, thereby avoiding low-weight turbo codewords.

Capacity-approaching performances for turbo codes are achieved using interleavers with lengths of order several thousand bits, and the best-performing interleavers are usually found by randomly choosing a permutation having the required length. For this reason random, or pseudo-random, interleavers are often used in simulations. For practical applications, however, some sort of structure is usually applied as an aid to implementation. Interleavers are discussed in more detail in Section 5.4.3. In the meantime we will use randomly chosen interleavers.

Code rate

If the first component code inputs k_1 bits per state transition with rate $r_1 = k_1/n_1$ then it will output $n_1 - k_1$ parity bits for each state transition. If the second code inputs k_2 bits per state transition with rate $r_2 = k_2/n_2$ then it will output $n_2 - k_2$ parity bits for each state transition. The set of message bits is the same for both component codes, thus $k_1 = k_2 = k$, so the rate of the turbo code will be

$$\begin{aligned} r &= \frac{k}{k + (n_1 - k) + (n_2 - k)} = \frac{k}{n_1 + n_2 - k} \\ &= \frac{k}{k/r_1 + k/r_2 - k} = \frac{r_1 r_2}{r_1 + r_2 - r_1 r_2}. \end{aligned}$$

To increase the code rate the output of one or both of the convolutional codes can be punctured (see e.g. Example 4.10 for how to puncture convolutional codes).

Example 5.2 In this example we consider the component encoders used to define the rate-1/2 turbo encoder presented in the original paper of Berrou, Glavieux and Thitimajshima [82]. The encoders both use the rate-1/2 systematic recursive convolutional code with generator matrix

$$G = \begin{bmatrix} 1 & \dfrac{1 + D^4}{1 + D + D^2 + D^3 + D^4} \end{bmatrix}.$$

The component codes are concatenated in parallel, as in Figure 5.1, producing a rate-$1/3$ turbo code. However, puncturing is used for both encoders, using the puncturing patterns

$$P_1 = \begin{bmatrix} 1 & 1 \\ 1 & 0 \end{bmatrix}, \quad P_2 = \begin{bmatrix} 0 & 0 \\ 0 & 1 \end{bmatrix},$$

and the final code rate is $1/2$. The top row of zeros in P_2 reflects that the two codes share the same message bits, which are only transmitted once. For this example we will use the length-10 interleaver

$$\Pi = [3, 7, 6, 2, 5, 10, 1, 8, 9, 4]$$

and so will encode 10 message bits at a time. The trellises are not terminated, so there are no padding bits. The 10-bit message

$$\mathbf{u} = [1\ 0\ 0\ 1\ 1\ 0\ 1\ 0\ 1\ 0]$$

is passed into the first encoder, which returns the parity bits

$$\mathbf{p}^{(1)} = [1\ 1\ 0\ 1\ 1\ 1\ 0\ 0\ 0\ 1].$$

An interleaved version of the message,

$$\Pi(\mathbf{u}) = [0\ 1\ 0\ 0\ 1\ 0\ 1\ 0\ 1\ 1],$$

is passed into the second encoder, which returns the parity bits

$$\mathbf{p}^{(2)} = [0\ 1\ 1\ 0\ 1\ 0\ 1\ 0\ 0\ 0].$$

After puncturing, the actual transmitted bits are

$$\begin{aligned} \mathbf{u} &= [1\ 0\ 0\ 1\ 1\ 0\ 1\ 0\ 1\ 0], \\ \mathbf{p}^{(1)} &= [1\ 0\ 1\ 0\ 0\ ], \\ \mathbf{p}^{(2)} &= [1\ 0\ 0\ 0\ 0\ ]. \end{aligned}$$

Since for every 10 message bits there are 20 codeword bits (10 message bits plus five parity bits for each encoder), the rate of this turbo code is $1/2$.

Code termination

A convolutional trellis can be terminated (returned to the zero state) by appending b padding bits to the message (see e.g. Figure 4.5). Both turbo component codes can be terminated in this way. However, since the order of the message bits is different for each encoder, the final states of the encoders can also differ and so the padding bits needed to terminate each encoder may not be the same. A standard approach is to terminate both encoders but transmit the padded message bits only for the first encoder, along with the extra parity bits generated by both encoders. The second component decoder can treat the missing padding bits as having been punctured and use the parity bits alone for decoding the corresponding trellis edges.

If the unterminated code encodes K message bits into N codeword bits then the terminated code will encode those same message bits into $N + 3b$ codeword bits, so the rate of the terminated turbo code will thus be reduced to $K/(N + 3b)$.

Example 5.3 Figure 5.2 shows the encoding circuit for the turbo code in Example 5.2 including the padding bit generator.

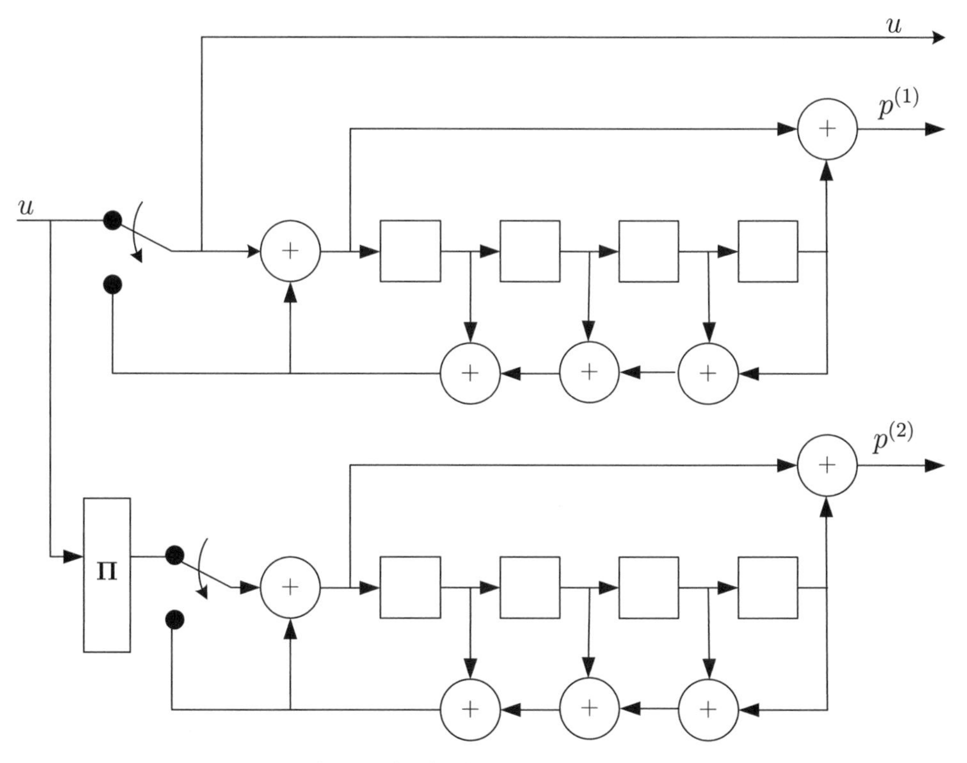

Figure 5.2 The encoder for the turbo code in Example 5.3.

5.3 Iterative turbo decoding

For *turbo decoding*, the soft decoding of each individual convolutional code is done by the BCJR algorithm in exactly the same way as described in Section 4.3.2. The main difference for the BCJR decoders inside a turbo decoder is the source of their a priori information. For the decoder of a stand-alone convolutional code, the a priori information about the message bits comes from knowledge of the source (see e.g. Example 4.12). In a turbo code the two component codes share the same message bits, and in the turbo decoder they share their information about the message bits as a priori information fed into each decoder. This extra information passed between the decoders is often referred to as extrinsic information or feedback information.

The second difference in a turbo decoder is that each BCJR decoder is used more than once. The extrinsic information is updated and passed back and forward between the decoders over many iterations. One iteration of turbo decoding consists of one application of the BCJR algorithm for each component code. A turbo decoder can use any number of iterations; however, increasing the iteration count increases the decoding time. Around 10–20 iterations is pretty standard but, generally, the longer the codewords the larger the number of iterations required to decode correctly and (as we will see in Chapter 7) codes operating close to their noise correction limits will require many more decoder iterations.

Figure 5.1 shows a block diagram for a turbo decoder. The decoder has as input $\mathbf{R}^{(u)}$, $\mathbf{R}^{(1)}$ and $\mathbf{R}^{(2)}$, the log likelihood ratios from the channel for $\mathbf{u}$, $\mathbf{p}^{(1)}$ and $\mathbf{p}^{(2)}$ respectively.

In the first iteration, $l = 1$, the Code 1 decoder has no extrinsic information available and so the input into the log BCJR algorithm is $\mathbf{R}^{(u)}$ and $\mathbf{R}^{(1)}$. The log BCJR decoder will output the a posteriori probabilities for the message bits, also as log likelihood ratios, $\mathbf{L}_1^{(1)}$ (see Algorithm 4.2). The new, extrinsic, information that the Code 1 decoder has created about the message bits in the first iteration, indicated by the subscript 1, is thus

$$\mathbf{E}_1^{(1)} = \mathbf{L}_1^{(1)} - \mathbf{R}^{(u)},$$

where $\mathbf{R}^{(u)}$ is the information that was already known about the message bits prior to the Code 1 decoding.

The inputs to the Code 2 decoder from the channel are $\Pi(\mathbf{R}^{(u)})$ and $\mathbf{R}^{(2)}$, with the $\mathbf{R}^{(u)}$ values interleaved to be in the same order as that of the message bits when they were input into the Code 2 encoder. For the lth iteration, the Code 2 decoder will use the extrinsic information from the first encoder, $\mathbf{E}_l^{(1)}$, as extra a priori information $\mathbf{A}_l^{(2)}$ about the message bits. The vector of LLRs from the Code 1 decoder will be interleaved so as to be in the same order as the Code 2 message bits:

$$\mathbf{A}_l^{(2)} = \Pi(\mathbf{E}_l^{(1)}).$$

The Code 2 log BCJR decoder will output the a posteriori probabilities it calculates for the message bits as log likelihood ratios $\mathbf{L}_l^{(2)}$ (see Algorithm 4.2). The new, extrinsic, information created by the Code 2 decoder about the message bits at iteration l is thus

$$\mathbf{E}_l^{(2)} = \mathbf{L}_l^{(2)} - \Pi(\mathbf{R}^{(u)}) - \mathbf{A}_l^{(2)}, \tag{5.1}$$

where $\Pi(\mathbf{R}^{(u)}) + \mathbf{A}_l^{(2)}$ is the information that was already known about the message bits prior to this iteration of Code 2 decoding.

In the second, and subsequent, iterations the Code 1 decoder will repeat the BCJR algorithm but now will have extra a priori information in the form of the extrinsic information from the Code 2 decoder created in the previous iteration. Thus in the lth iteration the extrinsic information from the Code 2 decoder is de-interleaved so as to be in the same order as the Code 1 message bits:

$$\mathbf{A}_l^{(1)} = \Pi^{-1}(\mathbf{E}_{l-1}^{(2)}).$$

The new, extrinsic, information about the message bits from the Code 1 decoder is thus

$$\mathbf{E}_l^{(1)} = \mathbf{L}_l^{(1)} - \mathbf{R}^{(u)} - \mathbf{A}_l^{(1)}, \tag{5.2}$$

since $\mathbf{R}^{(u)} + \mathbf{A}_l^{(1)}$ is the information that was already known about the message bits prior to this iteration of Code 1 decoding. Note that the log likelihood ratios from the channel remain unchanged throughout the turbo decoding; only the extrinsic information changes at each iteration.

The decoder can be halted after a fixed number of iterations or when predetermined criteria have been reached (we will consider these in Section 5.5.2). The final decision can be made on the basis of the output of either component decoder or an average of both outputs. To speed up decoding, the two BCJR decoders can also be implemented in parallel. In this case the extrinsic information into the Code 2 decoder will be the extrinsic information generated by the Code 1 decoder in the previous iteration.

Algorithm 5.1 completes $I_{\max}$ iterations of turbo decoding. We denote the trellis structure of the first component convolutional code as "trellis1" and the trellis

Algorithm 5.1 Turbo decoding

```
1:  procedure TURBO(trellis1,trellis2,Π,I_max,R^(u),R^(1),R^(2))
2:      E^(2) = zeros;
3:      for l = 1 : I_max do                          ▷ For each iteration l
4:          A^(1) = Π^{-1}(E^(2))
5:          L^(1) = logBCJR(trellis1,R^(u),R^(1),A^(1))
6:          E^(1) = L^(1) − R^(u) − A^(1)
7:          A^(2) = Π(E^(1))
8:          L^(2) = logBCJR(trellis2,Π(R^(u)),R^(2),A^(2))
9:          E^(2) = L^(2) − Π(R^(u)) − A^(2)
10:     end for
11:     output Π^{-1}(L^(2))
12: end procedure
```

structure of the second component convolutional code as "trellis2" (although in many cases the component codes will be identical). The interleaver used is Π. The received LLRs are split into three: $\mathbf{R}^{(u)}$, the LLRs of the transmitted message bits; $\mathbf{R}^{(1)}$, the LLRs of the parity bits belonging to Code 1; and $\mathbf{R}^{(2)}$, the LLRs of the parity bits generated by Code 2. Note that if puncturing has occurred then the received LLRs of the punctured bits will be zero.

Example 5.4 In this example we consider decoding for the turbo code presented in Example 5.2. The 20-bit codeword $\mathbf{c} = [\mathbf{u}\ \mathbf{p}^{(1)}\ \mathbf{p}^{(2)}]$ generated in Example 5.2 was transmitted over a BI-AWGN channel with $\mu = 1$ (for μ see (1.5)) and variance $\sigma^2 = 0.5012$ and the received values were

$$\begin{aligned}
\mathbf{y}^{(u)} &= [-0.4620\ \ -0.4126\ 0.2881\ \ -1.1397\ \ -0.2145\\
&\qquad 0.9144\ \ -0.7271\ 1.6833\ 0.6207\ 0.7262],\\
\mathbf{y}^{(1)} &= [-1.5556\ x\ 0.6315\ x\ \ -0.3619\ x\ 0.6077\ x\ 0.9563\ x],\\
\mathbf{y}^{(2)} &= [x\ \ -0.1177\ x\ 0.6122\ x\ 0.7948\ x\ 0.5888\ x\ 1.1217],
\end{aligned}$$

where x indicates a punctured bit. The received sequence is to be decoded with three iterations of turbo decoding.

To begin, the received LLRs for the codeword bits are calculated using (1.9) and the LLRs of the punctured bits are set to zero. Thus

$$\begin{aligned}
\mathbf{R}^{(u)} &= [-1.8436\ \ -1.6465\ 1.1497\ \ -4.5480\ \ -0.8560\\
&\qquad 3.6489\ \ -2.9015\ 6.7173\ 2.4769\ 2.8979],\\
\mathbf{R}^{(1)} &= [-6.2077\ 0\ 2.5200\ 0\ \ -1.4442\ 0\ 2.4250\ 0\ 3.8161\ 0],\\
\mathbf{R}^{(2)} &= [0\ \ -0.4697\ 0\ 2.4430\ 0\ 3.1717\ 0\ 2.3496\ 0\ 4.4762].
\end{aligned}$$

The received LLRs $\mathbf{R}^{(u)}$ and $\mathbf{R}^{(1)}$ are sent to the Code 1 decoder and the a priori LLRs $\mathbf{A}_1^{(1)}$ are set to zero. After running the log BCJR algorithm (see Algorithm 4.2), the Code 1 decoder returns

$$\begin{aligned}
\mathbf{L}_1^{(1)} &= [-8.0338\ \ -0.5330\ 0.0547\ \ -4.3310\ \ -1.0501\\
&\qquad 2.4858\ \ -1.6502\ 5.7755\ 1.3293\ 2.8979].
\end{aligned}$$

The extrinsic information $\mathbf{E}_1^{(1)} = \mathbf{L}_1^{(1)} - \mathbf{R}^{(u)}$ generated by the Code 1 decoder is given by

$$\begin{aligned}
\mathbf{E}_1^{(1)} &= [-6.1901\ 1.1134\ \ -1.0950\ 0.2170\ \ -0.1942\\
&\qquad -1.1631\ 1.2513\ \ -0.9417\ \ -1.1476\ 0.0000].
\end{aligned}$$

Next, the received LLRs $\Pi(\mathbf{R}^{(u)})$ and $\mathbf{R}^{(2)}$ are sent to the Code 2 decoder along with the a priori LLRs $\mathbf{A}_1^{(2)} = \Pi(\mathbf{E}_1^{(1)})$. The output of the Code 2 decoder, after it has run the log BCJR algorithm, is $\mathbf{L}_1^{(2)}$, where

$$\Pi^{-1}(\mathbf{L}_1^{(2)}) = [-8.6444\ 2.7847\ \ -0.5576\ \ -4.0711\ \ -2.7108$$
$$3.3102\ \ -3.3850\ 6.3897\ 0.8894\ 3.4507].$$

The extrinsic information output of the Code 2 decoder is $\mathbf{E}_1^{(2)} = \mathbf{L}_1^{(2)} - \mathbf{A}_1^{(2)} - \Pi(\mathbf{R}^{(u)})$. This is de-interleaved to give the a priori information $\mathbf{A}_2^{(1)} = \Pi^{-1}(\mathbf{E}_1^{(2)})$ to the Code 1 decoder for the second iteration,

$$\mathbf{A}_2^{(1)} = [-0.6106\ 3.3177\ \ -0.6123\ 0.2599\ \ -1.6607$$
$$0.8244\ \ -1.7348\ 0.6142\ \ -0.4399\ 0.5528],$$

while the received LLRs are unchanged. Running the log BCJR algorithm again, the Code 1 decoder now returns

$$\mathbf{L}_2^{(1)} = [-9.9305\ 2.4954\ 2.3209\ \ -4.7324\ \ -2.1302$$
$$4.5622\ \ -4.7249\ 5.6598\ \ -0.7831\ 3.4507].$$

The extrinsic information from the Code 1 decoder is consequently updated to

$$\mathbf{E}_2^{(1)} = [-7.4762\ 0.8242\ 1.7835\ \ -0.4443\ 0.3865$$
$$0.0889\ \ -0.0886\ \ -1.6717\ \ -2.8201\ 0.0000],$$

changing the a priori LLRs to $\mathbf{A}_2^{(2)} = \Pi(\mathbf{E}_2^{(1)})$ for the Code 2 decoder. The received LLRs fed into the Code 2 decoder are of course unchanged. Running the log BCJR algorithm again returns $\mathbf{L}_2^{(2)}$, from

$$\Pi^{-1}(\mathbf{L}_2^{(2)}) = [-10.5074\ 3.7493\ 3.6777\ \ -5.2659\ \ -2.8535$$
$$4.9604\ \ -5.0385\ 6.2522\ \ -2.4831\ 3.6064].$$

The extrinsic information output of the Code 2 decoder is $\mathbf{E}_2^{(2)} = \mathbf{L}_2^{(2)} - \mathbf{A}_2^{(2)} - \Pi(\mathbf{R}^{(u)})$. This is de-interleaved to give the a priori information $\mathbf{A}_3^{(1)} = \Pi^{-1}(\mathbf{E}_2^{(2)})$ to the Code 1 decoder to begin the third iteration:

$$\mathbf{A}_3^{(1)} = [-1.1876\ 4.5716\ 0.7445\ \ -0.2736\ \ -2.3841$$
$$1.2226\ \ -2.0485\ 1.2067\ \ -2.1399\ 0.7084].$$

The received LLRs are unchanged. The log BCJR algorithm is run again and the Code 1 decoder now returns

$$\mathbf{L}_3^{(1)} = [-12.5194\ 5.2123\ 5.8637\ \ -6.0724\ \ -4.3319$$
$$6.9691\ \ -7.0429\ 7.6092\ \ -3.0452\ 3.6064].$$

The extrinsic information from the Code 1 decoder is consequently updated to

$$\mathbf{E}_3^{(1)} = [-9.4882\ 2.2871\ 3.9695\ -1.2508\ -1.0919$$
$$2.0976\ -2.0929\ -0.3147\ -3.3821\ -0.0000],$$

changing the a priori LLRs for the Code 2 decoder to $\mathbf{A}_3^{(2)} = \Pi(\mathbf{E}_3^{(1)})$. Running the log BCJR algorithm again returns $\mathbf{L}_3^{(2)}$, from

$$\Pi^{-1}(\mathbf{L}_3^{(2)}) = [-13.4690\ 6.2159\ 6.4358\ -6.6448\ -4.6704$$
$$7.9641\ -9.1029\ 8.5536\ -4.3190\ 5.4488].$$

The three iterations are thus finished and the final hard decision on the message is given by the signs of the log likelihood ratios:

$$\begin{aligned}\hat{\mathbf{u}} &= \tfrac{1}{2}\left(\text{sign}\left(\Pi^{-1}(-\mathbf{L}_3^{(2)})\right) - 1\right)\\ &= [1\ 0\ 0\ 1\ 1\ 0\ 1\ 0\ 1\ 0].\end{aligned}$$

Three iterations of turbo decoding have successfully corrected the errors added by the channel.

While we have used a very short code for simplicity, the performance of a turbo code improves dramatically as the length of the interleaver is increased. Indeed, the best performances for turbo codes are achieved using interleavers with lengths of several thousand bits.

Example 5.5 In this example we consider again the performance of the code in Example 5.2 but this time looking at the average performance of the code over a number of signal-to-noise ratios. Using a randomly generated length-1024 interleaver, and varying the signal-to-noise ratio from 1 to 5 dB, the average performance of the decoder is calculated by simulating enough codewords that the decoder fails at least 50 times. The average word error rate (WER) and bit error rate (BER) of the decoder are plotted in Figure 5.3 versus the channel signal-to-noise ratio. This process was repeated for longer interleaver lengths, up to $K = 65\,536$, and more iterations. The average bit error rates of the decoder for these interleaver lengths are plotted in Figures 5.3 and 5.4.

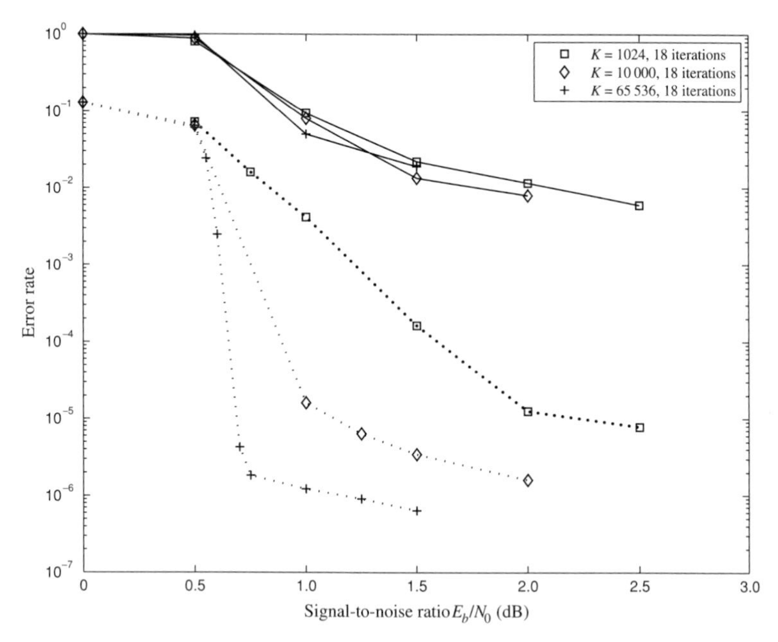

Figure 5.3 The BER (dotted curves) and WER (solid curves) performance of turbo decoding for the code from Example 5.5 for varying signal-to-noise ratios and interleaver lengths. In each case 18 turbo decoding iterations were carried out.

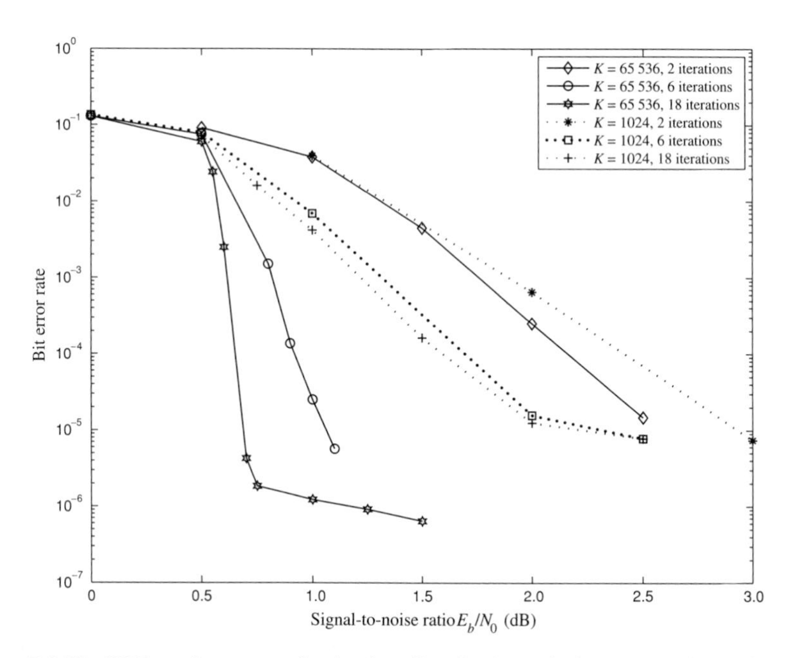

Figure 5.4 The BER performance of turbo decoding for the code from Example 5.5 for two interleaver lengths and three different numbers of iterations.

Figure 5.4 shows the characteristic BER curve for turbo-like codes, which can be divided into three regions. The first is the *non-convergence region*, where the BER is high and relatively flat as the SNR is increased. The second region shows a rapid decrease in BER for small increments in SNR and is commonly called the *waterfall region* (cf. Figure 2.5) or less commonly the *turbo cliff*. Lastly, in the *error floor region* the slope of the curve flattens out and the BER reductions are small as the SNR increases.

Figures 5.3 and 5.4 show three trends common to turbo codes in general. Firstly, increasing the interleaver length improves the code BER performance by decreasing the average SNR of the waterfall region. (The value to which the SNR of the waterfall region can be decreased simply by increasing K reaches a limit, the *threshold* SNR, which will be discussed further in Chapter 7.) Secondly, increasing the interleaver length seems to have little effect on the code's WER performance; this behavior will be discussed in Chapter 8. Thirdly, as is indicated by Figure 5.4, increasing the number of decoder iterations improves the decoder's performance. This improvement reaches a limit after which extra iterations provide little benefit. The performance of a turbo decoder, regardless of the number of iterations used, can be bounded by considering the performance of the ML decoder for that code and, again, is discussed in Chapter 8.

5.4 Turbo code design

In this section we will consider how the choice of component encoders, interleaver length and interleaver structure affect the performance of the resulting turbo code.

5.4.1 Interleaving gain

Consider a parallel concatenated turbo code with component encoders C_1 and C_2 and a length-K interleaver. Since the code is systematic, a weight-w message $\mathbf{u}$ will produces a turbo codeword with weight

$$d = w + p^{(1)} + p^{(2)},$$

where $p^{(1)}$ and $p^{(2)}$ are the weights of the parity-bit sequences from C_1 and C_2 respectively.

If $w = 1$ and the component encoders are finite impulse response (FIR) encoders then weight-1 messages will produce finite-weight outputs (i.e. the impulse response), whereas if the encoders are infinite impulse response (IIR) encoders for the same codes then weight-1 messages give infinite-weight outputs.

(See the end of Section 4.2 for an introduction to IIR and FIR encoders.) This difference is particularly important for turbo codes since a low-weight output for a weight-1 input results in a very low overall codeword weight, d. The choice of interleaver cannot improve an FIR encoder since a weight-1 input produces the encoder's impulse response regardless of its location (we will assume input sequences long enough that the occurrence of bits at the edge of the sequences, *edge effects*, can be ignored).

However, for a message with weight 2 or more the interleaver can help to improve the overall concatenated codeword weight. If $w \geq 2$ the output weight of the component code depends on the location of the 1 entries in $\mathbf{u}$. A "good" interleaver will permute the message $\mathbf{u}$ in such a way that if $\mathbf{u}$ has poorly placed 1 entries (in the sense that they lead to a low-weight parity sequence) then $\Pi(\mathbf{u})$ will not.

For a length-K interleaver and a weight-w message there are $\binom{K}{w}$ possible weight-w sequences output by an interleaver. We will call a "bad" interleaver one that just shifts the message, since in this case shifting the input to a convolutional encoder will just shift its output and will not improve its Hamming weight (ignoring edge effects). There are K possible shifted versions of a length-K message. Thus the probability that a randomly chosen interleaver will be bad is $K/\binom{K}{w}$. A good interleaver is thus chosen randomly with probability

$$1 - \frac{K}{\binom{K}{w}} \approx 1 - \frac{K}{K^w} = 1 - K^{1-w}.$$

Thus, as w is increased the probability that a good interleaver will be chosen increases, i.e. the probability that both $\mathbf{u}$ and $\Pi(\mathbf{u})$ are encoded to low-weight parity sequences decreases. Further, when $w \geq 2$, and K increases, the probability of a good interleaver also increases.

For weight $w = 1$ inputs there is zero chance that a good interleaver will be produced. This is reasonable since a single non-zero bit can only be shifted in position – there is no other option. Thus for FIR encoders, for which $w = 1$ inputs produce finite-weight outputs, there is no gain obtained by interleaving. However, for IIR encoders, for which weight-1 inputs produce infinite-weight outputs, there are no "bad" weight-1 inputs; the lowest-weight "bad" inputs occur for $w = 2$, where the probability that a good interleaver is obtained increases with K and there is a significant gain obtained by interleaving. (Of course the interleaver can also spread the weight-2 inputs for FIR encoders, but the performance of the FIR turbo code will be dominated by the low-weight codewords due to the weight-1 messages.)

More generally, FIR encoders will always return to state zero after a message sequence followed by b or more zeros, where b is the maximum memory order of the encoder. However, IIR encoders will only return to state zero for one in 2^b

message sequences followed by zeros. Because of this, IIR encoders are much less likely to map low-weight message sequences to low-weight outputs.

In Chapter 8 we will show that, when IIR encoders are used, the error correction improvement resulting from the consequent reduction in the probability of low-weight codewords decreases their contribution to the bit error rate by a factor $1/K$, which is the so-called *interleaving gain*. We will consider the interleaving gain in more detail in Chapter 8.

5.4.2 Choosing the component codes

The important property of the component codes will be both the number a_d (see (1.36)) of low-weight codewords generated and the message weight w that generates them. Two encoders for the same code will of course have the same a_d values since they have the same codeword sets. Where they differ will be in the total message weight B_d (see (1.40)) associated with all the weight-d outputs. Generally, the first two B_d terms for systematic IIR encoders are higher than those for the associated FIR encoders. However, for the next few codeword weights the trend is reversed and B_d grows more slowly for the IIR encoders, making them better performers in low-SNR channels where the larger-weight codewords play an increasing role in the overall error rate.

As we saw in the previous discussion, it can in fact be beneficial for a turbo component encoder to map low-weight inputs to high-weight outputs. Indeed, the choice of mapping from message sequences to codeword sequences by the component encoder can even play a more significant role in turbo code performance than the minimum distance of the component code.

Example 5.6 The following example from [88] highlights the difference between considering d_{min} as a design criterion for the component codes as opposed to considering the mappings from messages to codewords. Consider two possible IIR encoders, C_{A}, with generator matrix

$$\begin{bmatrix} 1 & \dfrac{1+D^2}{1+D+D^2} \end{bmatrix},$$

and C_{B}, with generator matrix

$$\begin{bmatrix} 1 & \dfrac{1+D+D^2}{1+D^2} \end{bmatrix},$$

for a rate-1/3 turbo code. The minimum parity sequence ($p = 2$) for the C_{A} code is generated by the weight-3 input sequence with three consecutive 1s. With worst-case interleaving, this input will produce a minimum-weight turbo codeword with $d_{\text{min}} = 7$. Meanwhile the minimum-weight parity sequence

($p = 3$) for the C_{B} code is generated by the weight-2 input sequence $0 \cdots 01010 \cdots 0$. With worst-case interleaving, this input will produce a minimum-weight turbo codeword with $d_{\min} = 8$. A turbo code employing the C_{B} encoder thus has a better minimum distance for the same rate, making it a better code in the classical sense. However, once we consider the interleaver the C_{A} code will show its strengths.

For a turbo encoder we want to consider not just the lowest-weight codewords but their multiplicities as well. Even though using the C_{A} encoder produces a weight-7 turbo codeword this codeword will only occur if the interleaver just shifts the weight-3 message sequence so that both component encoders produce weight-2 outputs. However, as we saw above, a randomly chosen length-K interleaver will do this with very low probability ($1/K^2$ where K is large). Of much more significance for the BER of the turbo code are the weight-2 messages, which are much more likely to be just shifted across (with probability $1/K$) making the production of low-weight codewords by weight-2 inputs much more likely.

For the weight-2 sequence $0 \cdots 010010 \cdots 0$ the C_{A} encoder will return to the all-zeros state after the second 1, producing low-weight outputs in the form $0 \cdots 011110 \cdots 0$. Such a sequence is called *self-terminating*. Without interleaving the codeword produced would be weight-10. In contrast, for the weight-2 sequence $0 \cdots 01010 \cdots 0$ the encoder never returns to the all-zeros state; such a sequence is called *non-self-terminating*, and the output parity sequence is of the form $0 \cdots 01111 \cdots 1$ where the all-ones sequence continues until the trellis is terminated. This sequence accumulates a very large weight unless it starts near the end of the block.

The weight-2 input sequence $0 \cdots 010010 \cdots 0$ is not the only self-terminating weight-2 sequence for this encoder. In general, weight-2 sequences with 1s separated by 3δ bit positions, where $\delta = 1, 2, \ldots, K-1$ are also self-terminating, producing outputs with weight $4 + 2\delta$.

For the C_{B} encoder, however, the weight-2 sequences with their 1s separated by 2δ bit positions, where again $\delta = 1, 2, \ldots, K-1$ are the self-terminating sequences producing outputs with weight $3 + \delta$. Thus, for a given message length the C_{B} encoder has both more possible self-terminating weight-2 sequences and also lower output weights associated with them. Simulation results confirm that turbo codes using the C_{A} encoder outperform turbo codes using the C_{B} encoder.

Example 5.6 shows that the different properties of the encoders for weight-2 sequences can play a significant role in reducing the *number* of low-weight codewords (but not the *minimum* codeword weight) produced by a turbo encoder with a random interleaver.

It also suggests a criterion for the design of the interleaver. If the interleaver can map all messages producing low-weight self-terminating sequences into non-self-terminating sequences for the second encoder then low-weight turbo codewords can be avoided. Of course the interleaver will have a better chance of doing this if there are fewer self-terminating sequences with low weight. Thus a sensible design criterion for component encoders could be the growth in output weight as the separation between two non-zero inputs is increased. A coarser, but still useful, measure of how a convolutional encoder will fare in a turbo code is given by its minimum output weight for any weight-2 input.

Effective free distance

The above discussion shows that, if IIR codes are used, weight $w = 1$ messages cannot produce low-weight outputs and, as the weight w of the message sequence $\mathbf{u}$ is increased, the probability that both $\mathbf{u}$ and $\Pi(\mathbf{u})$ result in a low-weight output sequence decreases dramatically. Thus, it makes sense to consider the case $w = 2$ when designing the component codes and interleaver so as to reduce the number of low-weight codewords. This observation is backed up by simulation results, which give clear evidence that, when using turbo codes with IIR component encoders, the codewords corresponding to weight-2 messages make an important contribution to the bit error rate. Consequently, a measure called the *effective free distance* d_{ef} is defined as the lowest-weight codeword for any weight-2 message.

Using the parallel concatenation of systematic IIR codes,

$$d_{\text{ef}} = 2 + p_{\text{ef}}^{(1)} + p_{\text{ef}}^{(2)}, \tag{5.3}$$

where $p_{\text{ef}}^{(i)}$ is the lowest-weight parity sequence generated by the ith component encoder for any weight-2 input and d_{ef} is thus the lowest-weight turbo codeword produced by any weight-2 input. The component codes are systematic, so we can also write

$$d_{\text{ef}} = d_{\text{ef}}^{(1)} + d_{\text{ef}}^{(2)} - 2 \tag{5.4}$$

since, for a systematic code,

$$d_{\text{ef}} = p_{\text{ef}} + 2.$$

Note that the effective free distance of the concatenated code may be greater than the minimum free distance of the code, since inputs with weight > 2 may produce lower overall codeword weights. Algorithms for finding the free distance of turbo codes can be found in [89]. The definition of d_{ef} assumes that both component codes produce the minimum-weight output possible for a weight-2 input, so it does not embrace the possibility that the interleaver is designed well enough to avoid this.

Consider an IIR encoder with generator function

$$f(D)/g(D),$$

where $f(D), g(D)$ are unspecified polynomials, and input $x(D)$. The output for this message,

$$x(D)f(D)/g(D),$$

will be finite if and only if $x(D)f(D)$ is divisible by $g(D)$. For a weight-2 message we have $x(D) = 1 + D^i$, and the smallest i for which $g(D)$ divides $x(D)$ is the *period* of $g(D)$. The period of a binary polynomial of degree v is less than or equal to $2^v - 1$. To reduce the number of possible weight-2 inputs that produce finite-weight outputs, we would like a polynomial $g(D)$ with the greatest possible period for a given v, i.e. $2^v - 1$. Such a polynomial is called *primitive*.

A rate-$1/n$ recursive convolutional encoder with memory order v and generator matrix

$$[1 \quad f_1(D)/g(D) \quad \cdots \quad f_{n-1}(D)/g(D)]$$

can achieve a maximum p_{ef} equal to

$$(n - 1)(2^{v-1} + 2).$$

Any generators $f_1(D)/g(D)$, where $g(D)$ is a primitive polynomial of degree v and $f_i(D)$ is any monic polynomial of degree v except $g(D)$, will achieve this maximum provided that the members of the set of $n - 1$ numerator polynomials

$$f_1(D), \ldots, f_{n-1}(D)$$

have greatest common divisor 1. With a primitive divisor polynomial, all weight-2 inputs are guaranteed to cause a trellis path that passes through all the code trellis states before any possible return to the all-zeros state. Of course, choosing a larger v can produce a larger d_{ef}; however, as the complexity of BCJR decoding scales with the state complexity 2^v, practical implementations currently use $v = 3$ or 4.

A search over the effective free distance of possible systematic rate-1/2 $[1, f/g]$ convolutional encoders (f and g are $f(D)$ and $g(D)$ in octal notation) in [90] found the set of encoders having the best effective free distance for each degree v; this is given in Table 5.1. (If two encoders have the optimal effective free distance then the one with the smaller multiplicity is chosen.)

Note that, with the exception of the case $v = 1$, the minimum free distance of the code is less than d_{ef} since inputs with weight greater than 2 produce the lowest-weight codewords for these encoders.

Table 5.1

v	f	g	p_{ef}
1	2	3	1
2	5	7	4
3	17	15	6
4	37	31	10
5	57	75	18
6	115	147	34

5.4.3 Interleaver design

The role of the interleaver is twofold. Firstly, it is chosen to improve the code's weight spectrum, i.e. it will permute the message $\mathbf{u}$ in such a way that if $\mathbf{u}$ has poorly placed non-zero entries (in the sense that they lead to a low-weight parity sequence) then $\Pi(\mathbf{u})$ does not have such entries. Secondly, the interleaver randomizes the location of the message symbols in each component codeword, which reduces correlation between the extrinsic information passed between component decoders.

As we saw earlier an improvement in the code's weight spectrum can be achieved if the self-terminating sequences input to one encoder are permuted to non-self-terminating sequences before being input to the second encoder.

If the interleaver could map all messages containing self-terminating sequences into non-self-terminating sequences, the turbo code would have a minimum weight that grows linearly with the block size K. Unfortunately, it is unlikely that this can be achieved (see Section 8.2.6). However, the principle can be applied to the most troublesome self-terminating sequences (i.e. those with the lowest output weight).

Of course the set of self-terminating sequences is different for every encoder, so new interleavers need to be designed if new component encoders are chosen. Nevertheless, some general observations can be made. In particular, for a weight-2 message the weight of the encoded output increases linearly with the distance apart of the 1 entries. If the 1 entries are very far apart then the encoder can be thought of as producing its weight-1 infinite impulse response until the second 1 arrives. Thus if the interleavers map weight-2 messages with closely spaced 1s to weight-2 sequences with 1s far apart, some lowest-weight codewords can be avoided.

Spreading factor

An interleaver Π has *spreading factor* (S, R) if any entries within S bits of each other at the input will be spread to at least R bits from each other at the output.

That is,

$$\text{if } |i - j| \leq S \text{ then } |\pi_i - \pi_j| \geq R,$$

where π_i is the ith element of Π. The *S-parameter* of an interleaver is the maximum S value such that $S \leq R$.

A common interleaver is the row–column interleaver, which is implemented by writing the sequence to be interleaved row-wise into an $M \times N$ matrix and reading the entries out column-wise.

Example 5.7 The length-16 row–column interleaver will write the message sequence

$$\mathbf{u} = [u_1\ u_2\ u_3\ u_4\ u_5 \cdots u_{16}]$$

into a matrix as

$$\begin{bmatrix} u_1 & u_2 & u_3 & u_4 \\ u_5 & u_6 & u_7 & u_8 \\ u_9 & u_{10} & u_{11} & u_{12} \\ u_{13} & u_{14} & u_{15} & u_{16} \end{bmatrix},$$

and read it out as

$$[u_1\ u_5\ u_9\ u_{13}\ u_2\ u_6\ u_{10}\ u_{14}\ u_3\ u_7\ u_{11}\ u_{15}\ u_4\ u_8\ u_{12}\ u_{16}].$$

As $M \times N$ row–column interleavers write into rows and read from columns, inputs on the same row (i.e. within N bits of each other) are spread to at least M bits apart at the output. Entries that wrap across rows are spread even further; for example, the Nth and $(N+1)$th bits are spread out to be $NM - 1$ bits apart at the output.

Variations on the row–column interleaver write entries column-wise and read row-wise, or write right-to-left and/or read bottom-to-top. While these block interleavers can have large spreading factors, their extreme regularity means that the low-weight codewords that do exist have very high multiplicities. In general, while interleaver regularity can guarantee the removal of all codewords below a certain weight, it can actually increase the number of codewords at the next lowest weights. Indeed, interleavers designed to break up all self-terminating weight-2 sequences actually increase the probability of producing "bad" sequences with weights 4 and higher.

Higher-weight input sequences are much more likely than weight-2 input sequences to be broken up by randomly chosen interleavers. However, a purely random interleaver is also likely to reproduce a few of the weight-2 input sequences with the lowest output weights.

One way to ensure a good spreading factor without requiring regularity is to constrain a random interleaver in such a way as to ensure a desired S-parameter. One such interleaver is called an *S-random interleaver*.

Example 5.8 To construct an S-random interleaver a set P is initialized as the set of all integers from 1 to K. Each interleaver digit π_i is an integer randomly selected, one at a time, from the set of valid integers in P, thus reducing the size of P by one. The selected integer is then compared with the S previously selected integers. If the value of the current integer is within $\pm S$ of the value of any of them, it is placed back in P and a new integer selected. This process is repeated until all K integers are selected.

Consequently, the S-random algorithm will ensure that if $|i - j| \leq S$ then $|\pi_i - \pi_j| > S$. The searching time for this algorithm increases with S, and the algorithm is not guaranteed to finish successfully. However, it is generally possible to find an interleaver for $S < \sqrt{(K/2)}$ in a reasonable time.

Modifications to the basic S-random interleaver can impose a bound on the sum of the input and output spacing between any two bits, i.e.

$$|i - j| + |\pi_i - \pi_j| \leq S,$$

or constrain the algorithm to ensure certain implementation features (see below) such as producing low-delay and prunable interleavers.

The key of S-random interleavers is that they achieve a particular S-value while maintaining their "randomness", a measure of which is given by an interleaver's dispersion ratio.

Dispersion

The set $\mathcal{D}$ of *displacement vectors* for an interleaver Π is the set of all possible (Δ_x, Δ_y) pairs such that

$$\Delta_x = j - i, \ \Delta_y = \pi_j - \pi_i, \quad \forall 0 \leq i < j < K.$$

The *dispersion* of an interleaver is the size $|\mathcal{D}|$ of the set of all displacement vectors.

The smallest set of displacement vectors belongs to the identity permutation $\pi_i = i$, i.e.

$$\mathcal{D} = \{(1, 1), (2, 2), \ldots, (K - 1, K - 1)\},$$

and $|\mathcal{D}| = K - 1$. Since there are $K(K-1)/2$ choices for i and j the largest possible set of displacement vectors is of size $K(K-1)/2$. Thus

$$K - 1 \leq |\mathcal{D}| \leq K(K-1)/2.$$

The dispersion value can be used to quantify the randomness of an interleaver. A purely random interleaver will have a large number of displacement vectors while a structured interleaver will have very few. The $N \times N$ row–column interleaver, for example, has dispersion $2N(N-1)$. For a given length-K interleaver the *dispersion ratio* ν is the ratio of the dispersion of that interleaver to the maximum possible dispersion:

$$\nu = \frac{|\mathcal{D}|}{K(K-1)/2}.$$

The row–column interleaver has a low value of ν:

$$\nu_{\mathrm{RC}} = \frac{2N(N-1)}{N^2(N^2-1)/2},$$

which goes to zero as the interleaver length grows. A randomly selected interleaver, however, has a dispersion ratio of around 0.8.

Figure 5.5 shows plots of π_i versus i plots for row–column, random and S-random interleavers. The row–column interleaver has a good spread but a low dispersion. The random interleaver, however, has a poor spread, with clumps of π_i values closely spaced. The S-random interleaver combines good dispersion with good spread.

Interleaver delay

An interleaver is causal if the output at time t depends only on the current or previous inputs, i.e. $\pi_i \leq i$ for all i. An interleaver can be made causal by introducing an appropriate *delay* at the input and/or output. If the input to the interleaver is serial, i.e. the bits arrive one at a time with u_1 at $t = 1$ and u_2 at $t = 2$ etc., and the output of the interleaver is serial then the delay is the number of time units for which the user will need to wait, over and above the K time units for reception of the the length-K input, before the full output sequence is provided by the interleaver.

Example 5.9 The length-16 row–column interleaver, which produces

$$\Pi(\mathbf{u}) = [u_1\ u_5\ u_9\ u_{13}\ u_2\ u_6\ u_{10}\ u_{14}\ u_3\ u_7\ u_{11}\ u_{15}\ u_4\ u_8\ u_{12}\ u_{16}],$$

requires a delay of 3 time units for the output bit at time $t = 2$ (since it must wait for the input at time $t = 5$), and a delay of 6 time units for the output bit at time $t = 3$ (since it must wait for the input at time $t = 9$), and so on. Inspecting all the output bits, the maximum required delay is 9 time units (for the output at $t = 4$).

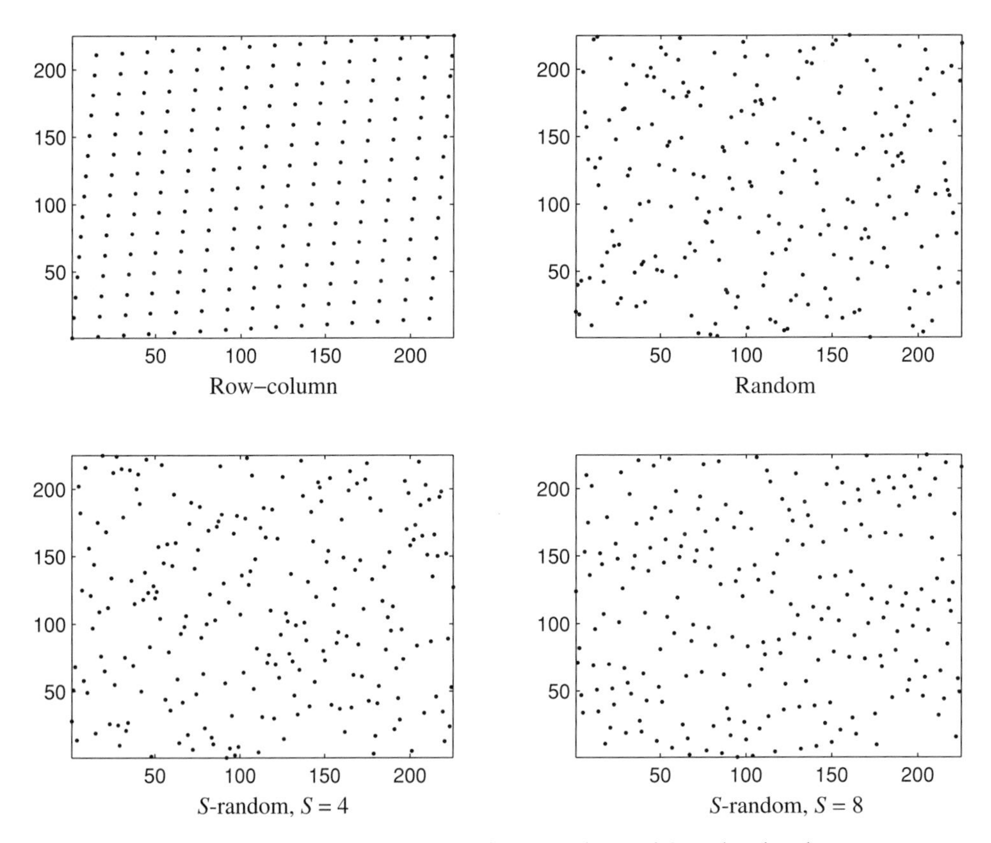

Figure 5.5 Plots of π_i versus i for row–column, random and S-random interleavers.

The total *latency* of the interleaver is the sum of the delays of both the interleaver and the de-interleaver. For the row–column interleaver the interleaver and de-interleaver delays are both $(N-1)(M-1)$ and so the latency is $2(N-1)(M-1)$.

The interleaver delay also determines how much memory is required to build a minimum-delay interleaver since the delayed inputs will need to be stored until the interleaver is ready for them.

Odd–even interleavers

When the parity bits at the output of an encoder are to be punctured, the interleaver can be designed to ensure that the parity bits associated with each message bit are evenly allocated across the message. For example, the rate-$1/2$ turbo code in Example 5.2 was formed by puncturing every second parity bit from each of the two rate-$1/2$ component codes. Unpunctured, each message bit is encoded to two parity bits, one from each encoder. If the message bits are randomly interleaved then the same message bit may have both its corresponding parity bits punctured while another will have both its parity bits unpunctured.

To obtain even error protection for this rate-$1/2$ turbo code simply requires that the message bits located in odd bit positions in the message (which will have their associated parity bits punctured by the first component code) remain in odd bit positions in the interleaved message (and so will not have their parity bits punctured by the second component code). An even–odd interleaver ensures this by permuting the even and odd entries in the message independently.

The same principle can be applied to punctured turbo codes of any rate. Of course, if unequal error protection is desired then the interleaver can be designed to deliberately allocate a selection of message bits in a way that decreases the number of parity bits punctured.

Prunable interleavers

Often a communications standard will call for flexibility in codeword length, i.e. the interleaver length will need to change depending on user requirements and channel conditions. Rather than store a new interleaver for each allowed codeword length it may be desirable to store a single interleaver (for the longest length) that can be modified to give shorter interleavers as required. Such an interleaver is called *prunable*. A purely random interleaver is prunable – every subset of entries is also a random interleaver. However, structured and pseudo-random designs (such as S-random interleavers) are not intrinsically prunable – the required properties can be lost in a shorter version of the interleaver.

Algorithmic interleavers

The chosen interleaver can be stored in memory, or the desired properties can be incorporated into algorithmic interleavers which can be described by a small algorithm and so generated in real time. The original Berrou–Glavieux interleaver is a good example of this (see [82, 92]).

Example 5.10 For $N = 2^n$ and $M = 2^m$, n, m integers, where $K = NM$, and define a set of prime numbers

$$p = \{17, 37, 19, 29, 41, 23, 13, 7\}.$$

Then for each $i = 1 : NM$ the Berrou–Glavieux interleaver is given by

$$\pi_i = c_i + Mr_i,$$

where

$$
\begin{aligned}
r_0 &= \text{mod}\,(i, M), \\
c_0 &= (i - r_0)/M, \\
l &= \text{mod}\,(r_0 + c_0, 8), \\
r_i &= \text{mod}\,(p_{l+1}(c_0 + 1) - 1, N), \\
c_i &= \text{mod}\,((M/2 + 1)(c_0 + r_0) - 1, M),
\end{aligned}
$$

and mod (a, b) means a mod b.

Interleavers for parallel decoders

Using windowed approaches to BCJR decoding (see Section 5.5.1) it is possible to implement partially parallel turbo decoders. For such decoders, the design of the interleaver can be particularly important in optimizing the decoder efficiency (in much the same way that edge routing is important in partially parallel LDPC decoder implementation). Indeed, the concept of protograph constructions (see Section 3.3.1) for LDPC codes with an efficient decoder implementation can also be used in the construction of parallelizable turbo interleavers (see e.g. [91]).

5.4.4 Design rules

We now summarize the design rules for turbo codes. Firstly, systematic IIR convolutional codes should be used for the component codes, since, unlike FIR encoders, IIR encoders map weight-1 messages to infinite-weight outputs, providing an interleaving gain when the interleaver spreads the weight-2 messages. In choosing a particular IIR encoder, the use of an encoder with maximum effective free distance can result in improved error floor performance owing to its avoidance of low-weight turbo codewords. Alternatively, see Chapter 7, we can choose component encoders that improve the performance of the turbo code at low SNRs by improving its threshold.

In theory, the designer of a turbo code should simultaneously consider the component codes and interleaver. For example, consider the scenario where a particular choice of encoder produces a turbo code with a very small d_{ef} but for which it is straightforward to design an interleaver that guarantees that at no point do both encoders produce their minimum weight outputs. If the next-lowest-weight output which is not avoided by the chosen interleaver was quite high, a good turbo code would be rejected if the interleaver and component code were not considered jointly. However, in practice, the simultaneous optimization of both the encoders and the interleavers together is usually too difficult a task.

The encoders are chosen first (perhaps by optimizing d_{ef}) and the interleaver is then designed for the chosen encoders.

Ideally, the interleaver would permute terminating sequences to non-terminating sequences, achieving a global optimum for the chosen component encoders by taking into account the input sequences of all weights simultaneously. This is not computationally possible, so we are left with some general design considerations. A "good" interleaver is one with a large S-parameter, for the prevention of very-low-weight codewords, a high dispersion ratio to keep the multiplicities small for the remaining low-weight codewords and a small delay. The S-random type of interleaver nicely combines a high S-parameter with a relatively high dispersion and makes a suitable de facto standard with which to compare new interleaver designs. Such an interleaver is still, however, a large pseudo-random device with inherent implementation issues; thus interleaver designs targeted directly to the component codes should be able to improve upon the performance of S-random interleavers. Overall, the best choice of interleaver will depend on the choice of the component encoders and the application area.

In summary:

- The component codes need to be IIR to provide an interleaving gain.
- The performance of turbo codes can be shown to be influenced by the codewords produced by weight-2 messages. These codewords, see (5.3), have weight

$$d_{\text{ef}} = d_{\text{ef}}^{(1)} + d_{\text{ef}}^{(2)} - 2.$$

  To maximize $d_{\text{ef}}^{(1)}$ and $d_{\text{ef}}^{(2)}$ we require that
  - the generator functions for the component codes should be of the form

$$f(D)/g(D)$$

    where $g(D)$ is a primitive polynomial of degree v, and
  - $f(D)$ is any monic polynomial of degree v except $g(D)$.
- There can be significant performance differences between codes with the same values of N, v and d_{ef}. Searches for good numerator polynomials $f(D)$ may take into account the following factors (in order of increasing complexity):
  - maximization of the minimum distance of the concatenated code (which is not necessarily the same since d_{ef} since the minimum distance is considered over all input weights, not just $w = 2$),
  - inclusion of more of the weight spectrum of the turbo code, i.e. $w = 3, 4$, etc., or, alternatively,
  - minimization of the BER in simulated results using iterative decoding.
- The design of an interleaver depends upon the chosen component codes and the application. A good interleaver design will ensure that low-weight outputs are

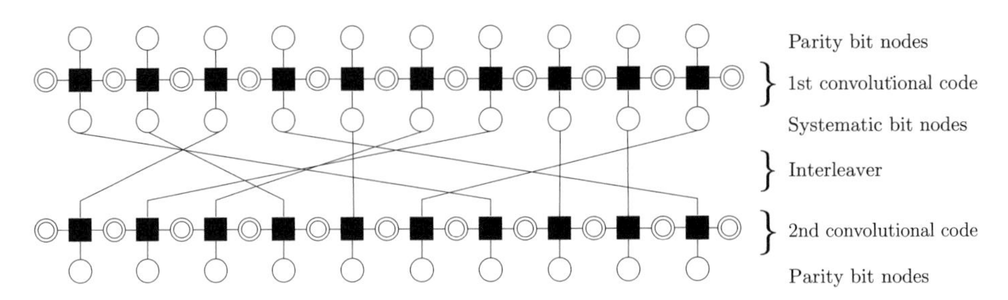

Figure 5.6 The factor graph for the turbo code in Example 5.2.

not simultaneously produced by both encoders. In general, important design features include

– a large S-factor,
– a high dispersion ratio,
– ease of implementation; depending on the application this may point to low-delay, prunable, parallelizable or algorithmic interleavers.

5.5 Factor graphs and implementation

Turbo codes can be represented graphically using *factor graphs*. Figure 5.6 shows the factor graph for the turbo code used in Example 5.2. A factor graph has three node types: bit nodes, shown as open circles, constraint nodes, shown as solid squares, and state nodes, shown as double circles. As turbo codes are systematic we can distinguish between the *systematic-bit nodes* corresponding to the K message bits in the codeword and the *parity-bit nodes* corresponding to the $N - K$ parity bits in the codeword.

The turbo code in Example 5.2 encodes a length $K = 10$ message so there are ten systematic-bit nodes. Similarly, each component convolutional encoder produces ten parity bits, so there are ten parity-bit nodes for the first convolutional encoder, at the top of the graph, and ten parity-bit nodes for the second convolutional encoder, at the bottom of the graph. Both encoders transition through 11 states while encoding the message from the starting state at $t = 0$ to the final state at $t = 10$. These states are represented in the factor graph as state nodes. Running from left to right, the first state node represents the state at time $t = 0$ while the last state node represents the state at time $t = 10$. Lastly, the constraint nodes show dependences between the bit and state nodes. Recall that the parity bit generated at time t depends on the state at time $t - 1$ and the message bit input at time t. Similarly the state at time t depends on the state at time $t - 1$ and the input bit at time t. Thus a constraint node connects the state at time $t - 1$ and the message bit input at time t with the parity bit at time t and the state at time t.

The factor graph representation of turbo codes shows how iterative turbo decoding can be considered as a form of message-passing decoding. In one iteration of the turbo decoder a number of different messages are generated, as follows.

- The bit nodes are loaded with the received LLRs from the channel and, in the case of the systematic-bit nodes, the extrinsic LLRs from the other half of the graph. Values of Γ are then generated using the messages from the bit nodes.
- For the BCJR decoder of the first code,
  - $\mathcal{A}$ messages are passed from left to right through the state nodes,
  - $\mathcal{B}$ messages are passed from right to left through the state nodes, and
  - once the constraint nodes have all three $\mathcal{A}$, $\mathcal{B}$ and Γ messages, extrinsic LLRs are calculated and passed to the systematic-bit nodes.
- The extrinsic LLR messages are passed through the interleaver to the second component code.
- For the BCJR decoder of the second code,
  - $\mathcal{A}$ messages are passed from left to right through the state nodes,
  - $\mathcal{B}$ messages are passed from right to left through the state nodes, and
  - once the constraint nodes have all three $\mathcal{A}$, $\mathcal{B}$ and Γ messages, extrinsic LLRs are calculated and passed to the systematic-bit nodes.

Using factor graphs we can see interesting parallels between turbo and LDPC codes. In fact the Tanner graph of LDPC codes is a factor graph with only two node types, bit nodes and constraint nodes. The constraint nodes in the factor graphs of LDPC codes are, of course, just single parity-check equations.

Generalizations of the original turbo codes include more than two convolutional encoders concatenated in parallel. As we will see in Chapter 8, this can improve the interleaving gain and result in asymptotically better turbo codes but will of course require more puncturing for reasonable code rates.

The concept of irregularity for LDPC codes can also be applied to these turbo codes, although in a less straightforward way. Like the generalized turbo codes, an irregular turbo code will have more than two convolutional encoders – but not all the message bits will be encoded by every one of them. Some message bits may be encoded by a convolutional code with a lower rate or with higher state complexity, affording the same sort of extra error protection that using a high-degree bit node provides to selected bits in an irregular LDPC code.

5.5.1 Implementation and complexity

When it comes to encoding turbo codes, the component convolutional encoders are particularly simple to implement, requiring components no more complex than shift registers and modulo-2 adders. The interleaver is usually implemented via a memory array with messages written into and read out of the memory according to the particular interleaver permutation. The interleaver permutation can be stored in the memory or, for an algorithmic interleaver (see Section 5.4.3), calculated as required.

While the first convolutional encoder can produce each output parity bit as it receives each new input bit, the second convolutional encoder will suffer a delay as it waits for the required input bits to be interleaved. For example, the second encoder of the turbo code in Example 5.2 must wait for the third message bit before it can output its first parity bit.

As in the case of message-passing decoding for LDPC codes, the complexity of the turbo decoder has three aspects: the complexity of the factor graph nodes, the complexity of routing for the interleaver and the number of iterations required for the decoder.

Node complexity

The node complexity for turbo decoding depends on the variant of the BCJR decoding algorithm employed. In practice the max log MAP decoder is usually used, with a lookup table for the correction term. Using a relatively small lookup table for the correction term can produce performances with a loss of only around 0.5 dB compared with the full log MAP algorithm.

Routing complexity

A high throughput interleaver can be implemented by directly mapping the corresponding edges in hardware; however, the long random-like interleavers that work best with turbo codes are particularly difficult to route. The interleaver can be implemented via a memory array with messages written into and read out of the memory according to the particular interleaver permutation.

Number of iterations

The maximum number of iterations required by a turbo decoder is usually around 10 to 20. However, the turbo decoder may arrive at a correct decoding long before the maximum number of iterations is completed. Ideally decoding would be stopped once the decoder is relatively sure that it has reached its final decision.[1]

A decoder is said to have *converged* when further iterations will not change the decoding result. A test to determine whether the decoder has converged is called a *stopping criterion* since decoding halts when the test is satisfied. For turbo codes a number of alternative stopping criteria have been proposed that reduce the complexity of turbo decoding, and we will consider a few of

[1] Unlike for block codes (where a quick syndrome check will tell whether a valid codeword has been found), the MAP convolutional decoder outputs message-bit LLRs and, of course, every length-K binary vector is a valid message.

these in Section 5.5.2. Employing stopping criteria to improve throughput of course assumes that the receiver is handling the variable amount of time taken to decode each codeword using decoder pipelining. However, even without this feature, stopping the decoder as early as possible has the advantage of reducing energy consumption, which can be very useful in some devices.

Serial versus parallel implementation

Unlike message passing for LDPC codes, in the decoding of turbo codes the nodes of the factor graph cannot be updated in parallel. In fact a half-iteration of the turbo decoder requires the $\mathcal{A}$ messages (see above) to be passed one at a time from left to right through the state nodes and, similarly, the $\mathcal{B}$ messages must be passed one at a time from right to left (although this can be done at the same time as the left to right pass). This significantly increases the delay associated with an iteration of turbo decoding. It also requires a large amount of memory, as all the $\mathcal{A}$ and $\mathcal{B}$ messages need to be stored until the final step, that of calculating the extrinsic LLRs.

A solution is to use a windowed version of the BCJR algorithm. In this case the BCJR algorithm is run on a sub-trellis representing a small subset of the trellis segments in the full code trellis. For example, we may decide to decode the factor graph in Figure 5.6 using two windows, the first covering the first five constraint nodes and the second covering the last five constraint nodes. The two sub-trellises corresponding to each set of constraint nodes are treated as independent trellises: $\mathcal{A}$ and $\mathcal{B}$ messages are calculated for the first trellis at the same time as completely independent $\mathcal{A}$ and $\mathcal{B}$ messages are calculated for the second. Because the sub-trellis can start and/or finish in the middle of the overall code trellis, the BCJR decoding algorithm for the sub-trellis assumes that every starting or ending state is equally likely. To reduce the errors associated with this approximation, the BCJR algorithm can be run on sub-trellises that include a few extra trellis sections at either side of the window, so that the $\mathcal{A}$ and $\mathcal{B}$ messages are relatively accurate before they are used to calculate bit LLRs. Alternatively, initial values for the $\mathcal{A}$ and $\mathcal{B}$ messages can come from the results generated by adjacent windows in the previous iteration. Windowed decoding results in a significantly higher throughput and significantly reduced memory requirement for the turbo decoder.

The turbo decoder can thus use a serial implementation (such as full BCJR decoding), a parallel implementation (where every factor graph node is implemented in hardware and each sub-trellis is run at the same time) or a compromise between the two (perhaps enough nodes for one or two sub-trellises to be implemented in hardware). Again parallel implementations achieve a better throughput at the expense of circuit size while serial implementations require less space but have a worse throughput.

5.5.2 Stopping criteria

The turbo decoder can be run for the same fixed number of iterations for every received vector. However, for many received vectors the decoder will find the correct codeword after only one or two iterations, and so unnecessary extra computations will be carried out by running the extra iterations. A number of stopping criteria have been proposed by which the decoder can recognize when it has run enough iterations.

Hard-decision and absolute-value stopping criteria

A very simple stopping criterion is the hard decision criterion. For this test a hard decision on the output of the first decoder, i.e. $\text{sign}(\mathbf{L}_l^{(1)})$, is compared with a hard decision on the output of the second decoder, i.e. $\text{sign}(\Pi^{-1}(\mathbf{L}_l^{(2)}))$, and decoding is halted if the two agree.

Example 5.11 In this example the hard decision stopping criterion will be applied to the decoder in Example 5.4. After the first iteration the signs of the LLRs are given by

$$\text{sign}(\mathbf{L}_1^{(1)}) = [-1, -1, 1, -1, -1, 1, -1, 1, 1, 1]$$

and

$$\text{sign}(\Pi^{-1}(\mathbf{L}_1^{(2)})) = [-1, 1, -1, -1, -1, 1, -1, 1, 1, 1].$$

As the two component codes have different hard decisions on the message, the decoder will continue. After the second iteration the signs of the LLRs are given by

$$\text{sign}(\mathbf{L}_2^{(1)}) = [-1, 1, 1, -1, -1, 1, -1, 1, -1, 1]$$

and

$$\text{sign}(\Pi^{-1}(\mathbf{L}_2^{(2)})) = [-1, 1, 1, -1, -1, 1, -1, 1, -1, 1].$$

The component decoders now agree, so, applying the hard decision stopping criterion, the turbo decoder would halt and return this hard decision as the decoded message. After the third iteration the signs of the LLRs are unchanged:

$$\text{sign}(\mathbf{L}_3^{(1)}) = [-1, 1, 1, -1, -1, 1, -1, 1, -1, 1]$$

and

$$\text{sign}(\Pi^{-1}(\mathbf{L}_3^{(2)})) = [-1, 1, 1, -1, -1, 1, -1, 1, -1, 1].$$

Thus, while the magnitude of the LLRs has changed in the third iteration, their sign, and hence the decoding decision, has not.

One variation on the hard-decision stopping criterion compares the sign of the outputs across two iterations, e.g. it compares sign($\mathbf{L}_l^{(2)}$) with sign($\mathbf{L}_{l-1}^{(2)}$); another variation can halt decoding only if the signs agree across some fixed number of iterations. Once the decoders have converged, the magnitudes of the LLRs will increase but their signs will not change.

An alternative stopping criterion compares the minimum absolute value of the decoded LLRs with a threshold value. If any decoded LLRs fall below that threshold then the result is considered unreliable and the decoder proceeds. Once the decoder has converged, the absolute values of the decoded LLRs will continue to increase and the threshold will be met.

Cross entropy stopping criterion

Once decoding has converged, the a posteriori probabilities for **u** from each decoder should agree and so a useful measure of decoder convergence is the similarity of these distributions.

One way to measure the similarity of two probability distribution functions is to compute their *Kullback–Leibler (KL) divergence* also known as their *cross entropy*. The KL divergence is a measure of the difference between two probability distributions $P = [p_1, p_2, \ldots, p_n]$ and $Q = [q_1, q_2, \ldots, q_n]$ for the random variable $X = [x_1, x_2, \ldots, x_n]$. When P and Q are discrete distributions the KL divergence is given by

$$D(P||Q) = \sum_j p_j \log_2 \frac{p_j}{q_j}.$$

If the distributions are the same then the KL divergence is zero; the closer they are, the smaller the value of $D(P||Q)$.

For the turbo decoder, the distributions of interest are the probability distributions for each bit of **u** output from the first and second BCJR decoders at the lth iteration:

$$\begin{aligned} D(P_l^{(1)}||P_l^{(2)}) &= \sum_{i=1}^{K} \sum_{x\in\{0,1\}} p_l^{(1)}(u_i = x) \log_2 \frac{p_l^{(1)}(u_i = x)}{p_l^{(2)}(u_i = x)} \\ &= \sum_{i=1}^{K} \left(p_l^{(1)}(u_i = 1) \log_2 \frac{p_l^{(1)}(u_i = 1)}{p_l^{(2)}(u_i = 1)} \right. \\ &\qquad \left. + \, p_l^{(1)}(u_i = 0) \log_2 \frac{p_l^{(1)}(u_i = 0)}{p_l^{(2)}(u_i = 0)} \right). \end{aligned}$$

Using (1.3) and (1.4) we can easily translate from the log likelihood ratios in the decoder, $\mathbf{L}_l^{(1)}$ and $\mathbf{L}_l^{(2)}$, to the required probabilities $p_l^{(1)}(u_i = 1)$

and $p_l^{(1)}(u_i = 0)$:

$$p_l^{(1)}(u_i = 1) = \frac{e^{-L_l^{(1)}(u_i)}}{1 + e^{-L_l^{(1)}(u_i)}},$$

$$p_l^{(1)}(u_i = 0) = \frac{e^{L_l^{(1)}(u_i)}}{1 + e^{L_l^{(1)}(u_i)}},$$

and similarly for $\mathbf{L}_l^{(2)}$. Thus the cross entropy becomes

$$D(P_l^{(1)}||P_l^{(2)}) = \sum_{i=1}^{K} \left(\frac{e^{-L_l^{(1)}(u_i)}}{1 + e^{-L_l^{(1)}(u_i)}} \log_2 \frac{e^{-L_l^{(1)}(u_i)}}{1 + e^{-L_l^{(1)}(u_i)}} \frac{1 + e^{-L_l^{(2)}(u_i)}}{e^{-L_l^{(2)}(u_i)}} \right.$$
$$\left. + \frac{e^{L_l^{(1)}(u_i)}}{1 + e^{L_l^{(1)}(u_i)}} \log_2 \frac{e^{L_l^{(1)}(u_i)}}{1 + e^{L_l^{(1)}(u_i)}} \frac{1 + e^{L_l^{(2)}(u_i)}}{e^{L_l^{(2)}(u_i)}} \right).$$

Using the relations $\log ab = \log a + \log b$ and $e^{-a} = 1/e^a$ gives

$$D(P_l^{(1)}||P_l^{(2)})$$
$$= \sum_{i=1}^{K} \left(\frac{1/e^{L_l^{(1)}(u_i)}}{1 + 1/e^{L_l^{(1)}(u_i)}} \left(\log_2 \frac{e^{-L_l^{(1)}(u_i)}}{e^{-L_l^{(2)}(u_i)}} + \log_2 \frac{1 + e^{-L_l^{(2)}(u_i)}}{1 + e^{-L_l^{(1)}(u_i)}} \right) \right.$$
$$\left. + \frac{e^{L_l^{(1)}(u_i)}}{1 + e^{L_l^{(1)}(u_i)}} \log_2 \frac{1/e^{-L_l^{(1)}(u_i)}}{1 + 1/e^{-L_l^{(1)}(u_i)}} \frac{1 + 1/e^{-L_l^{(2)}(u_i)}}{1/e^{-L_l^{(2)}(u_i)}} \right)$$
$$= \sum_{i=1}^{K} \left(\frac{\log_2 e^{-L_l^{(1)}(u_i)} - \log_2 e^{-L_l^{(2)}(u_i)}}{1 + e^{L_l^{(1)}(u_i)}} + \frac{1}{1 + e^{L_l^{(1)}(u_i)}} \log_2 \frac{1 + e^{-L_l^{(2)}(u_i)}}{1 + e^{-L_l^{(1)}(u_i)}} \right.$$
$$\left. + \frac{e^{L_l^{(1)}(u_i)}}{1 + e^{L_l^{(1)}(u_i)}} \log_2 \frac{1 + e^{-L_l^{(2)}(u_i)}}{1 + e^{-L_l^{(1)}(u_i)}} \right)$$
$$= \sum_{i=1}^{K} \left(\frac{L_l^{(2)}(u_i) - L_l^{(1)}(u_i)}{1 + e^{L_l^{(1)}(u_i)}} + \log_2 \frac{1 + e^{-L_l^{(2)}(u_i)}}{1 + e^{-L_l^{(1)}(u_i)}} \right).$$

5.6 Bibliographic notes

In many respects the introduction of turbo codes in 1993 was a turning point in the field of error control coding, when, after more than 50 years of striving to achieve Shannon's capacity limit, it was suddenly and unexpectedly achieved [92]. Although LDPC codes had been introduced many years earlier, their potential had not been realized, and it was only after Berrou, Glavieux and

Thitimajshima presented "Near Shannon limit error correcting and decoding: turbo codes" [82] that the combination of concatenation, random-like interleaving and iterative decoding became the new standard for capacity-approaching error correction.

The choice of component codes appears to have reached a consensus. There is only a finite number of primitive polynomials of a given degree and computer searches have narrowed down the set of possible numerator polynomials to those with the best effective free distance [93–96]. Turbo applications use IIR convolutional codes with $v = 3$ memory elements (such as in wireless telephony) or $v = 4$ memory elements (such as in space applications). When designing a turbo code, the choice comes down to a trade-off between extra complexity (eight versus 16 states) and around a dB of performance improvement [57].

The design of turbo interleavers, however, is still under discussion. The good BER performances of the S-random interleavers introduced in [88] and their extensions (see e.g. [97]) have made them a de facto standard for comparison with new interleaver proposals. The interleavers used in practice are tweaked, usually by trial and error, to fit the particular component codes being employed. New, promising, algorithmic interleavers are based on permutation polynomials that can be algebraically tailored to the given component codes [98].

Since their introduction, there has been an explosion of efforts to understand and generalize turbo codes. Factor graphs, introduced by Wiberg, Loeliger and Kötter [99] and Wiberg [100] provide a unification of turbo decoding and LDPC decoding showing both as message-passing decoders on graphs. Indeed, both decoding algorithms can be seen as instances of a belief-propagation algorithm used in artificial intelligence [34, 101]. A good overview of factor graphs can be found in [102, 103].

There has also been a corresponding explosion in interest for applications of turbo codes. A Consultative Committee for Space Data Systems (CCSDS) standard for turbo codes was issued in fewer than six years after turbo codes were introduced, and turbo codes were first sent into space in 2004. The MESSENGER spacecraft, Mars Reconnaissance Orbiter and New Horizons are all currently using turbo codes for their data transmissions. (See [57] for an interesting overview of this application area.) The 3G cellular standards are also employing turbo codes for their data services. The CDMA2000 standard uses a turbo code with two identical rate-1/3 eight-state systematic recursive convolutional encoders for data transmission supporting rates up to 307.4 kb/s per code channel, with interleaver lengths between 378 and 12 282 bits.

The original digital video broadcasting standard (DVB-RCS) employed turbo codes; however, they have been replaced by LDPC codes in the newest (DVB-S2) standard. The turbo versus LDPC debate continues to rage in a number of application areas and it is not yet clear whether one will emerge victorious or whether each has its own application areas. Certainly turbo codes seem to be favored for low-rate applications and LDPC codes for those with higher rates. Meanwhile, the turbo encoder has a distinct complexity advantage over the LDPC encoder, while the LDPC decoder can be simpler than the turbo decoder to implement. Perhaps both will eventually be replaced by repeat–accumulate codes, which we will consider in the following chapter; these have the low-complexity encoder of turbo-like codes and the decoder and capacity-achieving performance of LDPC codes.

There is a number of excellent publications devoted to explaining turbo codes, for example [86, 92, 104–107]. In Chapter 6 we will consider the extension of the "turbo principle" to serially concatenated codes, while in Chapters 7 and 8 we will present techniques to analyze the performance of the turbo decoder.

5.7 Exercises

5.1 The UMTS (Universal Mobile Telecommunications System) standard turbo encoder is formed using the concatenation of two copies of the convolutional encoder with generator matrix

$$\left[1 \quad \frac{1+D+D^3}{1+D^2+D^3}\right].$$

Assuming that the encoders are unterminated,
(a) determine the unpunctured code rate,
(b) determine the codeword produced by the message $\mathbf{u} = [1\ 0\ 1\ 1\ 0\ 0]$ if the interleaver is

$$\Pi = [1, 4, 2, 6, 3, 5],$$

(c) choose a puncturing pattern to produce a rate-$1/2$ code and then re-encode $\mathbf{u}$ using the punctured encoder.

5.2 The CDMA2000 standard turbo encoder is formed using the concatenation of two copies of convolutional encoder with generator matrix

$$\left[1 \quad \frac{1+D+D^3}{1+D^2+D^3} \quad \frac{1+D+D^2+D^3}{1+D^2+D^3}\right].$$

Assuming that the encoders are unterminated,
(a) determine the unpunctured code rate,
(b) determine the codeword produced by the message $\mathbf{u} = [1\ 0\ 1\ 1\ 0\ 0]$ if the interleaver is

$$\Pi = [1, 4, 2, 6, 3, 5],$$

(c) choose a puncturing pattern to produce a rate-$1/2$ code and then re-encode $\mathbf{u}$ using the punctured encoder.

5.3 Repeat Exercise 5.2 now assuming that padding bits are added to terminate the encoders.

5.4 If the turbo encoder in Exercise 5.2 used a length-1000 interleaver instead, what would be the rate difference between the terminated and unterminated cases?

5.5 A turbo code is formed using the concatenation of the convolutional encoders

$$\begin{bmatrix} 1 & \dfrac{1}{1+D} \end{bmatrix} \text{ and } \begin{bmatrix} 1 & \dfrac{1}{1+D^2} \end{bmatrix},$$

and a length-9 row–column interleaver. Assuming that the encoders are unterminated,
(a) determine the code rate,
(b) determine the codeword produced by the message

$$\mathbf{u} = [1\ 0\ 1\ 1\ 0\ 0\ 1\ 1\ 0].$$

5.6 The codeword generated in Example 5.2 was transmitted over a BSC with crossover probability 0.2 and the received bits were

$$\begin{aligned} \mathbf{y}^{(u)} &= [0\ 1\ 1\ 0\ 0\ 1\ 0\ 1\ 0\ 1], \\ \mathbf{y}^{(1)} &= [0\ x\ 0\ x\ 1\ x\ 0\ x\ 0\ x], \\ \mathbf{y}^{(2)} &= [x\ 1\ x\ 1\ x\ 0\ x\ 1\ x\ 1]. \end{aligned}$$

Decode $\mathbf{y}$ with turbo decoding with a maximum of 10 iterations, applying the hard-decision stopping criterion.

5.7 The turbo code in Example 5.2 was used with puncturing to encode a length-6 message using the interleaver

$$\Pi = [1, 3, 5, 2, 4, 6].$$

The resulting unterminated codeword was sent on a binary symmetric channel with crossover probability 0.1, and the vector

$$\mathbf{y} = [0\ 1\ 1;\ 1\ 0\ 0;\ 1\ 0\ 1;\ 0\ 1\ 1;\ 0\ 1\ 0;\ 1\ 1\ 1]$$

was received. Decode $\mathbf{y}$ using turbo decoding with a maximum of 10 iterations
(a) applying the hard-decision stopping criterion,
(b) applying the cross-entropy stopping criterion.

5.8 Generate bit error rate plots for turbo decoding operating on the turbo code in Exercise 5.3 with a randomly constructed interleaver of length 100 bits sent on a BI-AWGN channel with signal-to-noise ratios between 1 and 5 dB.

5.9 Repeat Exercise 5.8 using a row–column interleaver.

5.10 What is the effective free distance of the turbo encoder with the component codes in Exercise 5.2?

5.11 What is the effective free distance of the turbo encoder with the component codes in Exercise 5.5?

5.12 What is the actual minimum distance for the turbo code in Exercise 5.5?

5.13 Give one advantage and one disadvantage of choosing the memory-order-6 (i.e. $v = 6$) convolutional encoder in the table near the end of Section 5.4.2 rather than the memory-order-5 (i.e. $v = 5$) convolutional encoder in the same table for the component encoders of a turbo code.

6
Serial concatenation and RA codes

The idea of concatenating two or more error correction codes in series in order to improve the overall decoding performance of a system was introduced by Forney in 1966. Applying random-like interleaving and iterative decoding to these codes gives a whole new class of turbo-like codes that straddle the gap between parallel concatenated turbo codes and LDPC codes.

Concatenating two convolutional codes in series gives serially concatenated convolutional codes (SC turbo codes). We arrive at turbo block codes by concatenating two block codes and at repeat–accumulate codes by concatenating a repetition code and a convolutional (accumulator) code.

This chapter will convey basic information about the encoding, decoding and design of serially concatenated (SC) turbo codes. Most of what we need for SC turbo codes has already been presented in Chapter 4. The turbo encoder uses two convolutional encoders, from Section 4.2, while the turbo decoder uses two copies of the log BCJR decoder from Section 4.3. The section on design principles will refer to information presented in Chapter 5 and the discussion of repeat–accumulate codes will use concepts presented in Chapter 2. A deeper understanding of SC turbo codes and their decoding process is explored in Chapters 7 and 8.

6.1 Serial concatenation

The first serial concatenation schemes concatenated a high-rate block code with a short convolutional code. The first code, called the *outer* code, encoded the source message and passed the resulting codeword to the second code, called the *inner* code, which re-encoded it to obtain the final codeword to be transmitted. At the decoder the inner code decoded the received sequence from the channel and passed its decoded message to the outer code, which used it to decode the original source message. An interleaver was used between the two codes to spread out any burst errors produced by the inner decoder.

While in practice most classical concatenated systems used an outer block code and inner convolutional code, both codes could be either block or

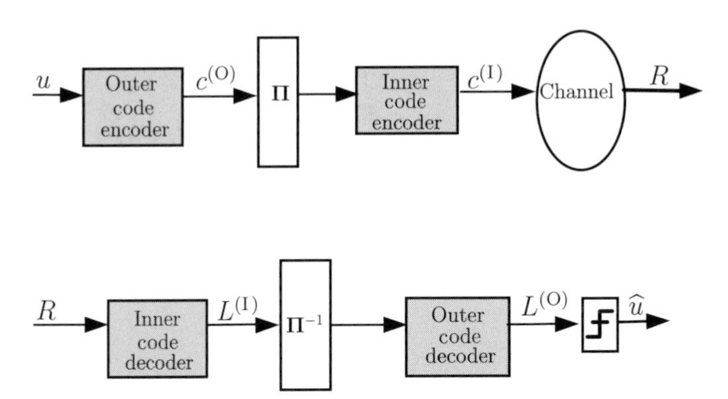

Figure 6.1 Serially concatenated error correction codes: encoding (top) and decoding (bottom).

convolutional codes or, in the case of the inner code, a trellis-coded modulation scheme could replace the error correction code.

As for a single code, the decoding performance of a concatenated coding system can be improved by using soft decisions. For the inner decoder this is straightforward; the soft information from the channel gives soft input probabilities. However, for the outer decoder to have access to soft information requires that the inner decoder provide soft outputs, which can be achieved by employing soft output Viterbi or BCJR decoding for the inner code.

Figure 6.1 shows a block diagram of a serially concatenated error correction code. The message to be encoded, $\mathbf{u}$, is passed to the outer encoder, so that $\mathbf{u}^{(\mathrm{O})} = \mathbf{u}$, and the codeword produced by the outer encoder, $\mathbf{c}^{(\mathrm{O})}$, once interleaved, becomes the message passed to the inner encoder: $\mathbf{u}^{(\mathrm{I})} = \Pi(\mathbf{c}^{(\mathrm{O})})$. The output from the inner encoder, $\mathbf{c}^{(\mathrm{I})}$, is then transmitted over the channel, so that $\mathbf{c} = \mathbf{c}^{(\mathrm{I})}$. For an $(N_\mathrm{O}, K_\mathrm{O})$ outer code and an $(N_\mathrm{I}, K_\mathrm{I})$ inner code we will assume that $N_\mathrm{O} = K_\mathrm{I}$ and that the interleaver has length K_I.[1] The concatenated code is thus an $(N_\mathrm{I}, K_\mathrm{O})$ code.

At the decoder this operation is reversed. The inner code takes the received LLRs from the channel, $\mathbf{R}^{(\mathrm{I})}$, and outputs a posteriori LLRs for its message, $\mathbf{L}^{(\mathrm{I})}$; the outer code takes these LLRs, once de-interleaved, as the received LLRs for its own codeword, $\mathbf{R}^{(\mathrm{O})} = \Pi^{-1}(\mathbf{L}^{(\mathrm{I})})$, and outputs a posteriori LLRs for the original message, $\mathbf{L}^{(\mathrm{O})}$.

If the inner code has rate $r_1 = k_1/n_1$ and the outer code has rate $r_2 = k_2/n_2$ then the concatenated code has rate

$$r = \frac{k}{n} = \frac{k_1}{n_2} = r_1 r_2.$$

[1] Although it is straightforward to concatenate several inner messages or outer codewords together to fit the interleaver length, or vice versa, it is simpler for our presentation to assume that $K_\mathrm{I} = N_\mathrm{O}$ is the interleaver length.

Code termination

A convolutional trellis with maximum memory b can be terminated (returned to the zero state) by appending b padding bits to the message (see e.g. Figure 4.5). The outer encoder can be terminated in this way by adding the appropriate b padding bits. The inner code has as input the codewords from the outer code but it can still be terminated in the standard way by adding the appropriate padding bits to the end of the interleaved outer codeword. However, for long codewords the performance loss due to an unterminated trellis is unnoticeable while for short codewords the rate loss due to the extra padded bits is not compensated by the improved performance due to terminating the trellis.

For a small range of code lengths, of around 100 bits, the performance of the concatenated code can be slightly improved, compared with that of both unterminated and standard termination of the inner code, by choosing padding sequences, slightly longer than b bits, which ensure a weight-2 output for the padded bits. However, again for short lengths the rate loss due to the extra padded bits is not compensated by the improved performance due to termination of the trellis and for longer codewords there is no noticeable performance improvement.

6.1.1 Serially concatenated turbo codes

When iterative "turbo" decoding is applied to serially concatenated codes, we will call such schemes SC turbo codes. The iterative decoder for SC codes uses the same principle as the decoder for parallel concatenated codes: the passing of soft information iteratively between component BCJR decoders. The difference for SC turbo codes is that the message bits of the inner decoder correspond to the entire codeword of the outer decoder. Thus the outer decoder generates extrinsic information about its codeword bits, $\mathbf{L}_{\mathrm{c}}^{(\mathrm{O})}$, as well as about its message bits, $\mathbf{L}^{(\mathrm{O})}$. Equation (4.32) showed the modification required to generate LLRs for the codeword bits.

6.1.2 Turbo decoding of SC codes

Figure 6.2 shows a turbo decoder for SC codes. The first iteration of the iterative decoder is identical to the traditional soft decoding of SC codes. The inner code takes the received LLRs from the channel, $\mathbf{R}^{(\mathrm{I})}$, and outputs a posteriori LLRs for its message, $\mathbf{L}_1^{(\mathrm{I})}$. In the first iteration the extrinsic information produced by the inner decoder is simply

$$\mathbf{E}_1^{(\mathrm{I})} = \mathbf{L}_1^{(\mathrm{I})}.$$

The outer code takes these LLRs, once de-interleaved, as the received LLRs for its own codeword. At iteration l,

$$\mathbf{R}_l^{(\mathrm{O})} = \mathbf{A}_l^{(\mathrm{O})} = \Pi^{-1}(\mathbf{E}_l^{(\mathrm{I})}).$$

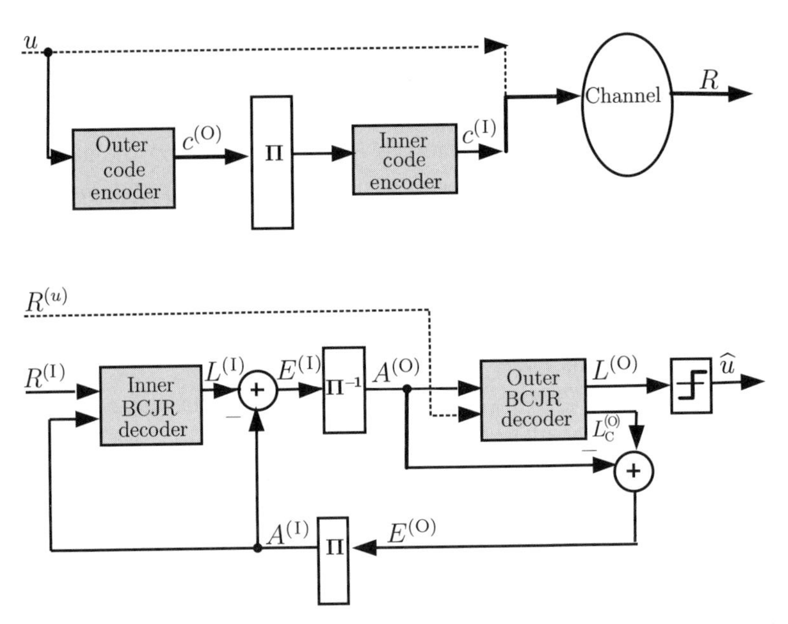

Figure 6.2 Serially concatenated convolutional codes: encoding (top) and turbo decoding (bottom). The broken lines show the additional components in a systematic SC code.

Since these LLRs come from the inner decoder rather than the channel, and so change at each iteration, they are labeled $\mathbf{A}_l^{(\mathrm{O})}$ to reflect this.

The difference from traditional soft decoding is that the outer BCJR decoder calculates a posteriori LLRs for its codeword bits, $\mathbf{L}_{\mathrm{c}}^{(\mathrm{O})}$, as well as for its message bits, $\mathbf{L}^{(\mathrm{O})}$. These LLRs are then used to calculate the extrinsic information generated by the outer decoder,

$$\mathbf{E}_l^{(\mathrm{O})} = \mathbf{L}_{\mathrm{c}l}^{(\mathrm{O})} - \mathbf{A}_l^{(\mathrm{O})}.$$

In the second and subsequent iterations the inner code decoder will repeat the BCJR algorithm but now will have extra a priori information about its message bits from the extrinsic information output by the outer code decoder in the previous iteration:

$$\mathbf{A}_l^{(\mathrm{I})} = \Pi(\mathbf{E}_{l-1}^{(\mathrm{O})}).$$

Recall that the message for the inner code is the codeword for the outer code; this is why $\mathbf{E}_{l-1}^{(\mathrm{O})}$ contains extrinsic information about all the codeword bits for the outer code. Since the inner decoder has now used extrinsic information produced by the outer decoder, this input information is subtracted from the LLRs produced by the inner decoder in order to create new extrinsic LLRs to pass to the outer decoder,

$$\mathbf{E}_l^{(\mathrm{I})} = \mathbf{L}_l^{(\mathrm{I})} - \mathbf{A}_l^{(\mathrm{I})}.$$

Algorithm 6.1 SCCC turbo decoding

```
 1: procedure SCCC TURBO(trellisI,trellisO,Π,I_max,R^(c),R^(u))
 2:     E^(O) = zeros;
 3:     for  l = 1 : I_max do                              ▷ For each iteration l
 4:         A^(I) = Π(E^(O))
 5:         [L^(I),not used] = logBCJR(trellisI,0,R^(c),A^(I))
 6:         E^(I) = L^(I) − A^(I)
 7:         A^(O) = Π^(-1)(E^(I))
 8:         [L^(O), L_c^(O)] = logBCJR(trellisO,R^(u),0,A^(O))
 9:         E^(O) = L_c^(O) − A^(O)
10:     end for
11:     output L^(O)
12: end procedure
```

The outer decoder's operation is unchanged. After all the decoder iterations have been completed, the outer decoder returns the a posteriori LLRs for its message bits, $\mathbf{L}^{(\mathrm{O})}$.

The performance of SC codes can be improved, albeit with a rate reduction, by also transmitting the original message bits over the channel. These codes are called *systematic SC* codes. The broken lines in Figure 6.2 show the changes required. The code rate of a systematic SC code with a rate $r_1 = k_1/n_1$ inner code and a rate $r_2 = k_2/n_2$ outer code is

$$\frac{k}{n} = \frac{k_1}{k_1 + n_2} = \frac{k_1}{k_1 + n_1/r_2} = \frac{k_1}{k_1 + k_1/(r_1 r_2)} = \frac{r_1 r_2}{r_1 r_2 + 1}.$$

The decoding of systematic SC codes is similar to that of non-systematic SC codes, the only difference being that the inner decoder now has extra received LLRs $\mathbf{R}^{(u)}$ for the message bits (see Figure 6.2).

Algorithm 6.1 completes $I_{\max}$ iterations of turbo decoding for serially concatenated codes. We will call "trellisO" the trellis structure of the outer component convolution code and "trellisI" the trellis structure of the inner component convolutional code. The interleaver is Π. The received LLRs are split into two, $\mathbf{R}^{(\mathrm{c})}$ for the LLRs of the transmitted codeword bits and $\mathbf{R}^{(\mathrm{u})}$ for the LLRs of the transmitted message bits if the code is systematic; otherwise $\mathbf{R}^{(\mathrm{u})}$ is all zeros. Note that the component encoders in an SC turbo code need not be systematic, so we will distinguish between the input and output LLRs rather than the input and parity LLRs. Thus we will write

$$[\mathbf{L}, \mathbf{L}_{\mathrm{c}}] = \mathrm{logBCJR}(\mathrm{trellis}, \mathbf{R}^{(\mathrm{u})}, \mathbf{R}^{(\mathrm{c})}, \mathbf{A})$$

for our BCJR decoder, which performs standard logBCJR decoding with the addition of (4.32) to produce codeword-bit LLRs.

Example 6.1 In this example we simulate the performance of a serially concatenated code with a rate-1 inner convolutional code $1/(1+D)$ and a rate-1/2 outer convolutional code $[1+D^2 \quad 1+D+D^2]$ from [108]. The average bit error rates of the decoder are plotted in Figure 6.3 using randomly generated interleavers. Next we repeat this process for longer interleaver lengths, up to $K = 5000$, and for more iterations. The average bit error rates of the decoder for these interleaver lengths are also plotted in Figure 6.3.

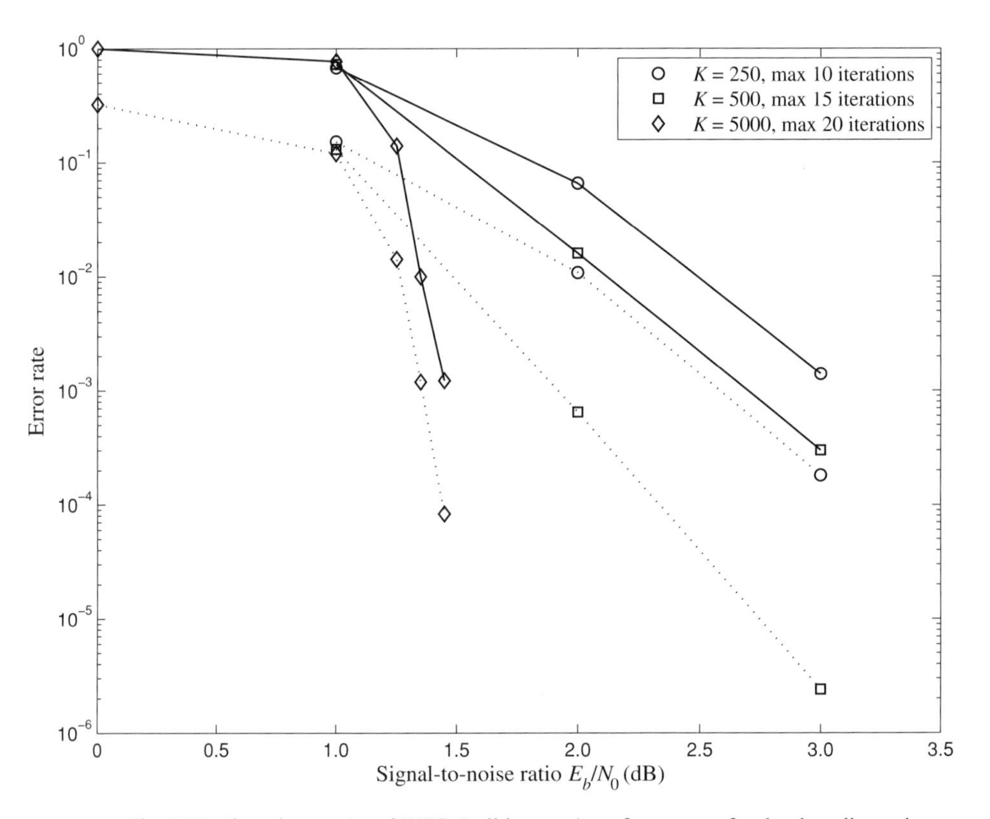

Figure 6.3 The BER (dotted curves) and WER (solid curves) performance of turbo decoding using a threshold stopping criterion for the code from Example 6.1, for various interleaver lengths.

Figure 6.3 shows the characteristic iterative decoding BER curve, with a non-convergence region where the BER is high and relatively flat as the SNR is increased, a waterfall region showing a rapid decrease in BER for small increments in SNR and an error floor region where the slope of the curve flattens out and the BER reductions are small as the SNR increases. Unlike the parallel concatenated turbo codes we saw previously, but like LDPC codes, SC turbo codes show an interleaving gain in their WER performance. We will see why this is so in Chapter 8.

In general, serially concatenated turbo codes do not have thresholds as good as those of parallel concatenated codes of the same rate, but they do usually have much lower error floors. However, careful design would produce SC turbo codes with a better threshold than the one shown here, which was chosen for its simplicity. See [109] for a comparison of serially concatenated turbo codes and parallel concatenated turbo codes.

6.1.3 Code design

For serial concatenation with a rate k_O/n_O outer convolutional code and rate k_I/n_I inner convolutional code, a fixed-length-K message is encoded. The outer code can thus be considered as an $(N_O = Kn_O/k_O, K_O = K, d_{\min}^{(O)})$ block code and the inner code as an $(N_I = K_I n_I/k_I, K_I = N_O, d_{\min}^{(I)})$ block code. Using a length $N_O = K_I$ interleaver, the concatenated code is thus an $(N = N_I, K = K_O, d_{\min})$ block code. Note that $d_{\min}$ is not necessarily $d_{\min}^{(I)}$ since the message sequences of the inner code that produce minimum-weight codewords may not be outputs of the outer code.

A weight-w message will produce a weight-d SC codeword for every i for which the outer code produces a weight-i output for a weight-w input and the inner code produces a weight-d output for a weight-i input. In Section 5.4 we saw that the interleaver has a higher probability of producing a "good" sequence (in the sense that it does not lead to low-weight outputs once encoded) for a higher-weight input sequence. For parallel concatenation, the interleaved sequences were simply the message and so could be of any weight. For serially concatenated codes, however, the interleaved sequences are the codewords of the outer code, which have minimum weight $d_f^{(O)}$. Thus a "good" SC turbo interleaver will permute weight-$d_f^{(O)}$ self-terminating sequences into non-self-terminating sequences, and a "good" interleaver will be more likely for larger $d_f^{(O)}$ values. Indeed, in Chapter 8 we will show that, when IIR inner encoders are used, the error correction improvement resulting from reducing the probability of low-weight codewords in this way decreases their bit error rate by a factor $K^{-\lfloor (d_f^{(O)}+1)/2 \rfloor}$, which is the SC turbo *interleaving gain*.

Thus the important property of the outer code will be its minimum free distance. The outer code need not be an IIR convolutional code and can even be a block code. Indeed, in the following section we will introduce very promising SC codes with a repetition code as the outer code. The main impediment to using block codes as outer codes, however, is the difficulty of obtaining low-complexity MAP decoders for them (i.e. the block code equivalent of the BCJR algorithm).

For the inner code, the codewords produced by weight-$d_f^{(O)}$ inputs, rather than the codewords produced by weight-2 inputs, will be the significant determinant of the low-weight codewords of the concatenated code. Finite impulse response encoders will always return to state zero after a message sequence followed by b or more zeros, where b is the maximum memory order of the encoder, so they will always output finite-weight sequences for finite-weight inputs. However, IIR encoders will only return to state zero for 1 in 2^b message sequences followed by zeros. Because of this, IIR encoders are much less likely to map low-weight message sequences to low-weight outputs and so are preferred for the inner codes of SC turbo codes.

Even though $d_{\mathrm{f}}^{(\mathrm{O})}$ may be greater than 2, the effective free distance $d_{\mathrm{ef}}^{(\mathrm{I})}$ of the inner code still has an important role to play. (Recall from Section 5.4 that d_{ef} is the minimum output weight produced by any weight-2 input.) If $d_{\mathrm{f}}^{(\mathrm{O})}$ is even, then low-weight codewords may include those composed of $d_{\mathrm{f}}^{(\mathrm{O})}/2$ weight-2 sequences that are self-terminating, so low-weight SC turbo codewords will include those of weight $d_{\mathrm{f}}^{(\mathrm{O})} d_{\mathrm{ef}}^{(\mathrm{I})}/2$. If $d_{\mathrm{f}}^{(\mathrm{O})}$ is odd then the low-weight codewords will include those produced by self-terminating weight-3 inputs. For an IIR encoder with generator

$$f(D)/g(D)$$

and input $x(D)$, the output for this message,

$$x(D)f(D)/g(D)$$

will be finite if and only if $x(D)f(D)$ is divisible by $g(D)$. For a weight-3 message $x(D) = 1 + D^i + D^j$ the output

$$(1 + D^i + D^j)f(D)/g(D),$$

will be finite only if $(1 + D^i + D^j)f(D)$ is divisible by $g(D)$, and finite outputs can be avoided for weight-3 inputs by including a factor $1 + D$ in $g(D)$.

As for parallel concatenated codes, the role of the interleaver for serially concatenated codes is twofold. Firstly, the interleaver is chosen to improve the code's weight spectrum, i.e. the interleaver will permute the codewords $\mathbf{c}^{(\mathrm{O})}$ from the inner code in such a way that $\Pi(\mathbf{c}^{(\mathrm{O})})$ minimizes poorly placed non-zero entries (in the sense that it leads to a low-weight parity sequence from the inner code). Secondly, the interleaver randomizes the location of the message symbols in each component codeword, which reduces the correlation between the extrinsic information passed between the component decoders. The discussion of interleavers for PC codes (see Section 5.4) is also relevant for SC codes.

In summary, standard design rules for serially concatenated turbo codes are as follows.

- The inner code needs to be IIR to ensure an interleaving gain.
- The outer code should have a large d_{f} or $d_{\min}$ if it is a block code (but it need not be IIR).

- The design rules for the inner component code follow those for the component codes of parallel concatenated codes, i.e. the code generator should be of the form

$$f(D)/g(D)$$

where $g(D)$ is a primitive polynomial of degree v and $f(D)$ can be any monic polynomial of degree v except $g(D)$.
- If the outer code has an odd $d_f^{(O)}$ value then choosing an inner code for which $g(D)$ contains a factor $1 + D$ can eliminate many low-weight codewords (those originated by self-terminating weight-3 inputs).
- Searches for good inner and outer codes comprise
  - maximizing the minimum distance of the concatenated code,
  - minimizing the BER in simulated results using iterative decoding.
- Interleaver design depends on the chosen component codes and application; however, in general important design features include
  - a large S-factor,
  - a high dispersion ratio, and
  - ease of implementation (depending on the application, this may point to low delay, prunable or algorithmic interleavers).

6.2 Repeat–accumulate codes

Repeat–accumulate (RA) codes are a specific class of serially concatenated codes in which the outer code is a rate-$1/q$ *repetition code* and the inner code is a convolutional code with generator $1/(1 + D)$. A $1/(1 + D)$ convolutional code simply outputs the modulo-2 sum of the current input bit and the previous output bit, i.e. it provides a running sum of all past inputs and so is often called an *accumulator*. These two component codes give repeat–accumulate codes their name. Figure 6.4 shows the encoder for RA codes.

Repeat–accumulate codes that transmit both the message bits and parity bits are called *systematic RA codes*. Often, systematic RA codes include a third component, called a *combiner*, placed before the accumulator to reduce the code rate. A rate-a combiner simply sums, modulo-2, each set of a bits input to it. In this case the repetition code can be thought of as the outer code and the combiner and accumulator together as the inner code. Again an interleaver is placed between the inner and outer codes. Figure 6.5 shows the encoder for systematic RA codes. Since RA codes are most often systematic, we will use the terminology "repeat–accumulate" when we mean "systematic repeat–accumulate" codes and specify references to non-systematic RA codes.The RA encoder can also be represented by using puncturing after the accumulator rather than placing a combiner before it. However, we will use the combiner as it will allow us later to represent, and decode, RA codes as LDPC codes.

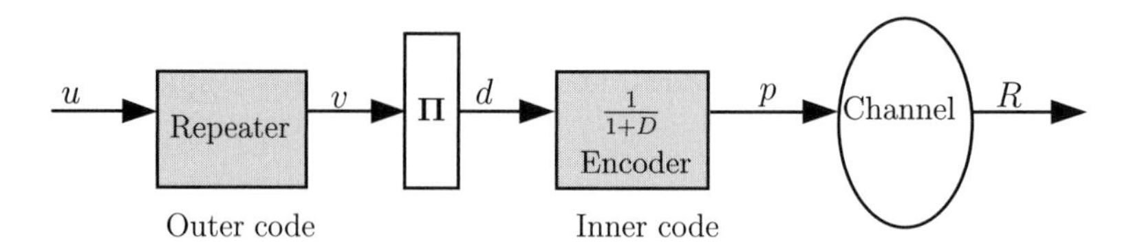

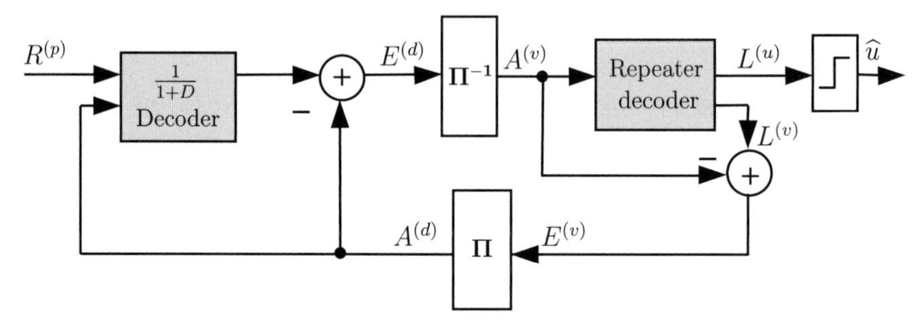

Figure 6.4 Non-systematic repeat–accumulate codes. Encoding (top) and turbo decoding (bottom).

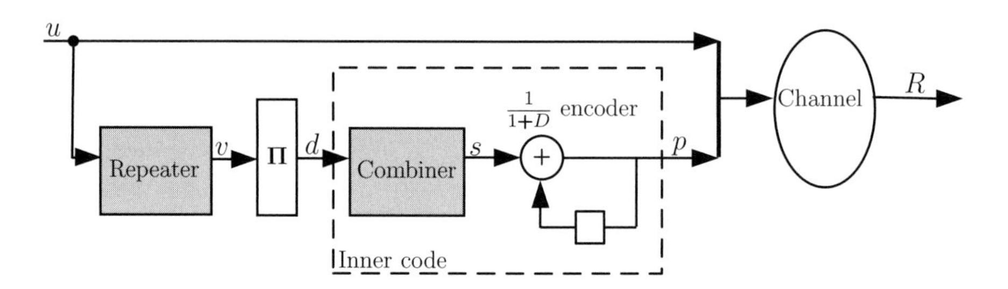

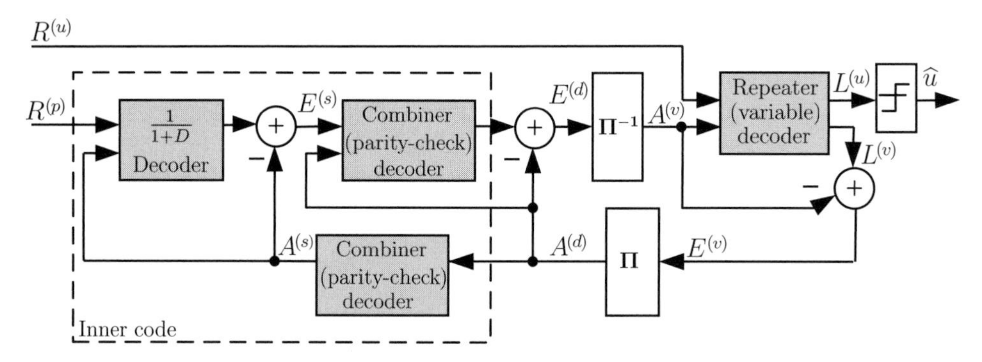

Figure 6.5 Systematic repeat–accumulate codes. Encoding (top) and turbo decoding (bottom).

6.2.1 Encoding RA codes

The simple component codes of RA codes lead to a particularly straightforward encoding process. The message bits are copied q times, interleaved, added modulo 2 in sets of a bits at a time and then passed through a memory-1 convolutional encoder.

More precisely, the encoding process is as follows. The input is the K message bits $\mathbf{u} = [u_1 \cdots u_K]$. The qK bits at the output of the repetition code are q copies of $\mathbf{u}$, in the form

$$\begin{aligned}\mathbf{v} &= [v_1\ v_2\ \cdots\ v_{qK}] \\ &= [\underbrace{u_1\ u_1\ \cdots\ u_1}_{q}\ \underbrace{u_2\ u_2\ \cdots\ u_2}_{q}\ \cdots\ \underbrace{u_K\ u_K \cdots u_K}_{q}]\end{aligned}$$

and so we have

$$v_i = u_{f(i)}, \qquad f(i) = \lceil i/q \rceil, \tag{6.1}$$

where $\lceil x \rceil$ denotes the smallest integer greater than or equal to x.

The interleaver pattern $\Pi = [\pi_1, \pi_2, \ldots, \pi_{qK}]$ defines a permutation of the bits in $\mathbf{v}$ to

$$\mathbf{d} = [d_1\ d_2 \cdots d_{qK}] = [v_{\pi_1}\ v_{\pi_2} \cdots v_{\pi_{qK}}]. \tag{6.2}$$

The bits at the output of the interleaver are summed, modulo 2, in sets of a bits by the combiner. The $m = Kq/a$ bits $\mathbf{s} = [s_1\ s_2\ \cdots\ s_m]$ at the output of the combiner are given by

$$s_i = d_{a(i-1)+1} \oplus d_{a(i-1)+2} \oplus \cdots \oplus d_{ai}, \qquad i = 1, 2, \ldots, m, \tag{6.3}$$

where $\oplus$ denotes modulo-2 addition.

Finally, the m parity bits $\mathbf{p} = [p_1\ p_2 \cdots p_m]$ at the output of the accumulator are defined by

$$p_i = p_{i-1} \oplus s_i, \qquad i = 1, 2, \ldots, m. \tag{6.4}$$

The encoder for an accumulator can also be called a *differential encoder*.

For systematic RA codes, $\mathbf{c} = [u_1\ u_2\ \ldots\ u_K\ p_1\ p_2\ \ldots\ p_m]$ is the final codeword and thus we have a code with length $n = K(1 + q/a)$ and rate $r = a/(a + q)$. For non-systematic RA codes, only the parity bits are sent to the receiver and so we have a code with length $N = Kq/a$ and rate $r = a/q$.

Example 6.2 The message

$$\mathbf{u} = [1\ \ 0\ \ 0\ \ 1]$$

is to be encoded using a length-10 RA code consisting of a $q = 3$ repetition code, an $a = 2$ combiner and the interleaver $\Pi = [1, 7, 4, 10, 2, 5, 8, 11, 3, 9, 6, 12]$.

Firstly, the message bits are each repeated three times to give

$$\mathbf{v} = [1\ \ 1\ \ 1\ \ 0\ \ 0\ \ 0\ \ 0\ \ 0\ \ 0\ \ 1\ \ 1\ \ 1].$$

Secondly, $\mathbf{v}$ is interleaved to give

$$\mathbf{d} = [1\ \ 0\ \ 0\ \ 1\ \ 1\ \ 0\ \ 0\ \ 1\ \ 1\ \ 0\ \ 0\ \ 1].$$

Thirdly, the combiner sums each set of two bits to give

$$\mathbf{s} = [1\ \ 1\ \ 1\ \ 1\ \ 1\ \ 1].$$

Lastly, the accumulator produces the running sum of $\mathbf{s}$ (from left to right),

$$\mathbf{p} = [1\ \ 0\ \ 1\ \ 0\ \ 1\ \ 0].$$

6.2.2 Turbo decoding of RA codes

The turbo decoding of non-systematic RA codes is identical to the turbo decoding of serial concatenated codes, where the outer code is the repetition code and the inner code is the accumulator. For systematic RA codes with a combiner the inner code is now a joint combiner–accumulator decoder and the repetition decoder has input LLRs from the channel, but the principle is still the same. Note that it is possible to specify (and even encode) a non-systematic RA code with a combiner, but such a code cannot be decoded. (Given only the parity bits from the channel there is no way in which the decoder can determine the values of the a bits that are summed by the combiner to produce each parity bit. Unless $a = 1$ there are multiple, equally valid, sets of a bits that sum to give the same output bit.) However, a class of capacity-approaching non-systematic RA codes has been proposed where a small number of message bits are transmitted to get the decoder started, a process called code doping; see [110].

Repetition code decoder

Recall that the repetition encoder repeats the message bits q times, so that $v_{q(i-1)+1}, v_{q(i-1)+2}, \ldots, v_{qi}$ are each equal to u_i. Given the LLRs of the bits in $\mathbf{v}$ as $\mathbf{A}^{(v)}$, the decoder for a repetition code simply calculates the LLRs for u_i as the sum of the LLRs of the corresponding repeated bits:

$$L_i^{(\mathrm{u})} = \sum_j A_j^{(v)} \tag{6.5}$$

for $j = q(i-1)+1, \ldots, qi$. That is, j indexes the q copies of u_i. The LLRs for the bits in v are simply

$$L_j^{(v)} = L_i^{(\mathrm{u})} \quad \forall i \in 1, \ldots, K, \ \ \forall j \in q(i-1)+1, \ldots, qi.$$

The extrinsic LLR for a bit in $\mathbf{v}$ is found by omitting the LLR of the selected bit from the sum, so that

$$E_j^{(v)} = \sum_{j' \neq j} A_{j'}^{(v)}, \tag{6.6}$$

where $j', j = q(i-1)+1, \ldots, qi$, that is, v_j, $v_{j'}$ are both copies of u_i. Or, equivalently,

$$E_j^{(v)} = L_j^{(v)} - A_j^{(v)}. \tag{6.7}$$

For a systematic repeat–accumulate code, the repetition decoder also has access to the received LLRs $\mathbf{R}^{(\mathrm{u})}$ for the message bits transmitted over the channel. In this case the repetition code decoder calculates

$$L_i^{(\mathrm{u})} = \sum_j A_j^{(v)} + R_i^{(\mathrm{u})}, \tag{6.8}$$

for $j = q(i-1)+1, \ldots, qi$, and (6.7) is unchanged.

Accumulator decoder

The accumulator is a $1/(1+D)$ convolutional code and so is decoded using standard log BCJR decoding (see Algorithm 4.2).

Combiner decoder

Recall that the combiner produces the modulo-2 sum of each set of a bits from the repeater:

$$s_i = d_{a(i-1)+1} \oplus d_{a(i-1)+2} \oplus \cdots \oplus d_{ai}.$$

Thus the combiner can be thought of as the parity-check equation

$$s_i \oplus d_{a(i-1)+1} \oplus d_{a(i-1)+2} \oplus \cdots \oplus d_{ai} = 0.$$

The decoder for a parity-check equation with input LLRs $\mathbf{R} = [R_1 \ \cdots \ R_i \ \cdots \ R_\rho]$ calculates the probability that bit i is zero or one on the basis of the probability that the parity-check equation will be satisfied if bit i is zero or one when the other bits have the probabilities given in $\mathbf{R}$, cf. (2.25):

$$\begin{aligned} E_i &= \log\left(\left(1 + \prod_{j \neq i} \frac{1 - e^{-R_j}}{1 + e^{-R_j}}\right) \bigg/ \left(1 - \prod_{j \neq i} \frac{1 - e^{-R_j}}{1 + e^{-R_j}}\right)\right) \\ &= \log \frac{1 + \prod_{j \neq i} f(R_j)}{1 - \prod_{j \neq i} f(R_j)} \end{aligned} \tag{6.9}$$

for each bit i, where

$$f(x) = \frac{1 - e^{-x}}{1 + e^{-x}}.$$

Joint combiner and accumulator decoder

Altogether, it requires three steps for joint log decoding of the combiner and accumulator. Firstly, the parity-check decoder computes the LLRs $\mathbf{A}^{(s)}$ for the bits in $\mathbf{s}$, see (6.3), based on the a priori LLRs $\mathbf{A}^{(d)}$ from the repetition decoder:

$$A_i^{(s)} = \log \frac{1 + \prod_j f(A_j^{(d)})}{1 - \prod_j f(A_j^{(d)})}, \tag{6.10}$$

where $j = a(i-1)+1, a(i-1)+2, \ldots, ai$ indexes the set of bits in $\mathbf{d}$ which were summed together to produce the ith bit in $\mathbf{s}$.

Secondly, log BCJR decoding for the accumulator with a priori LLRs $\mathbf{A}^{(s)}$ and received LLRs $\mathbf{R}^{(p)}$ returns updated LLRs $\mathbf{L}^{(s)}$ for the bits in $\mathbf{s}$:

$$\mathbf{L}^{(s)} = \text{logBCJR}\left(\frac{1}{1+D}, \mathbf{A}^{(s)}, \mathbf{R}^{(p)}\right).$$

The extrinsic information from the accumulator decoder,

$$\mathbf{E}^{(s)} = \mathbf{L}^{(s)} - \mathbf{A}^{(s)}, \tag{6.11}$$

is passed back to the parity-check decoder.

Thirdly, the parity-check decoder computes the LLRs $\mathbf{E}^{(d)}$ for the bits in $\mathbf{d}$:

$$E_j^{(d)} = \log \frac{1 + f(E_i^{(s)}) \prod_{j' \neq j} f(A_{j'}^{(d)})}{1 - f(E_i^{(s)}) \prod_{j' \neq j} f(A_{j'}^{(d)})}, \tag{6.12}$$

where $i = 1 : m$, and $j, j' \in \{a(i-1)+1, a(i-1)+2, \ldots, ai\}$. That is, j, j' index the set of bits in $\mathbf{d}$ that were summed together to produce the ith bit in $\mathbf{s}$.

Turbo decoding of RA codes

Figure 6.5 shows the turbo decoder for systematic RA codes.

At the lth iteration the repetition decoder receives the LLRs for the message bits from the channel, $\mathbf{R}^{(u)}$, and, in the second and subsequent iterations, the a priori LLRs

$$\mathbf{A}_l^{(v)} = \Pi^{-1}(\mathbf{E}_{l-1}^{(d)})$$

from the extrinsic information generated by the accumulator–combiner decoder, in the previous iteration.

The repetition decoder calculates the output LLRs $\mathbf{L}_l^{(u)}$ and $\mathbf{L}_l^{(v)}$ for the bits in $\mathbf{u}$ and $\mathbf{v}$ respectively. The extrinsic information from the repetition decoder,

$$\mathbf{E}_l^{(v)} = \mathbf{L}_l^{(v)} - \mathbf{A}_l^{(v)},$$

is interleaved to produce the a priori LLRs for the combiner–accumulator decoder,

$$\mathbf{A}_l^{(d)} = \Pi(\mathbf{E}_l^{(v)}).$$

The combiner–accumulator decoder receives the LLRs $\mathbf{R}^{(p)}$ for the parity bits from the channel and the a priori LLRs $\mathbf{A}_l^{(d)}$ from the extrinsic information generated by the repetition decoder. The combiner–accumulator decoder returns the LLRs $\mathbf{E}_l^{(d)}$.

At the final iteration, the a posteriori LLRs for the message bits from the repetition decoder are used to make a hard decision $\hat{\mathbf{u}}$.

Algorithm 6.2 completes $I_{\max}$ iterations of turbo decoding for an RA code having a rate-q repetition code for the outer component code, a combiner plus accumulator for the inner component code and an interleaver Π. The received LLRs are $\mathbf{R}^{(c)}$, giving the LLRs of the transmitted codeword bits, and $\mathbf{R}^{(u)}$, giving the LLRs of the transmitted message bits if the code is systematic; otherwise $\mathbf{R}^{(u)}$ is all zeros.

Example 6.3 Figure 6.6 shows the performance of turbo decoding of $q = 3$, $a = 3$ RA codes with randomly generated interleavers of various lengths.

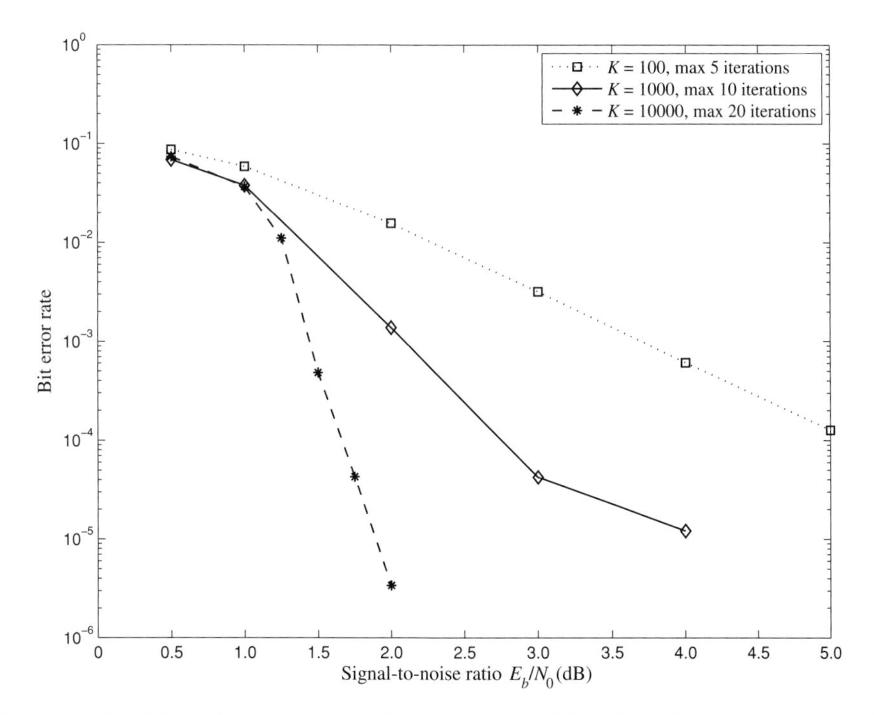

Figure 6.6 The BER performance of ($q = 3$, $a = 3$)-regular RA codes with randomly generated interleavers, using turbo decoding with a threshold stopping criterion.

Algorithm 6.2 RA turbo decoding

1: **procedure** RA TURBO(q,a,Π,$I_{\max}$,$\mathbf{R}^{(p)}$,$\mathbf{R}^{(\mathrm{u})}$)
2: $\mathbf{E}^{(d)}$ = zeros;
3: **for** $l = 1 : I_{\max}$ **do** ▷ For each iteration l
4: $\mathbf{A}^{(v)} = \Pi^{-1}(\mathbf{E}^{(d)})$
5: $\mathbf{L}^{(\mathrm{u})}$, $\mathbf{L}^{(v)}$ = logRdecoder(q,$\mathbf{R}^{(\mathrm{u})}$,$\mathbf{A}^{(v)}$)
6: $\mathbf{E}^{(v)} = \mathbf{L}^{(v)} - \mathbf{A}^{(v)}$
7: $\mathbf{A}^{(d)} = \Pi(\mathbf{E}^{(v)})$
8: $\mathbf{E}^{(d)}$ = logACdecoder(a,$\mathbf{R}^{(p)}$,$\mathbf{A}^{(d)}$)
9: **end for**
10: output $\mathbf{L}^{(\mathrm{u})}$
11: **end procedure**
12:
13: **procedure** LOGRDECODER(q,$\mathbf{R}^{(\mathrm{u})}$,$\mathbf{A}^{(v)}$)
14: **for** $i = 1 : \mathrm{length}(\mathbf{A}^{(v)})$ **do** ▷ For each bit i
15: **for** $j = q(i-1)+1, \ldots, qi$ **do**
16: $L_i^{(\mathrm{u})} = \sum_j A_j^{(v)}$
17: **end for**
18: **for** $j = q(i-1)+1, \ldots, qi$ **do**
19: $L_j^{(v)} = L_i^{(\mathrm{u})}$
20: **end for**
21: **end for**
22: output $\mathbf{L}^{(v)}$, $\mathbf{L}^{(\mathrm{u})}$
23: **end procedure**
24:
25: **procedure** LOGACDECODER(a,$\mathbf{R}^{(p)}$,$\mathbf{A}^{(d)}$)
26: $m = \mathrm{length}(\mathbf{R}^{(p)})$
27:
28: **for** $i = 1 : m$ **do**
29: $\mathrm{prod}_i = 1;$
30: **for** $j = a(i-1)+1 : ai$ **do**
31: $\mathrm{prod}_i = \mathrm{prod}_i \dfrac{1-\exp(-A_j^{(d)})}{1+\exp(-A_j^{(d)})}$
32: **end for**
33: $A_i^{(s)} = \log \dfrac{1+\mathrm{prod}_i}{1-\mathrm{prod}_i}$
34: **end for**
35:
36: $\mathbf{L}^{(s)} = \mathrm{logBCJR}\left(\dfrac{1}{1+D}, \mathbf{0}, \mathbf{R}^{(p)}, \mathbf{A}^{(s)}\right)$
37:
38: Continued over the page

Algorithm 6.2 RA Turbo Decoding continued

39: $\mathbf{E}^{(s)} = \mathbf{L}^{(s)} - \mathbf{A}^{(s)}$
40:
41: **for** $i = 1 : m$ **do**
42: $f_i = \dfrac{1 - \exp(-E_i^{(s)})}{1 + \exp(-E_i^{(s)})}$
43: **for** $j = a(i-1) + 1 : ai$ **do**
44: $g_j = 1;$
45: **for** $j' = a(i-1) + 1 : ai$ **do**
46: **if** $j' \neq j$ **then**
47: $g_j = g_j \dfrac{1 - \exp(-A_{j'}^{(d)})}{1 + \exp(-A_{j'}^{(d)})}$
48: **end if**
49: **end for**
50: $E_j^{(d)} = \log \dfrac{1 + f_i g_j}{1 - f_i g_j}$
51: **end for**
52: **end for**
53: output $\mathbf{E}^{(d)}$
54: **end procedure**

6.2.3 Sum–product decoding of RA codes

We have seen that systematic RA codes are a type of serially concatenated turbo code; however, importantly, they are also a special type of low-density parity-check code. The parity-check equation of an RA code comprises the combiner and accumulator equations (6.3) and (6.4) respectively:

$$\begin{aligned} p_i &= p_{i-1} \oplus s_i, \\ &= p_{i-1} \oplus d_{a(i-1)+1} \oplus d_{a(i-1)+2} \oplus \cdots \oplus d_{ai} \end{aligned}$$

or

$$p_i \oplus p_{i-1} \oplus d_{a(i-1)+1} \oplus d_{a(i-1)+2} \oplus \cdots \oplus d_{ai} = 0.$$

Recall that the d_i are just copies of the message bits (of which message bits is determined by the choice of interleaver) and p_i is just the ith parity bit, so it is straightforward to represent these parity-check equations by a parity-check matrix H. The first K columns of H correspond to the message bits and the last m columns correspond to the parity bits. For example, if we have $p_i = p_{i-1} \oplus u_{c_1} \oplus u_{c_2}$ then the ith row of H is 1 in columns c_1, c_2, $K + i$ and $K + i - 1$ and 0 elsewhere.

Example 6.4 In Example 6.2 we considered a $q = 3$, $a = 2$ RA code with $K = 4$ and $N = 10$ and interleaver $\Pi = [1, 7, 4, 10, 2, 5, 8, 11, 3, 9, 6, 12]$. Firstly, $q = 3$ so

$$\mathbf{v} = [u_1\ u_1\ u_1\ u_2\ u_2\ u_2\ u_3\ u_3\ u_3\ u_4\ u_4\ u_4].$$

Next, $\Pi = [1, 7, 4, 10, 2, 5, 8, 11, 3, 9, 6, 12]$ so

$$\mathbf{d} = [v_{\pi_1}\ v_{\pi_2} \cdots v_{\pi_{10}}] = [u_1\ u_3\ u_2\ u_4\ u_1\ u_2\ u_3\ u_4\ u_1\ u_3\ u_2\ u_4].$$

For the first parity bit we have

$$p_1 = d_1 \oplus d_2 = u_1 \oplus u_3,$$

so the first parity-check equation is

$$c_1 \oplus c_3 \oplus c_5 = 0$$

where $c_1, \ldots, c_K = u_1, \ldots, u_K$ and $c_{K+1}, \ldots, c_N = p_1, \ldots, p_m$. Similarly, the second parity bit is given by

$$p_2 = p_1 \oplus d_3 \oplus d_4 = p_1 \oplus u_2 \oplus u_4,$$

so the second parity-check equation is

$$c_2 \oplus c_4 \oplus c_5 \oplus c_6 = 0.$$

Repeating this for the remaining parity bits gives the RA parity-check matrix:

$$H = \begin{bmatrix} 1 & 0 & 1 & 0 & 1 & 0 & 0 & 0 & 0 & 0 \\ 0 & 1 & 0 & 1 & 1 & 1 & 0 & 0 & 0 & 0 \\ 1 & 1 & 0 & 0 & 0 & 1 & 1 & 0 & 0 & 0 \\ 0 & 0 & 1 & 1 & 0 & 0 & 1 & 1 & 0 & 0 \\ 1 & 0 & 1 & 0 & 0 & 0 & 0 & 1 & 1 & 0 \\ 0 & 1 & 0 & 1 & 0 & 0 & 0 & 0 & 1 & 1 \end{bmatrix}. \tag{6.13}$$

Generally, an $m \times N$ RA-code parity-check matrix H has two parts:

$$H = [H_1\ H_2], \tag{6.14}$$

where H_2 is an $m \times m$ matrix due to the accumulator,

$$H_2 = \begin{bmatrix} 1 & 0 & 0 & & 0 & 0 & 0 \\ 1 & 1 & 0 & \cdots & 0 & 0 & 0 \\ 0 & 1 & 1 & & 0 & 0 & 0 \\ & \vdots & & \ddots & & \vdots & \\ 0 & 0 & 0 & & 1 & 0 & 0 \\ 0 & 0 & 0 & \cdots & 1 & 1 & 0 \\ 0 & 0 & 0 & & 0 & 1 & 1 \end{bmatrix}, \tag{6.15}$$

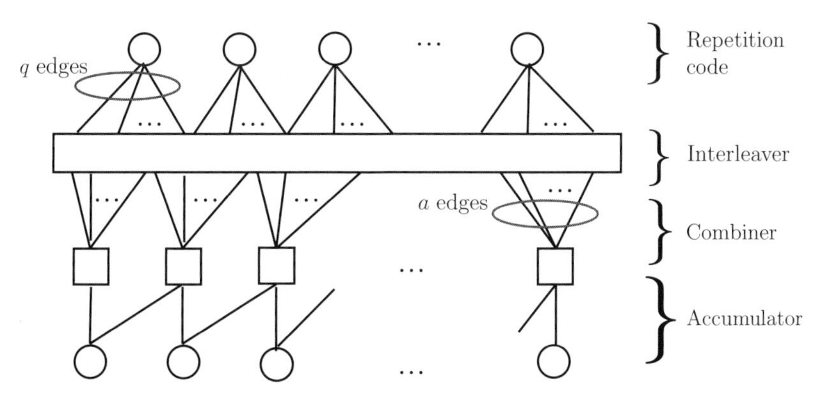

Figure 6.7 An RA-code Tanner graph.

and H_1 is an $m \times K$ matrix, with rows having weight a and columns having weight q, in which the choice of interleaver determines the positions of the entries in H_1. Two different interleaver patterns can describe the same RA code if the difference in the permutation pattern results in a difference in which copy of the same message bit is used (i.e. two 1 entries in H_1 are swapped with each other). Interleavers which produce repeated edges (see (6.23)) are not allowed.

Earlier we saw that an LDPC code can be put into an approximately lower triangular form to facilitate almost linear-time encoding. An RA code can be viewed as an LDPC code with a lower triangular form already built into the parity-check matrix during the code design.

As for LDPC codes, the Tanner graph of an RA code is defined by H. Figure 6.7 shows the Tanner graph representation of an RA code. Unlike for a general LDPC code, the message bits in the codeword of an RA code are easily distinguished from the parity bits. We distinguish between *systematic-bit nodes* corresponding to the K message bits in the codeword, shown at the top of the graph, and *parity-bit nodes* corresponding to the m parity bits in the codeword, which are shown at the bottom of the graph. The systematic-bit nodes have degree q while the parity-bit nodes have degree 2 except for the final parity-bit node, which has degree 1. The check nodes have degree $a + 2$ except the first, which has degree $a + 1$.

Example 6.5 Figure 6.8 shows the Tanner graph for the RA code from Examples 6.2 and 6.4.

The parity-check matrix of an RA code is called *(q, a)-regular* if the weights of the rows of H_1 all have the same value, a, and the weights of the columns of H_1 all have the same value, q. Note that a regular RA parity-check matrix has

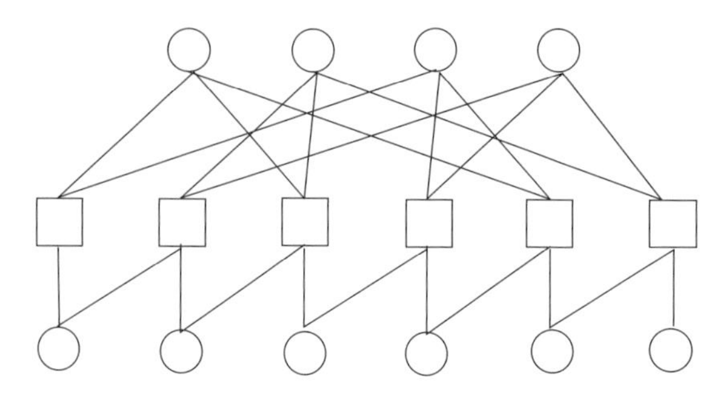

Figure 6.8 The Tanner graph for the RA parity-check matrix in Example 6.2.

columns of weight 2, and one column of weight 1, in H_2 and so is not regular in the sense of regular LDPC codes.

If the parity-bit nodes and systematic-bit nodes are treated as indistinguishable bit nodes then the sum–product decoding of an RA code is exactly the same as the sum–product decoding of an LDPC code with the same parity-check matrix. Alternatively, the sum–product algorithm can be scheduled slightly differently by updating systematic-bit node to check node messages, then check node to parity-bit node messages, then parity-bit node to check node messages and finally check node to systematic-bit node messages to complete one iteration of the decoder.

In a sum–product decoder for RA codes the parity-bit nodes are decoded in exactly the same way as the systematic-bit nodes. The turbo decoder for RA codes, however, uses BCJR decoding on a trellis for the parity bits. Because of this a single iteration of turbo decoding is more complex than a single iteration of sum–product decoding for the same code; however, turbo decoding usually requires fewer iterations overall.

Example 6.6 Figure 6.9 shows the BER performance of the sum–product decoding of ($q = 4, a = 4$)-regular RA codes with varying lengths and randomly constructed interleavers. The improvement in interleaving gain as the codeword length increases is very clear. Overall, regular RA codes show a waterfall and error floor performance similar to regular LDPC codes having the same average number of Tanner graph edges per codeword bit.

6.2.4 Irregular RA codes

Irregular repeat–accumulate (IRA) codes can be constructed using variable-rate repetition codes and/or combiners. To describe an IRA code we define as v_i the fraction of message bits encoded with each different-rate repetition

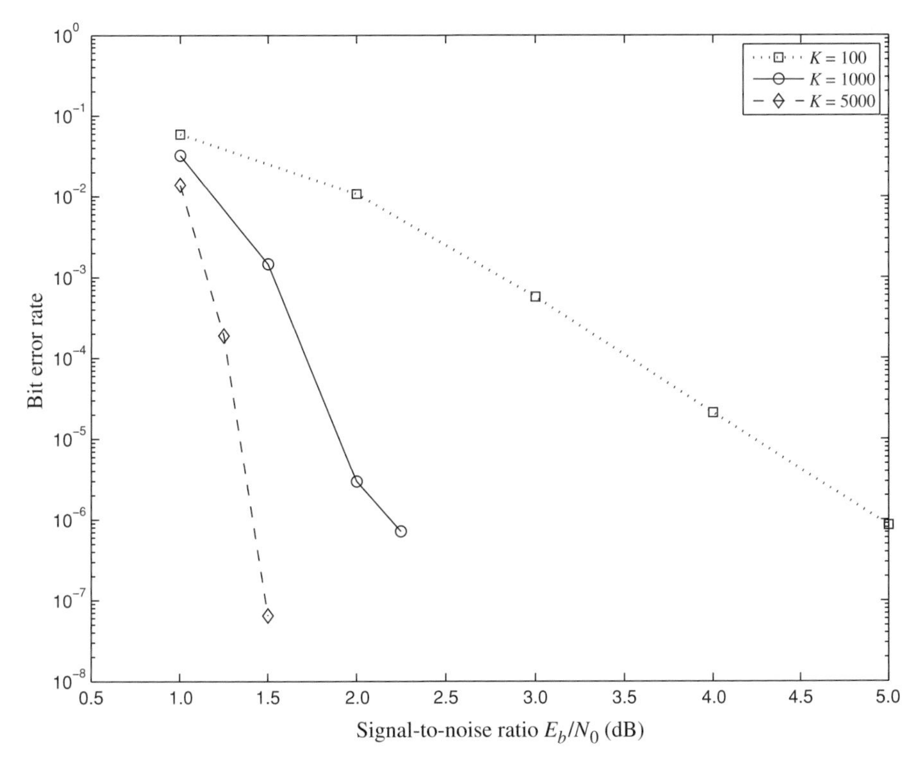

Figure 6.9 The BER performance of ($q = 4$, $a = 4$)-regular RA codes with randomly generated interleavers of various lengths, using sum–product decoding. After [115], © IEEE 2009.

code (equivalently the fraction of columns in H_1 of each weight or the fraction of systematic-bit nodes of each degree) and we define as h_i the fraction of the repeated bits encoded with each different-rate combiner (equivalently, the fraction of rows in H_1 of each weight or the fraction of check nodes of each degree). Note that the number of edges from check nodes to parity-bit nodes is fixed at 2, as H_2 remains the same as for a regular code. Collectively the set comprising $\mathbf{v}$ and $\mathbf{h}$ is the degree distribution of the code.

An alternative definition of the degree distribution denotes the fraction of edges that are connected to degree-i systematic-bit nodes by λ_i and the fraction of edges *from the systematic-bit nodes into check nodes* that are connected to i systematic-bit nodes by ρ_i. As for LDPC codes, we have

$$\sum_i \lambda_i = 1 \tag{6.16}$$

and

$$\sum_i \rho_i = 1, \tag{6.17}$$

and the functions

$$\lambda(x) = \sum_{i=2}^{\lambda_{\max}} \lambda_i x^{i-1} = \lambda_2 x + \lambda_3 x^2 + \cdots + \lambda_i x^{i-1} + \cdots, \tag{6.18}$$

$$\rho(x) = \sum_{i=2}^{\rho_{\max}} \rho_i x^{i-1} = \rho_2 x + \rho_3 x^2 + \cdots + \rho_i x^{i-1} + \cdots \tag{6.19}$$

can be defined. Note, however, that unlike for LDPC codes not all the edges are included in the definition of λ and ρ. The edges due to the accumulator (between the check nodes and parity-bit nodes) are not included.

A regular RA code will have

$$Kq = ma \tag{6.20}$$

non-zero entries in H_1 and a length-Kq interleaver. An IRA code will have

$$m\left(\sum_i h_i i\right) = K\left(\sum_i v_i i\right) \tag{6.21}$$

non-zero entries in H_1 and a length-$K(\sum_i v_i i)$ interleaver.

Example 6.7 The IRA code with parity-check matrix

$$H = \begin{bmatrix} 1 & 0 & 1 & 0 & 1 & 1 & 0 & 0 & 0 & 0 \\ 0 & 1 & 0 & 1 & 0 & 1 & 1 & 0 & 0 & 0 \\ 1 & 0 & 0 & 1 & 0 & 0 & 1 & 1 & 0 & 0 \\ 0 & 0 & 1 & 0 & 1 & 0 & 0 & 1 & 1 & 0 \\ 0 & 1 & 1 & 1 & 1 & 0 & 0 & 0 & 1 & 1 \end{bmatrix} \tag{6.22}$$

has degree distribution $v_2 = 0.4$, $v_3 = 0.6$, $h_2 = 0.6$, $h_3 = 0.2$, $h_4 = 0.2$ and 13 non-zero entries in H_1.

The sum–product decoding of IRA codes is exactly the same as the sum–product decoding of RA codes.

Example 6.8 Figure 6.10 shows the BER performance of the sum–product decoding of IRA codes with varying lengths and using randomly constructed interleavers. The IRA codes have an optimized degree distribution

$$\begin{aligned}\lambda(x) = {} & 0.04227x + 0.16242x^2 + 0.06529x^6 + 0.06489x^7 + 0.06207x^8 \\ & + 0.01273x^9 + 0.13072x^{10} + 0.04027x^{13} + 0.00013x^{24} \\ & + 0.05410x^{25} + 0.03031x^{35} + 0.13071x^{36} + 0.10402x^{99}\end{aligned}$$

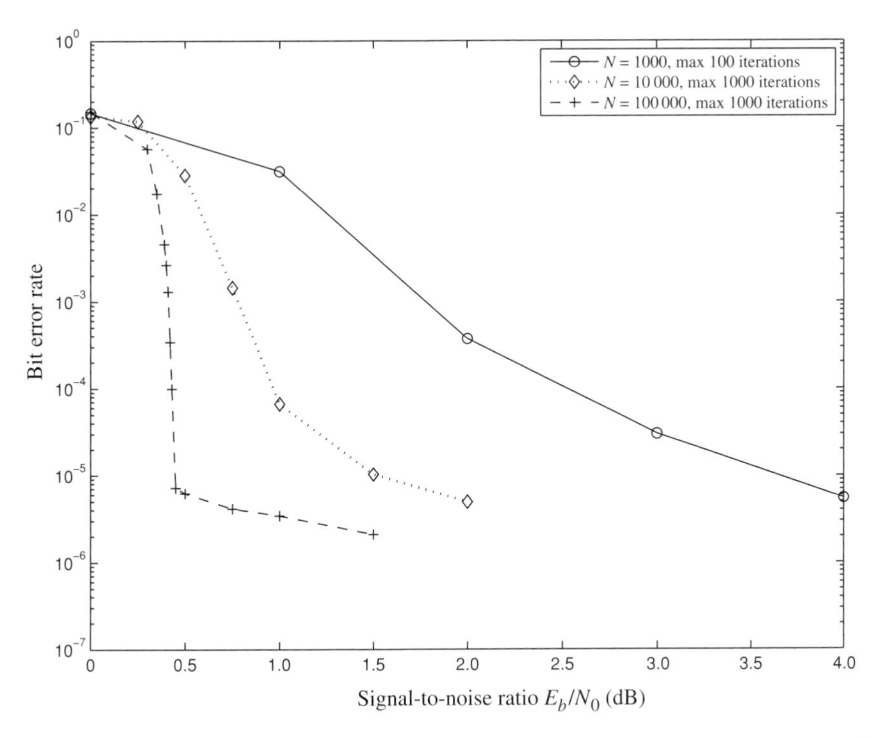

Figure 6.10 The BER performance of rate-1/2 IRA codes from Example 6.8 with varying lengths using sum–product decoding.

and $a = 8$, from [111]. The improvement in interleaving gain as the codeword length increases is very clear. Indeed, we will see in Chapter 7 that these codes have a threshold of 0.255 dB on the BI-AWGN channel, so the length $N = 100\,000$ code is getting close to the threshold for IRA codes with this degree distribution.

Again, irregular repeat–accumulate codes show a similar waterfall and error floor performance to that of irregular LDPC codes. That is, optimized (for threshold) irregular RA codes show a threshold improvement over regular codes but with a corresponding increase in the error rate of the error floor. We will see how to choose the degree distributions of IRA codes so as to optimize their threshold in Chapter 7 and consider their error floor performance in Chapter 8.

Encoding an IRA code requires that the encoder assigns the appropriate fraction of message bits to each different rate encoder. Similarly, the turbo decoder must keep track of which bits were encoded by which rate encoder so as to assign the appropriate LLRs to each decoder. For sum–product decoding this is incorporated in the description of the Tanner graph.

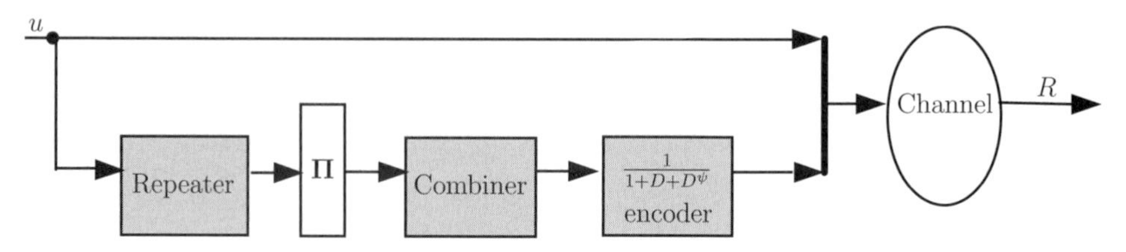

Figure 6.11 The encoder for w3IRA codes.

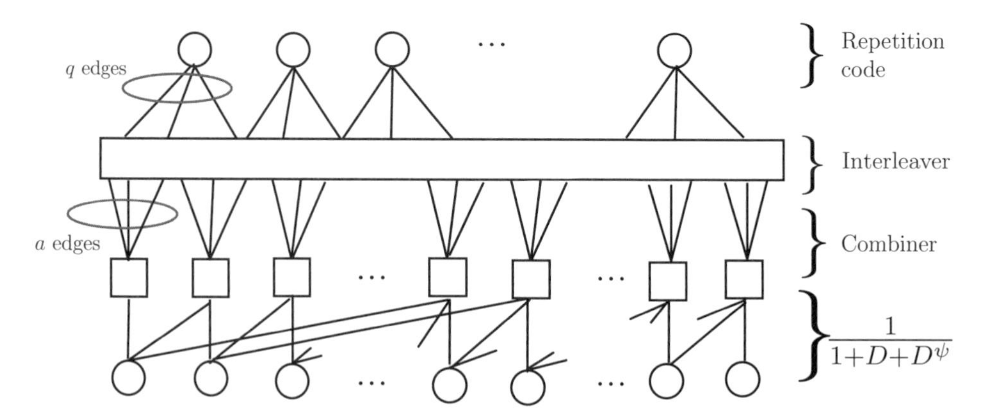

Figure 6.12 The Tanner graph representation of w3IRA codes.

6.2.5 Weight-3 IRA codes

The error floor performance of IRA codes can be improved by replacing the $1/(1 + D)$ convolutional encoder with a $1/(1 + D + D^{\psi})$ convolutional encoder, where ψ can be chosen to reduce the number of weight-2 columns in the parity-check matrix. The large number of weight-2 columns in the parity-check matrix, which is due to the accumulator, can hinder the error floor performance of IRA codes, particularly as higher-rate codes are considered (where the accumulator columns make up a greater proportion of the columns in H). The number of weight-2 columns remaining in H due to the parity bits is now ψ in a $1/(1 + D + D^{\psi})$ convolutional code, while the remainder of the parity bits contribute weight-3 columns. The codes are decoded using sum–product decoding in exactly the same way as RA and IRA codes.

Figure 6.11 shows the encoder for these *weight-3 IRA (w3IRA)* codes. Figure 6.12 shows the Tanner graph representation of w3IRA codes. Note that were a w3IRA code to be decoded with turbo decoding, choosing an encoder with a large ψ would dramatically increase the complexity of the BCJR decoder for the inner code. However, using sum–product decoding on the code's Tanner graph, the decoding complexity of w3IRA codes may be slightly increased over that of IRA codes owing to the $m - \psi$ extra edges in the Tanner graph. The extra edges in this part of the graph may result in fewer edges due to the repetition code once the degree distributions have been optimized. The encoder will require an extra (length-ψ) register to store the delayed parity bits, and an additional XOR summation for each parity bit. Overall the complexity difference is small.

Example 6.9 Figure 6.13 shows the BER performance of the sum–product decoding of the w3IRA codes with degree distribution

$$\begin{aligned} v(x) &= 0.7128x^2 + 0.0010x^6 + 0.0893x^7 + 0.090x^8 + 0.01236x^9 \\ &\quad + 0.01196x^{10} + 0.0504x^{13} + 0.0096x^{24} + 0.0514x^{25} \\ &\quad + 0.0311x^{35} + 0.00339x^{36} + 0.00877x^{99}, \\ h(x) &= 0.2832x^6 + 0.7168x^7 \end{aligned}$$

and

$$\Psi = 0.3583N.$$

Plots are given for varying lengths, with, for comparison, the IRA codes from Example 6.8. The w3IRA codes show a small loss in threshold for a gain in error floor performance.

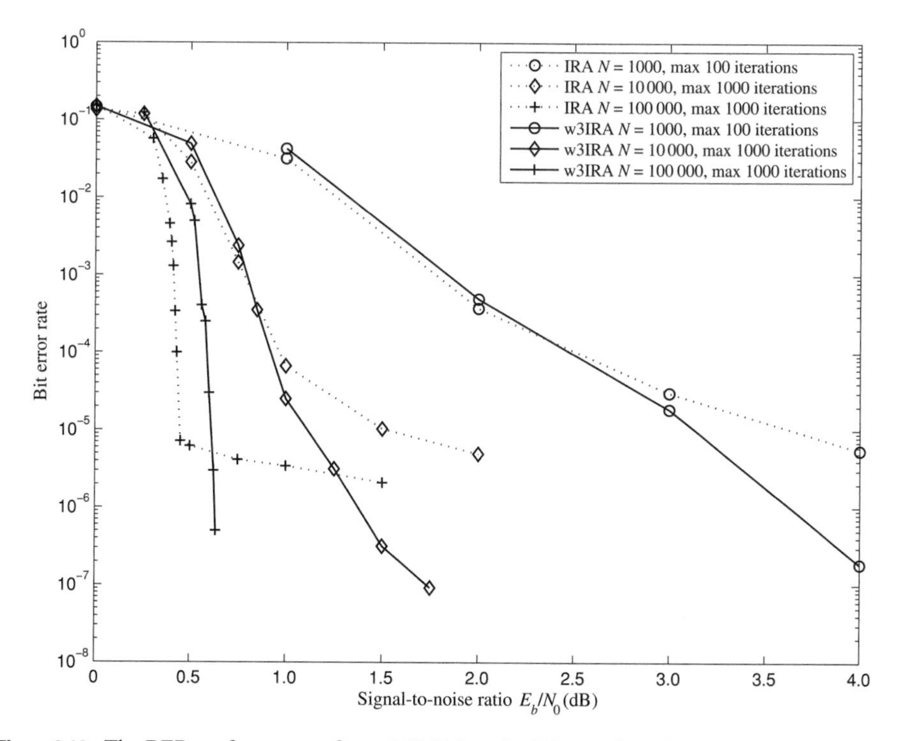

Figure 6.13 The BER performance of rate-1/2 IRA and w3IRA codes with varying lengths using sum–product decoding.

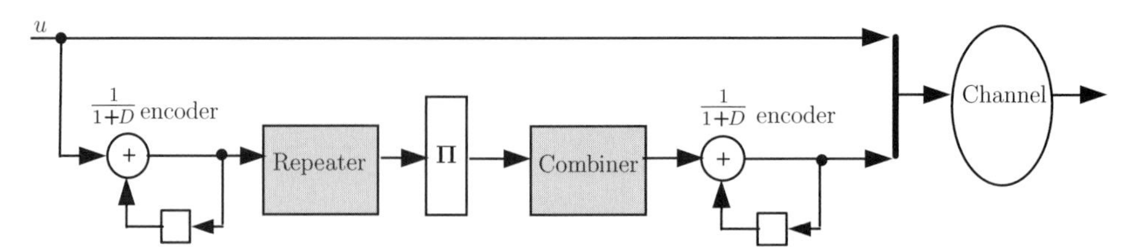

Figure 6.14 The encoder for accumulate–repeat–accumulate codes.

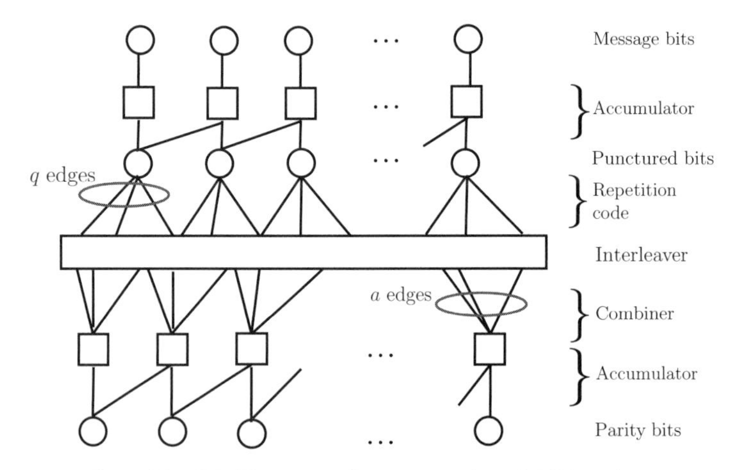

Figure 6.15 The Tanner graph representation of ARA codes.

6.2.6 Accumulate–repeat–accumulate codes

The thresholds of RA codes can be improved by precoding with an accumulator added before the repetition code in the encoder. Figure 6.14 shows the encoder for these *accumulate–repeat–accumulate* (ARA) codes; such codes are decoded using sum–product decoding as for LDPC codes. Figure 6.15 shows the Tanner graph representation of ARA codes. The systematic bits fed into the first accumulator are still transmitted through the channel; however, the bits at the output of the first accumulator, which will be passed into the repetition code, are not transmitted and so are often referred to as punctured bits.

The decoding complexity of ARA codes may be increased slightly over that of RA codes because of the extra edges in the Tanner graph. The added complexity in the encoder, however, simply results from the use of a single register and XOR block to implement the additional accumulator.

Like RA codes, ARA codes can be thought of as a particular type of LDPC code. Indeed the A4RA LDPC code protograph that we considered in Example 3.17 has its origins in an ARA code with repetition $q = 4$. The ARA codes offer both a simple encoder and capacity-approaching performance (capacity-achieving on the BEC).

Extensions to ARA codes include the addition of even more accumulators, as in *accumulate–repeat–accumulate–accumulate* codes, or they may only pass some fraction of the message bits through the first accumulator and send the remainder straight to the repetition encoder, in *multi-edge-type* ARA codes. The concept of precoding with an accumulator can also be extended to LDPC codes, producing accumulate LDPC codes, another class of capacity-approaching (again, capacity-achieving on the BEC) ensembles.

6.2.7 Code design

All the RA-like codes are easily encodable, either using turbo-like encoding as in Section 6.2.1 or using back substitution in H as in Section 2.3.1. The priority for code design is then the decoding performance of the code. Repeat–accumulate codes are typically decoded using sum–product decoding, so the same design rules as those outlined in Section 3.2 apply. However, the accumulator structure of H_2 somewhat constrains the design of the Tanner graph and the encoding of the RA codes requires that we specify an interleaver Π.

Choosing the degree distribution for IRA codes faces the same trade-off as for LDPC codes (threshold versus error floor). In Chapter 7 we will describe how to choose the degree distribution of IRA codes so as to optimize the code threshold, while for regular codes the average column weight of an RA code plays a similar role to the average column weight of LDPC codes (see Section 3.2.1). The average column weight of a regular RA code is

$$\frac{qK + 2m}{m + K},$$

where q is the repetition parameter, so an average column weight $\widehat{w_c}$ will require

$$q = \text{round}\frac{\widehat{w_c}(M + K) - 2M}{K}.$$

Thus for a rate-$1/2$ code $M = K$, so $\widehat{w_c} = 3$ when

$$q = \text{round}\left(\frac{6K - 2K}{K}\right) = 4.$$

Constructing a (q, a)-regular RA code interleaver requires a permutation $\Pi = [\pi_1, \pi_2, \ldots, \pi_{Kq}]$ such that

$$\pi_i \in \{1, 2, \ldots, Kq\} \quad \text{and} \quad \pi_i \neq \pi_j \quad \forall i \neq j.$$

The ith entry, π_i, specifies that the $\lceil i/a \rceil$th row of H has a 1 in the $\lceil \pi_i/q \rceil$th column, where as before $\lceil x \rceil$ denotes the smallest integer greater than or equal to x. Since H is binary repeated entries are not permitted, which is equivalent to requiring that

$$\lceil \pi_i/q \rceil \neq \lceil \pi_j/q \rceil \quad \forall i \neq j \text{ such that } \lceil i/a \rceil = \lceil j/a \rceil. \tag{6.23}$$

Since RA-like codes are usually decoded using sum–product decoding, the performance of the code is determined by the properties of the Tanner graph. In particular, the decoder performance is improved if short cycles in the code's Tanner graph are avoided (see Section 3.2.2). For an RA code we define two classes of 4-cycles. A *type*-1 4-*cycle* is said to occur if a column in H_1 contains two consecutive 1s (a second pair of 1s comprising the 4-cycle occurs in H_2). A *type*-2 4-*cycle* is formed if two columns of H_1 contain two entries in common.

Example 6.10 A type-1 4-cycle is shown below, in bold, in the parity-check matrix of a length-10 RA code,

$$H = \begin{bmatrix} 1 & 0 & 1 & 0 & 1 & 0 & 0 & 0 & 0 & 0 \\ 0 & \mathbf{1} & 0 & 1 & 1 & \mathbf{1} & 0 & 0 & 0 & 0 \\ 1 & \mathbf{1} & 0 & 0 & 0 & \mathbf{1} & 1 & 0 & 0 & 0 \\ 0 & 0 & 1 & 1 & 0 & 0 & 1 & 1 & 0 & 0 \\ 1 & 0 & 1 & 0 & 0 & 0 & 0 & 1 & 1 & 0 \\ 0 & 1 & 0 & 1 & 0 & 0 & 0 & 0 & 1 & 1 \end{bmatrix}.$$

Example 6.11 A type-2 4-cycle is shown in the parity-check matrix of another length-10 RA code,

$$H = \begin{bmatrix} \mathbf{1} & 0 & \mathbf{1} & 0 & 1 & 0 & 0 & 0 & 0 & 0 \\ 0 & 1 & 0 & 1 & 1 & 1 & 0 & 0 & 0 & 0 \\ 1 & 1 & 0 & 0 & 0 & 1 & 1 & 0 & 0 & 0 \\ 0 & 0 & 1 & 1 & 0 & 0 & 1 & 1 & 0 & 0 \\ \mathbf{1} & 0 & \mathbf{1} & 0 & 0 & 0 & 0 & 1 & 1 & 0 \\ 0 & 1 & 0 & 1 & 0 & 0 & 0 & 0 & 1 & 1 \end{bmatrix}.$$

To avoid type-1 4-cycles we require that

$$\lceil \pi_i/q \rceil \neq \lceil \pi_j/q \rceil \quad \forall i \neq j \text{ such that } \lceil i/a \rceil = \lceil j/a \rceil \pm 1,$$

while to avoid type-2 4-cycles we require that

$$\lceil \pi_j/q \rceil \neq \lceil \pi_i/q \rceil \quad \forall i \neq j \text{ such that}$$

$$\exists\ k,l \text{ where } \lceil l/a \rceil = \lceil j/a \rceil, \quad \lceil k/a \rceil = \lceil i/a \rceil, \quad \lceil \pi_l/q \rceil = \lceil \pi_k/q \rceil.$$

Cycles cannot be formed solely within the columns of H_2 and so type-1 and type-2 4-cycles cover all possible 4-cycles in an RA code.

Type-1 4-cycles can be avoided by using an S-random interleaver (see Section 5.2) and specifying that $S \geq \max(q-1, 2a-1)$. However, this can be a more stringent requirement than what is actually required to avoid type-1 4-cycles. More importantly, an S-random interleaver only acts within the S adjacent entries of the interleaver and so cannot be used to avoid the type-2 4-cycles formed by interleaver entries spaced widely apart.

The row–column interleaver is particularly bad for RA codes since it can add a large number of 4-cycles when $a > 1$.

Example 6.12 The parity-check matrix of a length-16 rate-1/2 (2, 2)-regular RA code with a row–column interleaver is given by

$$H = \begin{bmatrix} \mathbf{1} & 0 & \mathbf{1} & 0 & 0 & 0 & 0 & 0 & 1 & 0 & 0 & 0 & 0 & 0 & 0 & 0 \\ 0 & 0 & 0 & 0 & 1 & 0 & 1 & 0 & 1 & 1 & 0 & 0 & 0 & 0 & 0 & 0 \\ \mathbf{1} & 0 & \mathbf{1} & 0 & 0 & 0 & 0 & 0 & 0 & 1 & 1 & 0 & 0 & 0 & 0 & 0 \\ 0 & 0 & 0 & 0 & 1 & 0 & 1 & 0 & 0 & 0 & 1 & 1 & 0 & 0 & 0 & 0 \\ 0 & 1 & 0 & 1 & 0 & 0 & 0 & 0 & 0 & 0 & 0 & 1 & 1 & 0 & 0 & 0 \\ 0 & 0 & 0 & 0 & 0 & 1 & 0 & 1 & 0 & 0 & 0 & 0 & 1 & 1 & 0 & 0 \\ 0 & 1 & 0 & 1 & 0 & 0 & 0 & 0 & 0 & 0 & 0 & 0 & 0 & 1 & 1 & 0 \\ 0 & 0 & 0 & 0 & 0 & 1 & 0 & 1 & 0 & 0 & 0 & 0 & 0 & 0 & 1 & 1 \end{bmatrix}.$$

There are four type-2 4-cycles in H, one of which is shown in bold.

As for other classes of iterative codes, an RA-like code with a randomly constructed permutation will usually perform quite well, while a structured construction can improve the performance of short-to-medium-length codes and also aid in reducing implementation complexity. Suitably modified versions of most construction methods for LDPC codes can also be applied to RA-like codes. Indeed, RA-like codes can be thought of as a type of multi-edge-type LDPC code where the structure of the accumulator is incorporated into the edge definitions.

Bit-filling and progressive edge growth (PEG) techniques (see Example 3.15) can easily be applied to RA-like codes by fixing the accumulator portion of the Tanner graph and applying bit-filling or PEG algorithms to the remaining bit nodes.

It is also possible to construct RA-like codes using protographs (see Section 3.3.1). The edges between systematic-bit nodes and check nodes are permuted one set at a time, as for LDPC protograph constructions. However, the edges between the check nodes and parity-bit nodes are left as they are, in order to give the required accumulator pattern (i.e. the edge labeled 1 is permuted to link to the adjacent protograph but no other edge is moved). Protograph constructions are particularly useful for irregular codes, as the degree distribution of the protograph is maintained in the final code. Choosing circulant edge permutations can also aid in the implementation of the code.

Example 6.13 Figures 6.16 and 6.17 show the RA and ARA protographs defined in [113].

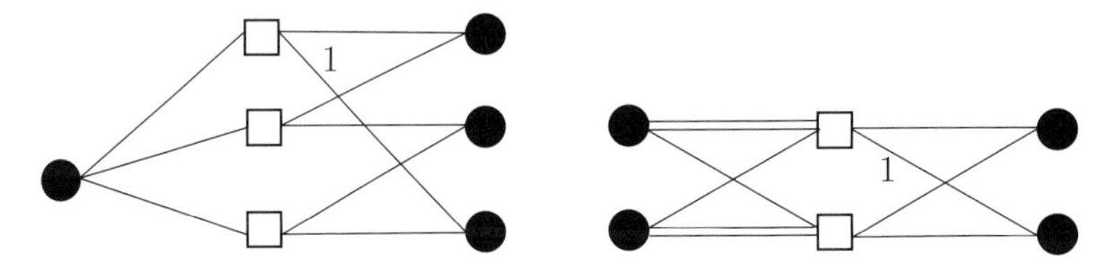

Figure 6.16 Protographs for repeat–accumulate codes. A protograph for a $(q = 3, a = 1)$ RA code (left) and for a $(q = 3, a = 3)$ RA code (right).

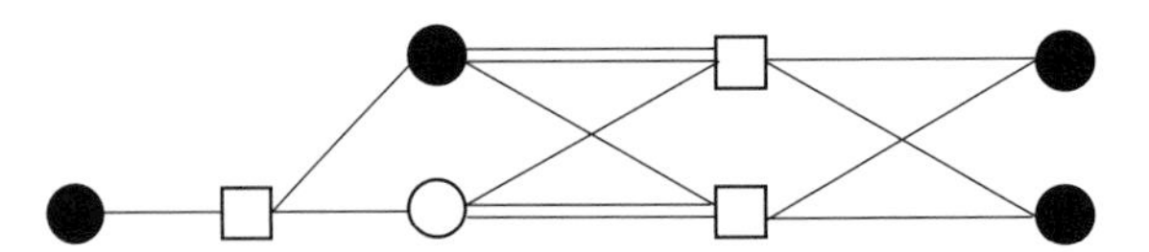

Figure 6.17 A protograph for a $(q = 3, a = 3)$ accumulate–repeat–accumulate code. The unfilled bit node represents a punctured bit.

Some algebraic constructions for LDPC codes can be modified for RA-like codes (see e.g. [114]). Also, new interleavers can be designed to target the repetition and accumulator component codes of the RA code. One example of this is L-type interleavers.

L-type interleavers

Repeat–accumulate codes can be constructed without small cycles by using L-type interleavers. An L-type interleaver is defined by the code parameters K and q and the integer-valued interleaver parameter L. The repetition code and interleaver can together be thought of as making a selection from the K message bits q times. First the K message bits are arranged in order in Π_1. Then message bits in Π_1 are selected to produce Π_2. After each selection, the process skips L bits ahead and starts again with the first unselected bit until all K message bits in Π_1 have been selected. This process is repeated $q - 1$ times; each time a selection is made from Π_{i-1} to form Π_i. In practice the set of bits $iK + 1$ to

$(i+1)K$ of Π, which we denote by Π_{i+1}, is a type of row–column permutation of the set of bits $(i-1)K+1$ to iK of Π, denoted Π_i, where the bits are written row-wise into a matrix with L columns and read out column-wise. Thus the L-type interleaver can be implemented as the concatenation of $q-1$ of these matrix row-write column-read operations.

Example 6.14 The parity-check matrix from [115] for a length-16 rate-1/2 (2, 2)-regular RA code with L-type ($L=2$) interleaver

$$\Pi = [1, 3, 5, 7, 9, 11, 13, 15, 2, 6, 10, 14, 4, 8, 12, 16]$$

is given by

$$H = \begin{bmatrix} 1 & 1 & 0 & 0 & 0 & 0 & 0 & 0 & 1 & 0 & 0 & 0 & 0 & 0 & 0 & 0 \\ 0 & 0 & 1 & 1 & 0 & 0 & 0 & 0 & 1 & 1 & 0 & 0 & 0 & 0 & 0 & 0 \\ 0 & 0 & 0 & 0 & 1 & 1 & 0 & 0 & 0 & 1 & 1 & 0 & 0 & 0 & 0 & 0 \\ 0 & 0 & 0 & 0 & 0 & 0 & 1 & 1 & 0 & 0 & 1 & 1 & 0 & 0 & 0 & 0 \\ 1 & 0 & 1 & 0 & 0 & 0 & 0 & 0 & 0 & 0 & 0 & 1 & 1 & 0 & 0 & 0 \\ 0 & 0 & 0 & 0 & 1 & 0 & 1 & 0 & 0 & 0 & 0 & 0 & 1 & 1 & 0 & 0 \\ 0 & 1 & 0 & 1 & 0 & 0 & 0 & 0 & 0 & 0 & 0 & 0 & 0 & 1 & 1 & 0 \\ 0 & 0 & 0 & 0 & 0 & 1 & 0 & 1 & 0 & 0 & 0 & 0 & 0 & 0 & 1 & 1 \end{bmatrix}.$$

A (3, a)-regular RA code can be constructed without 4-cycles whenever $K > a^3$, by using an L-type interleaver and setting $L = a$; L-type interleavers can also be used to produce RA codes guaranteed to be without 6-cycles in some cases. However, the low dispersion of these interleavers results in a progressively weakening performance for longer codes. Attempts have been made to modify L-type interleavers in order to improve this dispersion while maintaining an algorithmic construction.

In such modified L-type interleavers, Π_1 is unchanged and Π_i is formed starting with the same process as previously, i.e. the bits of Π_{i-1} are written row-wise into a matrix M_i with L columns and read out column-wise. Now, additionally, the bits from each column are written row-wise into another matrix, A_j, and read out column-wise. Thus the jth column of M_i is written into a matrix A_j having j columns.

The implementation of this second RA interleaver now requires $(q-1)L$ matrix write–read operations instead of $q-1$. However, the modified interleaver is still extremely simple to specify, requiring just three integer-valued parameters, L, q and K.

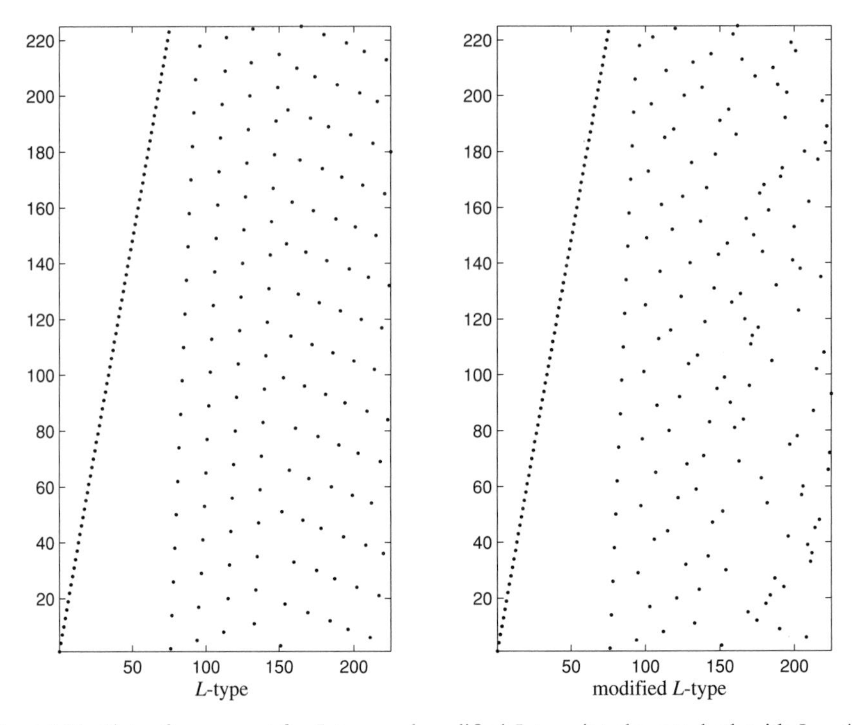

Figure 6.18 Plots of π_i versus i for L-type and modified L-type interleavers, both with $L = 4$.

Figure 6.18 shows π_i versus i plots for L-type and modified L-type interleavers. The extra row–column permutation has the effect of reducing the structure in the modified L-type interleavers.

Example 6.15 Figure 6.19 shows the BER performance of the sum–product decoding of $(q = 4, a = 4)$-regular RA codes from [115] with varying interleavers. At short lengths many of the structured interleavers perform on a par with randomly chosen interleavers. For longer codes, however, the low dispersion of the L-type interleaver becomes a problem, but this can be remedied by adding the extra steps in the modified L-type interleaver.

6.3 Bibliographic notes

The idea of concatenating two or more error correction codes in series in order to improve the overall decoding performance of a system was suggested long before turbo codes. In 1966 Forney proposed the serial concatenation of a high-rate block code with a short convolutional code. These "classical" concatenated

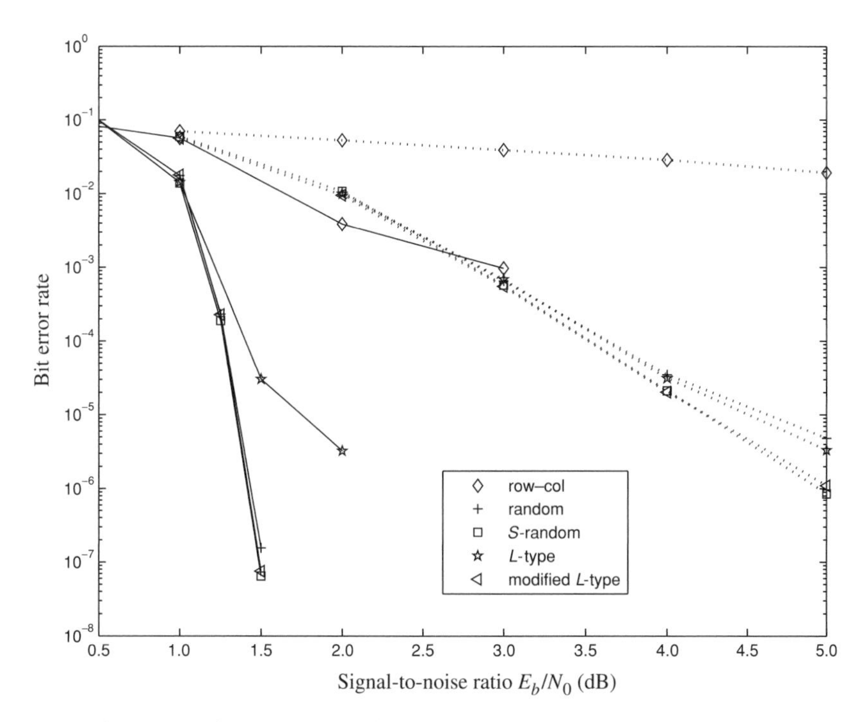

Figure 6.19 The BER performance of length-200 (dotted curves) and length 10 000 (solid curves) ($q = 4, a = 4$)-regular RA codes with varying interleavers. After [115], © IEEE 2009.

codes have been employed in applications where very good bit error rate performances are required, such as in deep space communications. The Voyager error correction scheme, for example, uses an inner convolutional code with an outer Reed–Solomon block code. This error correction system is still a standard in deep space communications. However, it was only once iterative turbo decoding was applied to serially concatenated codes by Benedetto, Montorsi and coworkers [109, 116, 117] that capacity-approaching performances were obtained.

While less research effort has been dedicated to serially concatenated turbo codes than to the parallel version, many results derived for parallel turbo codes apply when concatenating serially. The choice of the inner component code, and to some extent the interleaver, can be made using the same general principles as for turbo codes. For the outer code, the important property is its minimum distance, and there is no shortage of constructions from classical coding theory for codes with good $d_{\min}$ values. The only stumbling block is obtaining low-complexity MAP decoders for them.

Repeat–accumulate codes were first presented by Divsalar, Jin and McEliece as a simple class of serially concatenated turbo codes [118] for which coding theorems could be developed. However, it was soon realized that, although simple, RA codes and their extension to IRA codes were powerful codes in their own right (see e.g. [111, 119]). From another angle, designers of LDPC codes had also been considering RA-like codes in their attempts to find easily encoded LDPC codes [120, 121]. For more on RA codes and their constructions see [114, 115, 121–123].

The combination of simplicity with excellent decoding performance has made repeat–accumulate type codes a hot topic for the coding community. Indeed, the accumulate-repeat–accumulate codes introduced by Abbasfar, Divsalar and Yao [113, 124] achieve capacity on the BEC and do this with bounded complexity per message bit [125].

When decoded as LDPC codes, RA codes offer the advantages of LDPC codes at the decoder and simple encoding at the transmitter. For this reason it is likely that RA codes will be used in place of LDPC codes (and possibly even turbo codes) in future applications. Irregular RA codes (with column weights 8 and 3) are already in use in Second Generation digital video broadcasting (DVB-S2) and the LDPC codes adopted for the WiFi (IEEE 802.11n) and WiMAX (IEEE 802.16e) wireless LAN standards use RA base matrices.

We will consider serially concatenated codes in more detail in Chapters 7 and 8.

6.4 Exercises

6.1 A serially concatenated (SC) turbo code is formed using the concatenation of the convolutional encoder

$$\left[1 \quad \frac{1+D^2}{1+D+D^2}\right],$$

for the outer (convolutional) encoder,

$$\left[1 \quad \frac{1}{1+D^2}\right],$$

for the inner encoder and the length-6 interleaver

$$\Pi = [1, 3, 5, 6, 4, 2].$$

Given that the encoders are unterminated,
(a) determine the code rate,
(b) determine the codeword produced by the message $\mathbf{u} = [1\ 1\ 0]$.

6.2 An SC turbo code is formed using the concatenation of the convolutional encoders

$$\left[\frac{1}{1+D}\right] \text{ and } \left[1 \quad \frac{1}{1+D^2}\right],$$

for the inner and outer codes respectively and a row–column interleaver with two rows and three columns. Given that the encoders are unterminated,

(a) determine the code rate,

(b) determine the codeword produced by the message $\mathbf{u} = [1\ 0\ 1]$.

6.3 A codeword from the SC turbo code in Exercise 6.1 was sent on a binary symmetric channel with crossover probability 0.1, and the vector

$$\mathbf{y} = [1\ 1\ 0\ 1\ 0\ 1\ 1\ 1\ 0\ 1\ 0\ 1]$$

was received. Decode $\mathbf{y}$ using turbo decoding with a maximum of three iterations.

6.4 Generate a bit error rate plot for turbo decoding operating on the turbo encoder in Exercise 6.1 with a randomly constructed interleaver of length 100 sent on a BI-AWGN channel with signal-to-noise ratios between 1 and 5 dB.

6.5 Repeat Exercise 6.4 using a row–column interleaver.

6.6 Design a length-100 interleaver for the code in Exercise 6.4 and compare its BER performance with that of the random and row–column interleavers.

6.7 The message

$$\mathbf{u} = [1\ 0\ 1\ 0\ 1]$$

is to be encoded using a systematic RA code consisting of a $q = 2$ repetition code, an $a = 2$ combiner and the interleaver $\Pi = [1, 7, 4, 10, 2, 5, 8, 3, 9, 6]$. Give the resulting codeword.

6.8 The message

$$\mathbf{u} = [1\ 0\ 1\ 0\ 1]$$

is to be encoded using a non-systematic RA code consisting of a $q = 2$ repetition code, an $a = 1$ combiner and the interleaver $\Pi = [1, 7, 4, 10, 2, 5, 8, 3, 9, 6]$. Give the resulting codeword.

6.9 The message

$$\mathbf{u} = [1\ 0\ 1\ 0]$$

is to be encoded using an IRA code in which every second bit is encoded by a $q = 2$ repetition code and every other bit by a $q = 3$ repetition code. The combiner has $a = 2$ and the interleaver is $\Pi = [1, 7, 4, 10, 2, 5, 8, 3, 9, 6]$. Give the resulting codeword.

6.10 Give the node degree distribution of the IRA code considered in Example 6.9.

6.11 Find a parity-check matrix and draw the corresponding Tanner graph for the RA code in Exercise 6.7. What is the minimum girth and minimum stopping set size of the Tanner graph?

6.12 Find a parity-check matrix and draw the corresponding Tanner graph for the IRA code in Exercise 6.9. Give the edge-degree distribution of the code and calculate the minimum girth and minimum stopping set size of the Tanner graph.

6.13 Find a parity-check matrix and draw the corresponding Tanner graph for the RA code in Exercise 6.7, but replace the given interleaver with a row–column interleaver. What is the minimum girth and minimum stopping set size of the Tanner graph?

6.14 Find a parity-check matrix and draw the corresponding Tanner graph for the RA code in Exercise 6.7 but now replace the interleaver with an L-type interleaver having four columns ($L = 2$). What is the minimum girth and minimum stopping set size of the Tanner graph?

6.15 Repeat Example 6.6 for a $q = 3, a = 3$ RA code.

6.16 You are to choose a regular rate-1/2 RA code with $q \leq 5$ to transmit a length-100 message over a BI-AWGN channel. Use BER plots to determine which value of q is best for each range of SNR values.

6.17 Repeat Exercise 6.16 but now precode each RA code with an accumulator. For each value of q, decide whether you would choose the RA or ARA code.

6.18 Consider again the RA and ARA codes in Exercises 6.16 and 6.17 but now assume that they have the same approximate decoding complexity rather than the same repetition parameter. (Hint: recall that decoding complexity scales roughly with the number of Tanner graph edges per codeword bit.)

7 Density evolution and EXIT charts

7.1 Introduction

In Chapters 2–6 we introduced turbo, LDPC and RA codes and their iterative decoding algorithms. Simulation results show that these codes can perform extremely closely to Shannon's capacity limit with practical implementation complexity. In this chapter we analyze the performance of iterative decoders, determine how close they can in fact get to the Shannon limit and consider how the design of the codes will impact on this performance.

Ideally, for a given code and decoder we would like to know for which channel noise levels the message-passing decoder will be able to correct the errors and for which it will not. Unfortunately this is still an open problem. Instead, we will consider the set, or *ensemble*, of all possible codes with certain parameters (for example, a certain degree distribution) rather than a particular choice of code having those parameters.

For example, a turbo code ensemble is defined by its component encoders and consists of the set of codes generated by all possible interleaver permutations while an LDPC code ensemble is specified by the degree distribution of the Tanner graph nodes and consists of the set of codes generated by all possible permutations of the Tanner graph edges.

When very long codes are considered, the extrinsic LLRs passed between the component decoders can be assumed to be independent and identically distributed. Under this assumption the expected iterative decoding performance of a particular ensemble can be determined by tracking the evolution of these probability density functions through the iterative decoding process, a technique called *density evolution*.

Density evolution (DE) can be used to find the maximum level of channel noise which is likely to be corrected by a particular ensemble using the message-passing algorithm, called the *threshold* for that ensemble. This then enables the code designer to search for the ensemble with the best threshold, from which a specific code may then be chosen.

Density evolution can be approximated using *extrinsic information transfer* (EXIT) charts, which track the mutual information of extrinsic LLRs rather than their densities by assuming that the extrinsic LLR densities are Gaussian. The advantage of such tracking of mutual information is an easy-to-visualize graphical representation of the decoder.

7.2 Density evolution

If the channel is memoryless, under the assumption that the Tanner graphs are all cycle-free it is possible to determine the expected behavior of an iterative decoding algorithm over a given ensemble. Here the expectation is taken not only over all members of the ensemble but also over all possible realizations of the channel noise.

The derivation of the density evolution requires a number of properties of iterative ensembles. These are as follows.

- **Symmetric distributions** If the channel is symmetric, i.e. if $p(y|x = 1) = f(y)$ then $p(y|x = -1) = f(-y)$, the LLRs output by the iterative decoder are also symmetric.
- **All-zeros codeword** Using symmetric distributions, the iterative decoding performance can be shown to be independent of the codeword transmitted. This result allows the performance of a code–decoder pair to be modeled by sending only the all-zeros codeword.
- **Concentration** With high probability, a randomly chosen code from an ensemble will have an iterative decoding performance close to the ensemble average performance. This result makes the ensemble average performance a useful predictor of the behavior of any particular code from the ensemble.
- **Cycle-free graphs** As the codeword length goes to infinity the ensemble average performance of the iterative decoder approaches that of decoding on a cycle-free graph. If there are no cycles in the Tanner graphs of length $2l$ or less we can assume that the iterative decoder messages are independent for up to l iterations, which makes density evolution possible.

See [1] for the derivation of these results for a number of iterative ensembles, including LDPC and turbo ensembles.

We open this section by considering density evolution on the binary erasure channel (BEC). For this channel a transmitted bit is either correctly received or completely erased, so the density to be tracked is simply the erasure probability. Thus for the BEC the probability density function is effectively one-dimensional, resulting in a greatly simplified analysis. We then consider density evolution and the corresponding thresholds of regular codes on the BI-AWGNC, before presenting techniques for obtaining *irregular* degree distributions which enable the optimization of thresholds.

7.2.1 Density evolution on the BEC

Recall from Algorithm 2.1 that, for message-passing decoding on the BEC, the messages hold either the current value of the bit, which is 1 or 0 if known or e if the bit is erased. The densities to be tracked are thus the expected fraction of erased message bits.

Regular LDPC codes

Given an ensemble $\mathcal{T}(w_c, w_r)$ that consists of all regular LDPC Tanner graphs with bit nodes of degree w_c and check nodes of degree w_r, we want to know how the message-passing decoder will perform on the binary erasure channel using codes from this ensemble. We define q_l to be the probability that at iteration l a check-to-bit message is an e and p_l to be the probability that at iteration l a bit-to-check message is an e (i.e. p_l is the probability that a codeword bit remains erased at iteration l).

For message-passing decoding on the BEC, a parity-check equation can correct an erased bit if that bit is the only erased bit in the parity-check equation. Thus the check-to-bit message on an edge is e if one or more of the incoming messages on the other $w_r - 1$ edges into that check node is an e. To calculate the probability q_l that a check-to-bit message is e at iteration l we make the assumption that all the incoming messages are independent of one another. That is, we assume firstly that the channel is memoryless, so that none of the original bit probabilities were correlated, and secondly that there are no cycles in the Tanner graphs of length $2l$ or less, as a cycle will cause the messages to become correlated. With this assumption, the probability that none of the other $w_r - 1$ incoming messages to the check node is e is simply the product of the probabilities $(1 - p_l)$ that an individual message is not e . So the probability that one or more of the other incoming messages is e is one minus this product:

$$q_l = 1 - (1 - p_l)^{w_r - 1}. \tag{7.1}$$

At iteration l the bit-to-check message will be e if the original message from the channel was an erasure, which occurs with probability ε, and all the incoming messages at iteration $l - 1$ are erasures, each of which occurrences has probability q_{l-1}. Again we make the assumption that the incoming messages are independent of one another, and so the probability that the bit-to-check message is an e is the product of the probabilities that the other $w_c - 1$ incoming messages to the bit node, and the original message from the channel, were erased:

$$p_l = \varepsilon (q_{l-1})^{w_c - 1}. \tag{7.2}$$

Substituting for q_{l-1} from (7.1) gives

$$p_l = \varepsilon \left(1 - (1 - p_{l-1})^{w_r - 1}\right)^{w_c - 1}. \tag{7.3}$$

Prior to decoding, the value of p_0 is the probability that the channel erased a codeword bit:

$$p_0 = \varepsilon.$$

Thus, for a (w_c, w_r)-regular ensemble,

$$p_0 = \varepsilon, \quad p_l = \varepsilon \left(1 - (1 - p_{l-1})^{w_r - 1}\right)^{w_c - 1}. \tag{7.4}$$

The recursion in (7.4) describes how the erasure probability of message-passing decoding evolves as a function of the iteration number l for (w_c, w_r)-regular LDPC codes. Applying this recursion we can determine for which erasure probabilities the message-passing decoder is likely to correct the erasures.

Example 7.1 A code from the (3, 6)-regular ensemble is to be transmitted on a binary erasure channel with erasure probability $\varepsilon = 0.3$ and decoded with message-passing iterative decoding. The probability that a codeword bit will remain erased after l iterations of message-passing decoding (if the code's Tanner graph is free of cycles of size $2l$ or less) is given by the recursion

$$p_0 = 0.3, \quad p_l = p_0 \left(1 - (1 - p_{l-1})^5\right)^2.$$

Applying this recursion for seven iterations gives the following sequence of bit erasure probabilities:

$$p_0 = 0.3000, \quad p_1 = 0.2076, \quad p_2 = 0.1419, \quad p_3 = 0.0858,$$

$$p_4 = 0.0392, \quad p_5 = 0.0098, \quad p_6 = 0.0007, \quad p_7 = 0.0000.$$

Thus the bit erasure probability of a codeword from a cycle-free (3, 6)-regular LDPC code, transmitted on a BEC with erasure probability 0.3, will approach zero after seven iterations of message-passing decoding.

Irregular LDPC codes

Recall that an irregular parity-check matrix has columns and rows with varying weights (in the Tanner graph, bit nodes and check nodes with varying degrees). We designated the fraction of columns of weight i by v_i and the fraction of rows of weight i by h_i. The fraction of edges connected to degree-i bit nodes is denoted λ_i, and the fraction of edges connected to degree-i check nodes is denoted ρ_i.

From (7.1) we know that, at the lth iteration of message-passing decoding, the probability that a check-to-bit message is e, if all the incoming messages are independent, is

$$q_l = 1 - (1 - p_l)^{w_r - 1},$$

for an edge connected to a degree-w_r check node. For an irregular Tanner graph the probability that an edge is connected to a degree-w_r check node is ρ_{w_r}.

Thus, averaging over all the edges in an irregular Tanner graph gives the average probability that a check-to-bit message is in error:

$$q_l = \sum_i \rho_i \left(1 - (1 - p_l)^{i-1}\right) = 1 - \sum_i \rho_i (1 - p_l)^{i-1}.$$

Using the function

$$\rho(x) = \sum_{i=2}^{\rho_{\max}} \rho_i x^{i-1},$$

the equation for q_l can be written

$$q_l = 1 - \rho(1 - p_l).$$

From (7.2) we know that the probability that a bit-to-check message is e at the lth iteration of message-passing decoding, if all incoming messages are independent, is

$$p_l = \varepsilon \left(q_{l-1}\right)^{w_c - 1}$$

for an edge connected to a degree-w_c bit node. For an irregular Tanner graph the probability that an edge is connected to a degree-w_c bit node is λ_{w_c}. Averaging over all the edges in the Tanner graph gives the average probability that a bit-to-check message is in error:

$$p_l = \varepsilon \sum_i \lambda_i (q_{l-1})^{i-1}.$$

By defining the function

$$\lambda(x) = \sum_{i=2}^{\lambda_{\max}} \lambda_i x^{i-1},$$

the equation for p_l can be written

$$p_l = \varepsilon \lambda(q_{l-1}).$$

Finally, substituting for q_{l-1} we have

$$p_l = \varepsilon \lambda \left(1 - \rho\left(1 - p_{l-1}\right)\right).$$

Prior to decoding, the value of p_0 is the probability that the channel has erased a codeword bit:

$$p_0 = \varepsilon,$$

and so for irregular LDPC codes we have the recursion

$$p_0 = \varepsilon, \qquad p_l = p_0 \lambda \left(1 - \rho\left(1 - p_{l-1}\right)\right). \tag{7.5}$$

Algorithm 7.1 outlines the application of density evolution on the binary erasure channel with erasure probability ε for an LDPC ensemble with degree distribution (λ, ρ). In theory, density evolution considers the limit as the number

Algorithm 7.1 BEC density evolution

1: **procedure** $p = \text{DE}(\lambda, \rho, I_{\max}, \varepsilon)$
2: $\quad p_0 = \varepsilon$;
3: $\quad$ **for** $l = 1 : I_{\max}$ **do** $\quad \triangleright$ For each iteration l
4: $\quad\quad p_l = p_0 \lambda (1 - \rho (1 - p_{l-1}))$
5: $\quad$ **end for**
6: $\quad$ output p_l
7: **end procedure**

of iterations goes to infinity but in practice some large fixed number of iterations is specified.

Example 7.2 An irregular LDPC ensemble with degree distribution

$$\lambda(x) = 0.1x + 0.4x^2 + 0.5x^{19}$$

and

$$\rho(x) = 0.5x^7 + 0.5x^8$$

has rate

$$1 - \frac{\sum_i \lambda_i / i}{\sum_i \rho_i / i} \approx 0.5.$$

We apply the recursion from (7.5) over four different values of ε. Figure 7.1 shows the evolution of erasure probability with the number of iterations. We see

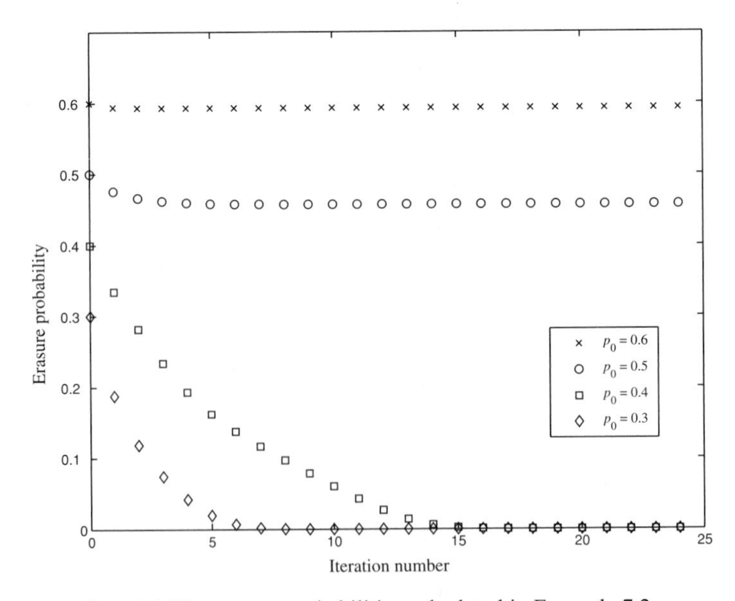

Figure 7.1 The erasure probabilities calculated in Example 7.2.

that for values of ε equal to 0.5 and above the probability that erasures remain does not decrease to zero even as l gets very large, whereas for values of ε equal to 0.4 and below the probability of error goes to zero.

7.2.2 Ensemble thresholds

The aim of density evolution is to determine for which channel erasure probabilities ε the message-passing decoder is likely to correct all the erased bits. Using (7.5) we have a means of approximating this as an average over all Tanner graphs with a given degree distribution (λ, ρ), by assuming that the graphs are cycle-free.

To examine the influence of ε on p_l we define the function

$$f(p, \varepsilon) = \varepsilon\lambda\left(1 - \rho\left(1 - p\right)\right).$$

The erasure probability at iteration l is then

$$p_l(\varepsilon) = f(p_{l-1}, \varepsilon),$$

where p and ε are probabilities and so can take values between 0 and 1. Here $f(p, \varepsilon)$ is a strictly increasing function in p for $\varepsilon > 0$. Thus, if $p_l > p_{l-1}$ then

$$p_{l+1} = f(p_l, \varepsilon) \geq f(p_{l-1}, \varepsilon) = p_l$$

for $\varepsilon \in [0, 1]$, so $p_l(\varepsilon)$ is a monotone sequence that is lower bounded at $p = 0$ by

$$f(0, \varepsilon) = \varepsilon\lambda\left(1 - \rho\left(1\right)\right) = \varepsilon\lambda\left(1 - 1\right) = 0$$

and upper bounded at $p = 1$ by

$$f(1, \varepsilon) = \varepsilon\lambda\left(1 - \rho\left(1 - 1\right)\right) = \varepsilon\lambda\left(1 - 0\right) = \varepsilon.$$

Since $f(p, \varepsilon)$ is a strictly increasing function in p,

$$0 \leq f(p, \varepsilon) \leq \varepsilon$$

for all $p \in [0, 1]$ and $\varepsilon \in [0, 1]$. Thus p_l converges to an element $p_\infty \in [0, \varepsilon]$. Further, for a degree distribution pair (λ, ρ) and an $\varepsilon \in [0, 1]$, it can be proven that if $p_l(\varepsilon) \to 0$ then $p_l(\varepsilon') \to 0$ for all $\varepsilon < \varepsilon'$. Indeed, there is a value ε^* called the *threshold* such that, for values of ε below ε^*, p_l approaches zero as the number of iterations goes to infinity while for values of ε above ε^* it does not. The threshold ε^* for a (λ, ρ) ensemble is defined as the supremum of ε for which $p_l(\varepsilon) \to 0$:

$$\varepsilon^*(\lambda, \rho) = \sup\{\varepsilon \in [0, 1] : p_l(\varepsilon)_{l\to\infty} \to 0\}.$$

In practice we can search for the threshold of an ensemble with a given degree distribution using Algorithm 7.2. Here p_{limit} is the erasure rate considered to be

Algorithm 7.2 Threshold

1: **procedure** $\varepsilon^* = \text{THRESHOLD}(\lambda,\rho,I_{\max},p_{\text{limit}},\varepsilon_H, \varepsilon_L,\delta\varepsilon)$
2: **repeat**
3: $\varepsilon = (\varepsilon_{\text{L}} + \varepsilon_{\text{H}})/2$
4: $p = \text{DE}(\lambda,\rho,I_{\max},\varepsilon)$
5: **if** $p < p_{\text{limit}}$ **then**
6: $\varepsilon_{\text{L}} = \varepsilon$
7: **else**
8: $\varepsilon_{\text{H}} = \varepsilon$
9: **end if**
10: **until** $(\varepsilon_{\text{H}} - \varepsilon_{\text{L}}) < \delta\varepsilon$
11: Return $\varepsilon^* = \varepsilon_{\text{L}}$
12: **end procedure**

close enough to zero, ε_{L} and ε_{H} are lower and upper bounds on the threshold (set to 0 and 1 respectively if one has no idea what the threshold might be) and $\delta\varepsilon$ sets the tolerance on the threshold result.

Example 7.3 Repeated application of the recursion (7.3) demonstrates that the threshold for a (3, 6)-regular ensemble on the binary erasure channel is between 0.4293 and 0.4295. Figure 7.2 shows the evolution of the erasure probability

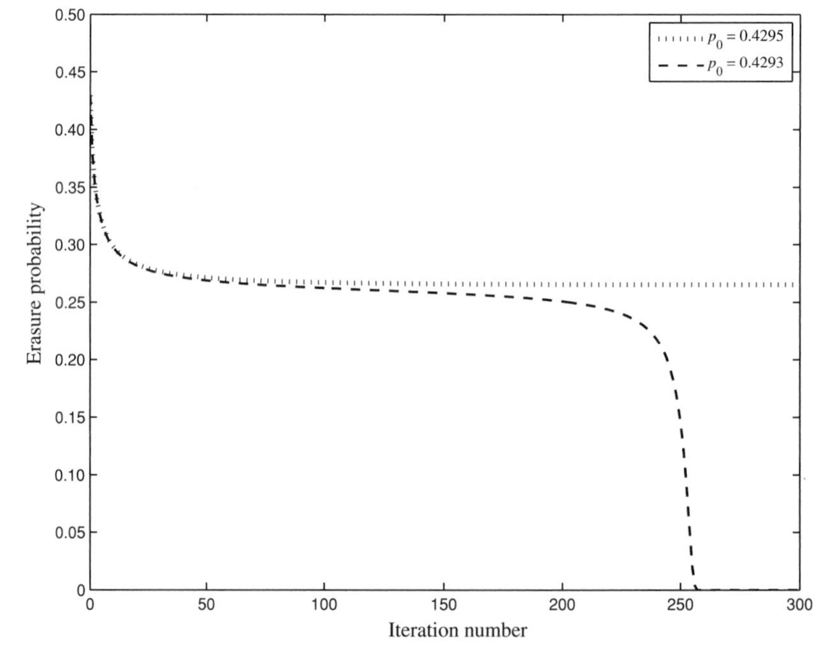

Figure 7.2 Finding the threshold of the (3, 6)-regular LDPC code considered in Example 7.3.

with iteration number for channel erasure probabilities just above and just below the code's threshold. Here we can see the mechanism behind the waterfall region of the BER curve from another perspective. At an erasure probability equal to 0.4295 the decoder returns a very high decoded erasure probability (around 0.26) but for a decrease in channel erasure probability by just 0.0002 the decoder corrects every one of the erased bits.

We see in Figure 7.2 one problem of decoding on channels with noise levels very close to the ensemble threshold: a very large number of iterations is required.

Example 7.4 We now extend Example 7.2 to find the threshold of an irregular LDPC ensemble with degree distribution

$$\lambda(x) = 0.1x + 0.4x^2 + 0.5x^{19},$$
$$\rho(x) = 0.5x^7 + 0.5x^8.$$

We saw in Figure 7.1 that for values of $\varepsilon \geq 0.5$ the probability that erasures remain does not decrease to zero even as l gets very large, whereas for values of $\varepsilon \leq 0.4$ the probability of error does go to zero. To close on the threshold even further, in Figure 7.3 we see that for values of $\varepsilon \geq 0.475$ the probability that erasures remain does not decrease to zero even as l gets very large, whereas for values of $\varepsilon \leq 0.465$ the probability of error does go to zero. For $\varepsilon = 0.47$

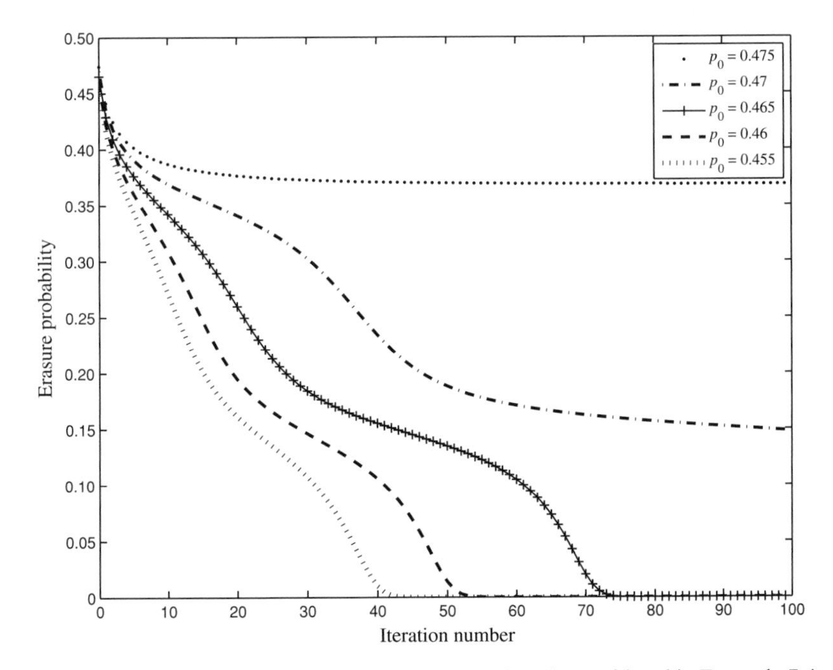

Figure 7.3 Finding the threshold of the irregular LDPC code considered in Example 7.4.

we would need to consider a larger number of decoder iterations to determine whether the decoder will converge to zero. We can conclude, however, that the threshold for this ensemble is an erasure probability between 0.465 and 0.475.

The examples above explain our observation in earlier chapters that iterative codes can require a larger value for $I_{\max}$ before their capacity-approaching potential is realized. Density evolution shows that as the channel erasure probability approaches the threshold of the ensemble, a larger number of iterations is required to decode with zero errors.

Stability

The recursion in (7.5) quickly results in very high order polynomials as the iteration number is increased. However, to understand its behavior when p_l is small we can approximate it using a Taylor series expansion of the right-hand side around 0:

$$p_l = f(p_{l-1}, \varepsilon) \approx f'(p, \varepsilon) p_{l-1}. \tag{7.6}$$

A function $f(x) = g(h(x))$ has a derivative with respect to x given by

$$\frac{df}{dx} = \frac{dg}{dh}\frac{dh}{dx}.$$

Thus for

$$f(p, \varepsilon) = \varepsilon\lambda(h(p)) \quad \text{where} \quad h(p) = 1 - \rho(1 - p)$$

the derivative of f with respect to p is

$$\frac{df(p, \varepsilon)}{dp} = \frac{d\lambda}{dh}\frac{dh}{dp}.$$

Evaluating this derivative at $p = 0$ we have that

$$h(p = 0) = 1 - \rho(1) = 0$$

and so

$$\left.\frac{d\lambda}{dh}\right|_{p=0} = \left.\frac{d\lambda}{dh}\right|_{h=0} = \lambda_2 + 2\lambda_3 h + \cdots + (i-1)\lambda_i h^{i-2} + \cdots \bigg|_{h=0} = \lambda_2$$

and

$$\left.\frac{dh}{dp}\right|_{p=0} = \left.\frac{d(1 - \rho(1 - p))}{dp}\right|_{(1-p)=1} = \rho'(1).$$

Substituting back into (7.6),

$$p_l \approx \varepsilon\lambda_2\rho'(1)p_{l-1} \tag{7.7}$$

as $p_l \to 0$.

For p_l to converge to zero as $l \to \infty$ requires that $p_l < p_{l-1}$ and thus, from (7.7),

$$\varepsilon\lambda_2\rho'(1) < 1. \tag{7.8}$$

Thus for a degree distribution pair (λ, ρ) to converge to zero on a binary erasure channel with erasure probability ε, λ_2 must be upper bounded by

$$\lambda_2 < \frac{1}{\varepsilon\rho'(1)}. \tag{7.9}$$

Equation (7.9) is called the *stability constraint* of density evolution and is usually written as

$$\lambda'(0)\rho'(1) < \frac{1}{\varepsilon},$$

where $\lambda'(0) = \lambda_2$.

7.2.3 Density evolution and repeat–accumulate codes

As systematic codes, repeat–accumulate codes have two types of bit nodes: *systematic-bit nodes*, which correspond to the message bits, and *parity-bit nodes*, which correspond to the parity bits. We will consider the sum–product algorithm, and thus the DE equations, scheduled slightly differently from LDPC codes. First, systematic-bit node to check node messages are sent, then check node to parity-bit node messages, then parity-bit node to check node messages and finally check node to systematic-bit node messages, to complete one iteration of the decoder.

For the sum–product decoding of RA codes there are thus four different types of messages, those from systematic-bit nodes to check nodes, denoted p_l, those from check nodes to parity-bit nodes, denoted s_l, those from parity-bit nodes to check nodes, denoted t_l, and finally those from check nodes to systematic-bit nodes, denoted q_l.

Recall that for an IRA code we designated the fraction of systematic-bit nodes of degree i (the fraction of columns of weight i in H_1) by v_i and the fraction of check nodes with i edges to systematic-bit nodes (the fraction of rows of weight i in H_1) by h_i. Alternatively, the fraction of edges from the check nodes that are connected to degree-i systematic-bit nodes is denoted by λ_i, and the fraction of edges *from the systematic-bit nodes* that are connected to degree-i check nodes by ρ_i. Importantly, edges between the check nodes and the parity-bit nodes play no role in the degree distribution.

The systematic-bit nodes perform the same function as the bit nodes for LDPC codes. Thus the message p_l from the systematic-bit nodes to the check nodes is an erasure if the incoming messages q_l, from the other check nodes generated in the previous iteration, and the original message from the channel, are all erasures:

$$p_l = \varepsilon\lambda\left(q_{l-1}\right).$$

The RA check nodes perform the same function as the check nodes for LDPC codes, except that along with the messages p_l from the systematic-bit nodes there are two messages t_l from the parity-bit nodes. Thus a message q_l from a check node to a systematic-bit node is an erasure if any incoming messages p_l from the other systematic-bit nodes or the two messages t_l from the two connected parity-bit nodes are erasures:

$$q_l = 1 - (1 - t_l)^2 \rho(1 - p_l).$$

A message s_l from the check nodes to the parity-bit nodes is an erasure if any incoming messages p_l from the systematic-bit nodes or the message t_l from the other parity bit node are erasures:

$$s_l = 1 - (1 - t_{l-1}) \sum_i h_i (1 - p_l)^i.$$

Finally, the parity-bit nodes have exactly two edges, so the message t_l from a parity-bit node to a check node will be an erasure if the original message from the channel and the message s_l from the other check node are both erasures:

$$t_l = \varepsilon s_l.$$

Example 7.5 Using density evolution, Figure 7.4 shows that the BEC threshold of a $q = 4$, $a = 4$ RA code is between 0.445 and 0.447. Thus a regular RA

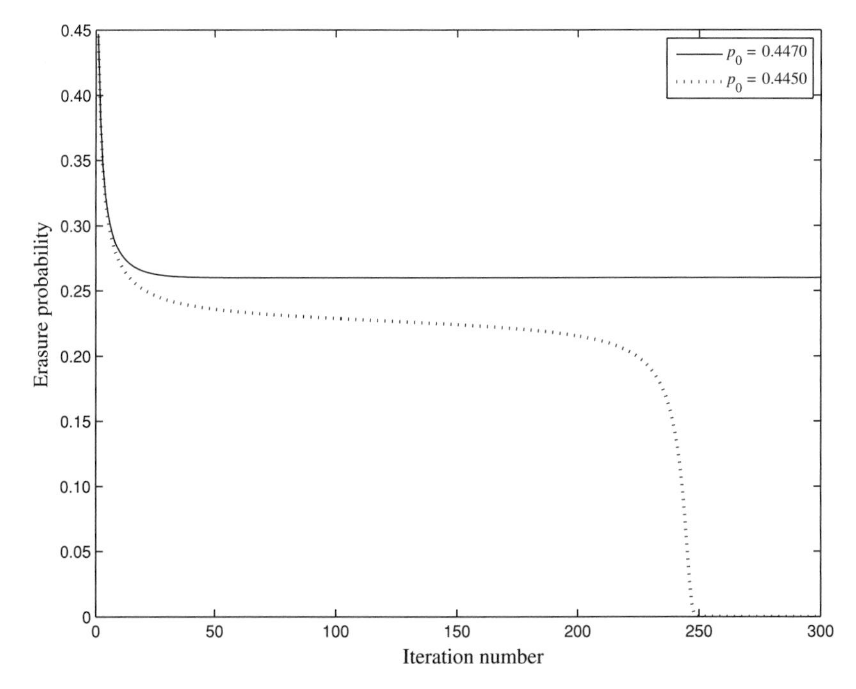

Figure 7.4 Finding the threshold of a $q = 4$ regular RA code in Example 7.5.

ensemble with the same average column weight as the (3, 6)-regular LDPC ensemble has a slightly better threshold. The reason may be that the regular RA ensemble is actually a bi-regular LDPC ensemble owing to the weight-2 columns in the accumulator.

7.2.4 Density evolution on general binary input memoryless channels

For message-passing decoding on general binary input memoryless channels, the bit-to-check messages are the log likelihood ratios (LLRs) of the probabilities that a given bit is 1 or 0. As these LLR values are continuous, the probability that a message has a particular LLR value is described by a probability density function (pdf).

Recall that the LLR of a random variable x is

$$L(x) = \log \frac{p(x=0)}{p(x=1)},$$

and so $L(x)$ will be positive if $p(x=0) > p(x=1)$ and negative otherwise. Consequently the probability that the corresponding codeword bit is a 1 is the probability that the LLR is negative.

To analyze the evolution of these pdfs in the message-passing decoder we define $p(M_l)$ to be the probability density function for a bit-to-check message at iteration l, $p(E_l)$ to be the probability density function for a check-to-bit message at iteration l and $p(R)$ to be the probability density function for the LLR of the received signal.

Again we make the assumption that all the incoming messages are independent of one another. That is, we are assuming firstly that the channel is memoryless, i.e. that none of the original bit probabilities were correlated, and secondly that there are no cycles in the Tanner graphs of length $2l$ or less, since a cycle would cause the messages to become correlated.

The outgoing message at a bit node is the sum of the incoming LLRs on the other edges into that node, (2.29):

$$M_{j,i} = \sum_{j' \in A_i,\ j' \neq j} E_{j',i} + R_i.$$

Since the incoming messages are independent, the pdf of the random variable formed from this summation can be obtained by the convolution [126, equation (6.39)] of the pdfs of the $w_c - 1$ incoming messages from the check nodes and the pdf of the incoming message from the channel:

$$p(M_l) = p(R) \otimes p(E_l)^{\otimes(w_c - 1)},$$

where $p(E_l)^{\otimes(w_c-1)}$ is the convolution of $w_c - 1$ copies of $p(E_l)$. Averaging over the bit degree distribution $\lambda(x)$:

$$p(M_l) = p(R) \otimes \sum_i \lambda_i p(E_l)^{\otimes(i-1)}.$$

The decoded LLRs L_l are, similarly,

$$p(L_l) = p(R) \otimes p(E_l)^{\otimes(w_c)},$$

and averaging over the bit-degree distribution $\lambda(x)$ gives:

$$p(L_l) = p(R) \otimes \sum_i \lambda_i p(E_l)^{\otimes(i)}.$$

The convolution operation can be evaluated numerically using fast Fourier Transforms (FFTs).

The function to be evaluated at each check node is (2.26):

$$E_{j,i} = \log \frac{1 + \prod_{i' \in B_j, i' \neq i} \tanh(M_{j,i'}/2)}{1 - \prod_{i' \in B_j,\, i' \neq i} \tanh(M_{j,i'}/2)}$$

where

$$\tanh(x/2) = \log \frac{e^x - 1}{e^x + 1}.$$

Thus to sum over two messages x and y requires the calculation of the probability density function

$$\begin{aligned} f(x, y) &= \log \frac{1 + \tanh(x/2)\tanh(y/2)}{1 - \tanh(x/2)\tanh(y/2)} \\ &= \log \frac{(e^x + 1)(e^y + 1) + (e^x - 1)(e^y - 1)}{(e^x + 1)(e^y + 1) - (e^x - 1)(e^y - 1)} \\ &= -\log \frac{e^x + e^y}{1 + e^{x+y}}. \end{aligned} \quad (7.10)$$

A method for finding the pdf of a function of two random variables is given in [126, equation (6.36)]. Briefly, given two random variables x and y and the function $z = f(x, y)$, the density of z can be found as follows:

$$f(z)dz = \iint_{\Delta D_z} f(x, y)dxdy$$

where D_z is the region of the xy-plane such that $z < g(x, y) < z + dz$.

Given a variable y with symmetric pdf, i.e. $p(y|x = 1) = p(-y|x = -1)$, if f is the probability density of the log likelihood ratio $L(y)$ then

$$z = \log \frac{f(z)}{f(-z)}$$

or, equivalently, $f(z) = e^z f(-z)$.

Since E, M, L all correspond to symmetric densities, fixing an all-zeros message gives an estimate of $p(L|x = 1)$, from which $p(L|x = -1)$ can be easily calculated, since

$$p(L|x = -1) = p(-L|x = 1).$$

Consequently the probability that a bit is in error, Pc, is the probability that the LLR is negative:

$$Pc = p(L < 0).$$

Example 7.6 Figure 7.5 shows the evolution of $p(M_l)$ for a (3, 6)-regular ensemble on a BI-AWGN channel with signal-to-noise ratio (E_b/N_0) equal to 1.12. On a BI-AWGN channel the pdf of the original received signal will be Gaussian with variance σ^2, reflecting the pdf of the noise. As the iteration number is increased,

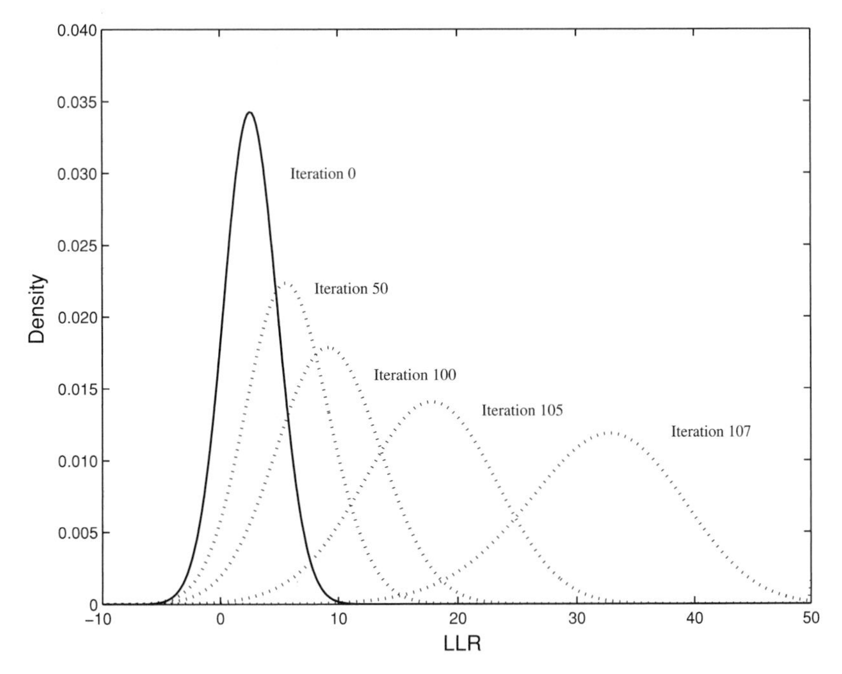

Figure 7.5 The evolution of the probability density functions $p(M_l)$ with iteration number, using density evolution. See Example 7.6.

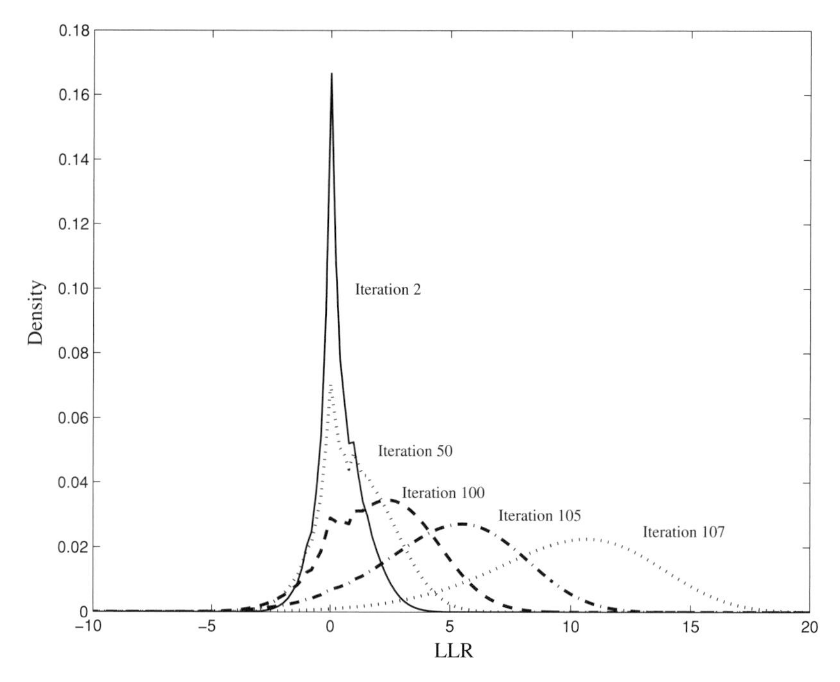

Figure 7.6 The evolution of probability density functions $p(E_l)$ with iteration number using density evolution. See Example 7.6.

the area under the curve for negative LLRs decreases and so the probability of error decreases. Figure 7.6 shows the evolution of $p(E_l)$ for the same ensemble. While the distributions of the check-to-bit messages are not Gaussian for the first few iterations, they do tend towards a Gaussian distribution as the decoding progresses.

Thresholds calculated using density evolution

The sum–product decoding algorithm preserves the order implied by the degradation of a binary input memoryless symmetric channel; i.e. as the channel improves, the performance of the sum–product decoder will also improve (see [127]). For example, on a BI-AWGN channel with noise variance σ^2, for all $\sigma' < \sigma$ the expected bit error probability of sum–product decoding, Pc, satisfies $Pc(\sigma') < Pc(\sigma)$.

The threshold of a given degree distribution for sum–product decoding is again the supremum of the channel noise values for which the probability of decoding error goes to zero as the iteration number is allowed to go to infinity. For a BI-AWGN channel with noise variance σ^2 the threshold is denoted σ^*:

$$\sigma^* = \sup\{\sigma : Pc(\sigma)_{l\to\infty} \to 0\}.$$

For the BI-AWGN channel with noise variance σ^2 the stability condition is

$$\lambda'(0)\rho'(1) < e^{1/2\sigma^2} \tag{7.11}$$

(see e.g. [127]).

Example 7.7 We would like to know at which BI-AWGN signal-to-noise ratios the codes from a (3, 6)-regular ensemble are likely to be able to correct the noise. At each iteration of density evolution the bit error rate of the message if decoding were halted is the probability that the log likelihood ratio L is negative. Applying density evolution for different channel noise levels, in Figure 7.7 we plot the sum–product decoding BER as a function of iteration number. We see that the BI-AWGN threshold for the (3, 6)-regular ensemble is between 0.881 and 0.879.

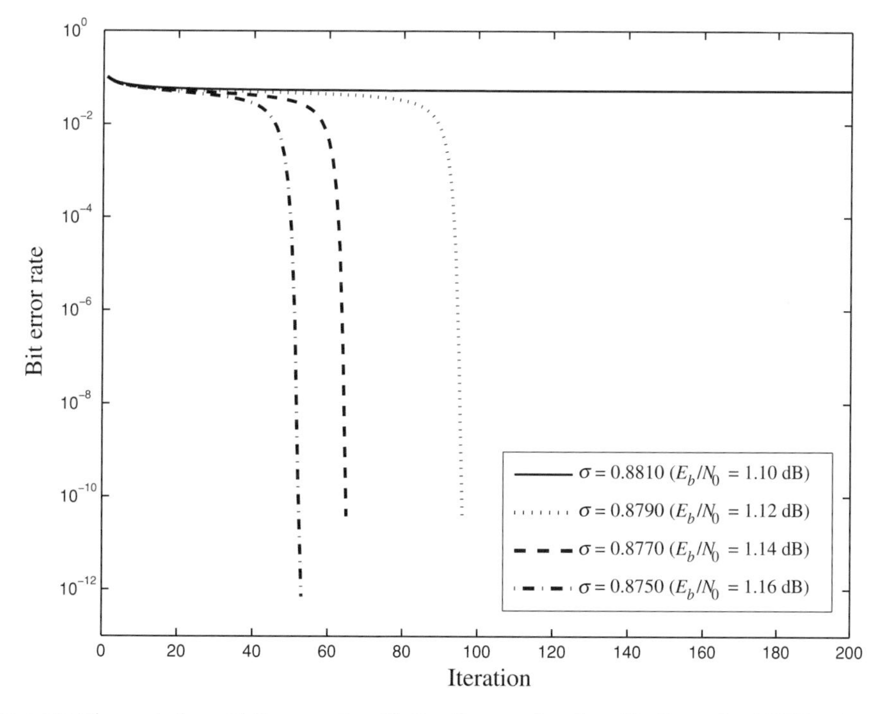

Figure 7.7 The evolution of bit error rate with iteration number for a (3, 6)-regular LDPC code on a BI-AWGN channel calculated using density evolution.

Calculating the BER as in Example 7.7, we can see that the average number of iterations required for correct decoding increases as the noise value approaches the threshold.

Example 7.8 The threshold performance of regular LDPC codes on a BI-AWGN channel is shown for a range of rates in Figure 7.8. For most code rates a column weight 3 gives the lowest threshold value for regular LDPC codes; however, for rates ≥ 0.95 the threshold for column-weight-4 regular codes is slightly higher than the threshold for column-weight-3 regular codes.

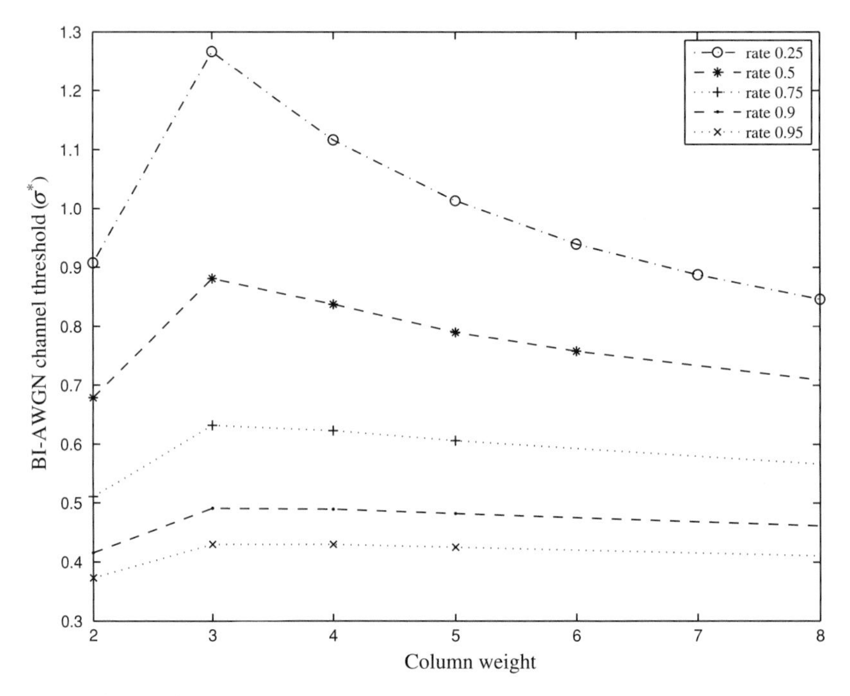

Figure 7.8 The thresholds σ^* of regular LDPC codes on a BI-AWGN channel as the column weight is varied, for five different rates.

Thus our analysis in the limit of infinite code length indicates that for regular LDPC codes on a BI-AWGN channel a column weight of 3 will give the best thresholds. Comparing the thresholds of regular LDPC codes with the Shannon limit (see Algorithm 1.1), see Figure 7.9, shows that the higher the rate of regular LDPC codes the smaller the gap to Shannon's capacity limit.

7.2.5 Density evolution and turbo codes

Density evolution can be applied to turbo code ensembles by tracking the evolution of the probability densities along the turbo code trellis. An analog of the concentration theorem for LDPC codes also applies to turbo codes, and turbo code thresholds have been proved to exist (see [128]).

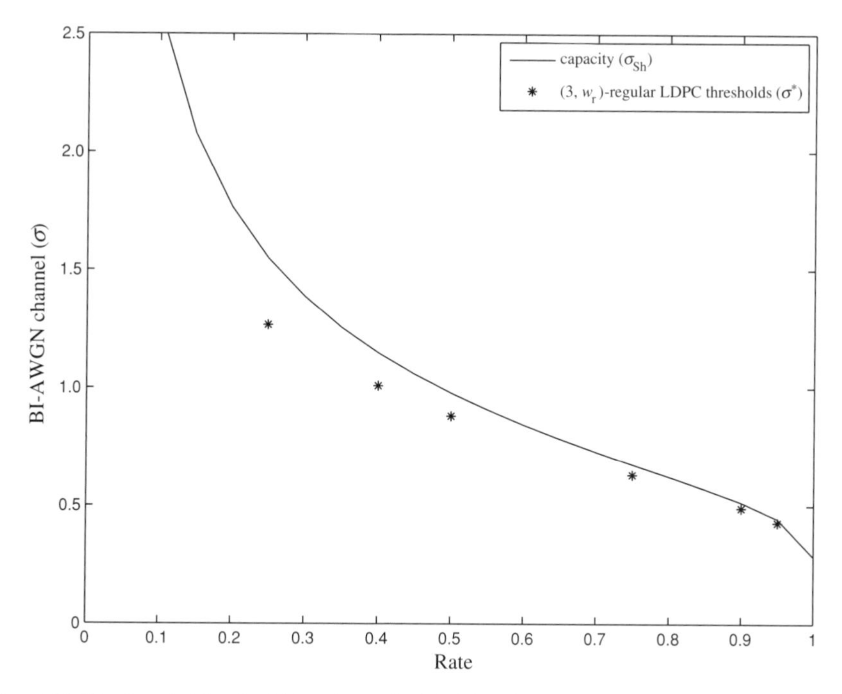

Figure 7.9 The thresholds σ^* of $(3, w_r)$-regular LDPC codes on a BI-AWGN channel and the BI-AWGN channel capacity σ_{Sh} for comparison.

However, the actual densities cannot be tracked through the BCJR decoder in the same way as they can for the sum–product decoder. In practice, Monte-Carlo simulation of the BCJR decoder is used to estimate the density functions of the extrinsic LLRs, using a histogram of the simulated LLRs. Unlike for turbo decoding, the actual extrinsic LLRs are not exchanged between the component BCJR decoders. Instead, the estimated pdf of the output LLRs for **E** from one component decoder is used to generate new input LLRs **A** for the other component decoder by sampling the estimated distribution.

By generating the input LLRs **A** in this way, we are assuming that they are an independent, noisy, observation of the source. That is, we are assuming that the extrinsic information passed between component decoders is uncorrelated with the received LLRs from the channel. This assumption can be reasonable if windowed BCJR decoding is used, so that the LLRs calculated in one iteration are independent in each window.

Because the actual component codes are simulated, a fixed code length must be chosen; however, this length must be sufficient for our assumption of uncorrelated extrinsic information to be valid. The all-zeros property of turbo decoding allows us to send only the all-zeros codeword.

The steps of density evolution for a turbo code are as follows.

(i) Choose a code length N and generate an all-ones vector $\mathbf{x}$ corresponding to the all-zeros codeword.

(ii) Fix the channel signal-to-noise ratio E_b/N_0 and create the received LLR vectors $\mathbf{R}^{(u)}$, $\mathbf{R}^{(1)}$ and $\mathbf{R}^{(2)}$:

$$R_i = \frac{2}{\sigma^2} y_i = \frac{2}{\sigma^2}(x_i + z_i),$$

where z_i is Gaussian with zero mean and variance $\sigma^2 = (2rE_b/N_0)^{-1}$ and r is the rate of the turbo code. Thus the vector $\mathbf{R}$ is generated in exactly the same way as for a Monte-Carlo simulation.

(iii) Initialize the input extrinsic information vector $\mathbf{A}^{(1)}$ to zero.

(iv) For I_{max} iterations,

(a) Run the BCJR decoder for $C^{(1)}$, with inputs $\mathbf{R}^{(u)}$, $\mathbf{R}^{(1)}$ and $\mathbf{A}^{(1)}$ and output $\mathbf{L}^{(1)}$. Calculate $\mathbf{E}^{(1)}$:

$$\mathbf{E}^{(1)} = \mathbf{L}^{(1)} - \mathbf{R}^{(u)} - \mathbf{A}^{(1)}.$$

(b) Estimate the probability distribution $p(E^{(1)})$ using a histogram of $\mathbf{E}^{(1)}$.

(c) Create an input extrinsic information vector $\mathbf{A}^{(2)}$ by sampling from the distribution $p(E^{(1)})$.

(d) Run the BCJR decoder for $C^{(2)}$, with inputs $\mathbf{R}^{(u)}$, $\mathbf{R}^{(2)}$ and $\mathbf{A}^{(2)}$ and output $\mathbf{L}^{(2)}$. Calculate $\mathbf{E}^{(2)}$:

$$\mathbf{E}^{(2)} = \mathbf{L}^{(2)} - \mathbf{R}^{(u)} - \mathbf{A}^{(2)}.$$

(e) Estimate the probability distribution $p(E^{(2)})$ using a histogram of $\mathbf{E}^{(2)}$.

(f) Create an input extrinsic information vector $\mathbf{A}^{(1)}$ by sampling from the distribution $p(E^{(2)})$.

(v) Calculate the probability of error based on the final message bit LLRs.

7.2.6 Designing ensembles with good thresholds

We have seen that the threshold of an ensemble of codes with a given degree distribution, or choice of component codes, can be found using density evolution. The question for code designers is then which degree distribution will produce the best threshold.

For turbo codes there is only a finite number of potential IIR convolutional encoders for each memory value v. An exhaustive search using density evolution can find the component codes for a given rate that return the best threshold.

Example 7.9 Density evolution applied to turbo codes in [128] on the BI-AWGN channel gives the component encoders, shown in the following table for each memory order from 2 to 6, that give the best threshold for a rate-1/2 turbo code.

Component code	σ^*
$(5, 7)$ $(v = 2)$	0.883
$(11, 13)$ $(v = 3)$	0.93
$(17, 31)$ $(v = 4)$	0.94
$(31, 45)$ $(v = 5)$	0.94
$(41, 107)$ $(v = 6)$	0.941

For LDPC and RA codes, however, trying every possible degree distribution is of course not practical and so optimization techniques are used to find the best degree distribution subject to the desired constraints. Optimizing over the density evolution algorithm is not straightforward, in particular because a gradient for the cost function is not defined. Nevertheless, two general optimization algorithms have been applied to finding the degree distributions of LDPC codes, *iterative linear programming* and the confusingly named (in our present context) *differential evolution*.

Algorithm 7.3 gives the general principle of an optimization procedure. Note that the method of generating a new candidate degree distribution is not specified. One option is a small perturbation of the best distribution found so far; another

Algorithm 7.3 Optimization

```
 1: procedure λ,ρ = DEGREE(I_max,p_limit,σ_H,σ_L,Trials)
 2:     Choose an initial λ,ρ
 3:     σ* = σ_H
 4:     λ*,ρ* = λ,ρ
 5:     for  i = 1 :Trials do
          σ = Threshold(λ,ρ,I_max,p_limit,σ_H,σ_L)
 6:       if σ* < σ then
 7:           σ* = σ
 8:           λ*,ρ* = λ,ρ
 9:       end if
10:       New λ,ρ = some perturbation of λ*,ρ*
11:     end for
12:     return λ*,ρ*,σ*.
13: end procedure
```

option is to average over the past few distributions. The channel parameter is given as σ; however, the technique extends to other binary input memoryless channels.

In practice a set of allowed degrees is usually specified (i.e. those i, j for which λ_i and ρ_j may be non-zero). Obviously, the larger the number of allowed variables in λ and ρ the longer the optimization will take. However, the possible degree distributions λ, ρ are constrained by (3.1), (3.2):

$$\sum_i \lambda_i = 1,$$

$$\sum_i \rho_i = 1,$$

and by

$$r = 1 - \frac{\sum_i \lambda_i / i}{\sum_i \rho_i / i}.$$

Thus three of the allowed λ, ρ parameters are dependent variables. Furthermore the stability constraint

$$\lambda'(0)\rho'(1) < e^{1/2\sigma^2}$$

must be satisfied, and so any candidate degree distributions that do not satisfy this constraint must be discarded.

Example 7.10 Capacity-approaching LDPC and RA codes are very long and very irregular. The famous LDPC ensemble with threshold 0.0045 dB from capacity on the BI-AWGN channel, found using density evolution, has bit degrees varying from 2 to 8000 [16]:

$$\begin{aligned}\lambda(x) = {} & 0.096294x + 0.095393x^2 + 0.033599x^5 + 0.091918x^6 \\ & + 0.031642x^{14} + 0.086563x^{19} + 0.0.0938962x^{49} + 0.006035x^{69} \\ & + 0.018375x^{99} + 0.086919x^{149} + 0.089018x^{399} + 0.057176x^{899} \\ & + 0.085816x^{1999} + 0.006163x^{2999} + 0.003028x^{5999} + 0.118165x^{7999}.\end{aligned}$$

Its average check edge degree is 18.5 and its threshold σ^* is 0.9781869.

The IRA ensemble from [111], given by

$$\begin{aligned}\lambda(x) = {} & 0.04227x + 0.16242x^2 + 0.06529x^6 + 0.06489x^7 + 0.06207x^8 \\ & + 0.01273x^9 + 0.13072x^{10} + 0.04027x^{13} + 0.00013x^{24} + 0.05410x^{25} \\ & + 0.03031x^{35} + 0.13071x^{36} + 0.10402x^{99}\end{aligned}$$

and $a = 8$, has a threshold 0.059 dB from capacity for the BI-AWGN channel.

In general, when using irregular Tanner graphs, the subset of the bit nodes having very high degrees will very quickly converge to their solution and, once correct, will pass high LLR values to their many connected nodes. Since the overall density of H needs to be low, a large proportion of degree-2 bit nodes is also required in order to reduce the average node degree while allowing some very-high-degree nodes. Thus a degree distribution with a good threshold will contain a few very-high-degree bit nodes, many degree-2 nodes, but no more than is allowed for by stability, and some nodes with degrees in between these. Irregularity in the check node degrees is not as essential, and generally one or two check degrees, chosen to achieve the required average row weight, are sufficient.

The BI-AWGN thresholds for some common iterative ensembles with rate approximately $1/2$ are given in the following table:

Code	σ^*	E_b/N_0
capacity	0.97869	0.1871
(3, 6)-regular LDPC	0.88	1.1
optimized irregular LDPC (see Example 7.10)	0.9781869	0.1916
turbo (see Example 7.9)	0.94	0.537
RA ($q = 3, a = 3$)	0.879	1.116
IRA (see Example 7.10)	0.971	0.255

Capacity-achieving ensembles

A rate-r ensemble can be called capacity achieving on the BEC if its threshold ε^* is arbitrarily close to $1 - r$ (the Shannon capacity limit of the BEC) as the code length goes to infinity.

The first capacity-achieving sequences, called heavy-tail Poisson sequences or more commonly tornado codes, have bit edge-degree distributions defined by

$$\lambda_i = \begin{cases} \dfrac{1}{H(D)(i-1)} & i = 2, \ldots, D+1, \\ 0 & \text{otherwise,} \end{cases}$$

where D is a fixed parameter, and

$$H(D) = \sum_{i=1}^{D} \frac{1}{i}.$$

The check edge degrees are defined for all $i \geq 1$ using a Poisson distribution for the function $\rho(x)$ and in practice will need to be truncated at some sufficiently large edge degree: thus

$$\rho(x) = e^{\varpi(x-1)},$$

where ϖ is the unique solution to

$$\frac{1}{\varpi}(1 - e^{-\varpi}) = \frac{1-r}{H(D)}\left(1 - \frac{1}{D+1}\right)$$

and r is the code rate. For more on these sequences, and the proof that ε^* does indeed come arbitrarily close to $1-r$ for them, see [14, 129].

Ensembles of IRA codes can also achieve the capacity of the binary erasure channel (see e.g. [119]) if the code length is allowed to increase indefinitely. While the complexity of decoding any particular ensemble is linear in the block length, the complexity of the LDPC and IRA ensembles required to reach capacity increases as capacity is approached. More specifically, if the ensemble threshold is a fraction $1-\delta$ of the BEC capacity, the decoding complexity can be shown [130] to scale with $\log(1/\delta)$. Ensembles of non-systematic doped IRA codes, accumulate–repeat–accumulate (ARA) codes and accumulate LDPC codes, however, can be shown to achieve capacity on the BEC with bounded decoding complexity (see [110] and [125]). A list of all the currently known capacity-achieving ensembles, and a comparison of their complexities, is given in [125].

It remains an open problem to find capacity-achieving ensembles for more general memoryless channels, or indeed to prove whether such ensembles can exist. Nevertheless, as we saw above, thresholds have been found for more general binary-input memoryless channels that are within a fraction of a decibel of the capacity limit. These results have earned iterative codes the title capacity-approaching for these channels.

7.2.7 Approximations to density evolution

As we saw in Example 7.6, the LLR densities of sum–product decoding on the BI-AWGN channel settle to a Gaussian-like distribution as the number of iterations increases. This trend is observed for all the iterative code ensembles and decoding algorithms that we have considered in this text.

Using this observation the complexity of finding an ensemble threshold can be greatly reduced by approximating the distributions of the iterative decoding extrinsic LLRs for the BI-AWGN channel by a Gaussian density function. Since a Gaussian pdf is completely described by its mean and variance this approximation greatly simplifies the application of density evolution, as only the mean and variance need be tracked through the message-passing decoding, not the entire pdf.

A non-Gaussian message U can be approximated by a Gaussian message U_g in a number of ways:

- the expected value of $\tanh(U_g/2)$ is taken as equal to the expected value of $\tanh(U/2)$;
- U_g is taken to have the same BER as U;
- U_g is taken to have the same signal-to-noise ratio as U;
- U_g is taken to have the same mutual information as U.

The Gaussian approximation U_g can be generated using the mean and variance measured for the actual distribution U or by measuring the mean of U and using the fact that U is symmetric, in which case the mean μ and variance σ^2 of U_g are related by

$$\sigma^2 = 2\mu.$$

Gaussian approximations can be used to find degree distributions for LDPC codes that are very close to capacity with far less computational cost. Indeed, the irregular LDPC code in Example 7.10 was found using a Gaussian approximation to DE. Once an ensemble has been found, its actual threshold can be computed using full DE.

Representing the extrinsic information transferred between component codes by a single parameter also allows an excellent visual representation of the decoder: the plot of its trajectory over each iteration. The approach of tracking the extrinsic information densities using mutual information, called extrinsic information transfer, has proven to be the most popular DE approximation technique, and so we will focus on this technique in the following.

7.3 EXIT charts

In the previous section, we saw how density evolution (DE) enabled the expected behavior of an ensemble of iteratively decoded codes to be tracked. This in turn enabled the calculation of noise thresholds for given code ensembles and paved the way for obtaining irregular degree distributions of codes capable of capacity-approaching performance. Density evolution, however, is very complex and does not easily offer a great deal of insight into the operation of the message-passing decoder. A solution to each of these issues is provided by approximating the extrinsic information passed between component decoders by a single parameter, the mutual information. A chart that tracks the mutual information at each iteration, i.e. an extrinsic information transfer (EXIT) chart, gives an excellent visual representation of the decoder.

Such charts predict the behavior of the full iterative decoder on the basis of only the simulated behavior of the individual component decoders, by making assumptions about how they will interact. Using this approach it is only necessary to simulate the behavior of each component decoder once, rather than that of the whole decoder for every iteration. As well as reducing the computations needed to examine a particular code, an EXIT chart explicitly shows the role of each component code in the overall performance, making it a valuable tool for code design.

7.3.1 Mutual information

Recall from Section 1.2.2 that the mutual information,

$$I(X;Y) = H(X) - H(X|Y),$$

between two random variables X and Y gives the amount of uncertainty in X that is removed by knowing Y. In our case X is the transmitted sequence and Y is the decoded sequence, so the more successful the decoder the more uncertainty is removed by knowing Y.

The mutual information can be calculated as

$$I(X;Y) = \sum_{x \in X} \sum_{y \in Y} p(x, y) \log_2 \frac{p(x, y)}{p(x)p(y)},$$

where $p(x, y)$ is the joint probability distribution of X and Y and $p(x)$ and $p(y)$ are the marginal probability distributions of X and Y respectively.

The mutual information of two continuous variables X and Y is

$$I(X;Y) = \iint p(x, y) \log_2 \frac{p(x, y)}{p(x)p(y)} dxdy,$$

where $p(x, y)$ is the joint probability density function of X and Y and $p(x)$ and $p(y)$ are the marginal probability density functions of X and Y respectively. When $p(x, y) = p(x)p(y)$, knowing Y reveals nothing about X and so the mutual information is zero.

For measuring the mutual information between the inputs and outputs of a BI-AWGN channel, we will be interested in a discrete-valued random variable $X \in \{-1, 1\}$ and a continuous random variable Y. The mutual information $I(X;Y)$ is then

$$I(X;Y) = \sum_{x \in \pm 1} \int p(x, y) \log_2 \frac{p(x, y)}{p(x)p(y)} dy.$$

Using Bayes' rule,

$$I(X;Y) = \sum_{x \in \pm 1} \int p(y|x)p(x) \log_2 \frac{p(y|x)p(x)}{p(x) \sum_{x' \in \pm 1} p(y|x')p(x')} dy$$

and thus

$$I(X;Y) = \sum_{x\in\pm 1} \int p(y|x)p(x)\log_2 \frac{p(y|x)}{p(y|x=1)p(x=1)+p(y|x=-1)p(x=-1)} dy. \quad (7.12)$$

In particular, for

$$y_i = \mu x_i + z_i \quad (7.13)$$

where z_i is Gaussian with mean zero and variance $\sigma^2 = 2\mu$, the conditional pdf of $\mathbf{y}$ is (1.7);

$$p(y|x) = \frac{1}{\sqrt{2\pi}\sigma} \exp\left(\frac{-1}{2\sigma^2}(y - \frac{\sigma^2}{2}x)^2\right).$$

Substituting into (7.12), we obtain

$$\begin{aligned} I(X;Y) &= J(\sigma) \\ &= 1 - \int \frac{1}{\sqrt{2\pi}\sigma} \exp\left(\frac{-1}{2\sigma^2}(y - \frac{\sigma^2}{2})^2\right) \log_2(1+e^{-y})dy. \end{aligned} \quad (7.14)$$

The expression in (7.14) can be solved by integrating numerically, or it can be approximated. Here we use the approximation from [131]:

$$J(\sigma) \approx \begin{cases} a_{J,1}\sigma^3 + b_{J,1}\sigma^2 + c_{J,1}\sigma, & 0 \le \sigma \le 1.6363, \\ 1 - \exp(a_{J,2}\sigma^3 + b_{J,2}\sigma^2 + c_{J,2}\sigma + d_{J,2}), & 1.6363 < \sigma < 10, \\ 1, & \sigma \ge 10, \end{cases} \quad (7.15)$$

where

$$\begin{aligned} &a_{J,1} = -0.0421061, \quad b_{J,1} = 0.209252, \quad c_{J,1} = -0.00640081, \\ &a_{J,2} = 0.00181491, \quad b_{J,2} = -0.142675, \quad c_{J,2} = -0.0822054, \\ &d_{J,2} = 0.0549608. \end{aligned}$$

Here $J(\sigma) = I(X;Y)$ when $y_i = \mu x_i + z_i$ and z_i is sampled from a Gaussian random variable with mean zero and variance $\sigma^2 = 2\mu$. The inverse function

$$\sigma = J^{-1}(I)$$

can be approximated by the function [131]:

$$\sigma = J^{-1}(I) \approx \begin{cases} a_{\sigma,1}I^2 + b_{\sigma,1}I + c_{\sigma,1}\sqrt{I}, & 0 \le I \le 0.3646, \\ -a_{\sigma,2}\log_e b_{\sigma,2}(1-I) - c_{\sigma,2}I, & 0.3646 < I < 1, \end{cases} \quad (7.16)$$

where

$$\begin{aligned} &a_{\sigma,1} = 1.09542, \quad b_{\sigma,1} = 0.214217, \quad c_{\sigma,1} = 2.33727, \\ &a_{\sigma,2} = 0.706692, \quad b_{\sigma,2} = 0.386013, \quad c_{\sigma,2} = -1.75017. \end{aligned}$$

7.3.2 EXIT charts for turbo codes

Recall from Section 5.3 that over each iteration of turbo decoding the extrinsic information is updated while the channel LLRs remain fixed. Then it is obvious that, for the decoder to converge, the extrinsic information should be providing improved information about the transmitted bits at each iteration. The usefulness of the extrinsic information can be measured by the mutual information $I(X; E)$ between the extrinsic information $\mathbf{E}$ and the transmitted symbols $\mathbf{x}$. That is, $I(X; E)$ quantifies the information $\mathbf{E}$ tells us about $\mathbf{x}$. For a turbo decoder to converge, we expect that each component decoder will produce extrinsic information $\mathbf{E}$ at its output with greater mutual information than that provided by the extrinsic information $\mathbf{A}$ at its input.

To measure the performance of a component decoder and determine whether it does indeed produce an increase in mutual information, an input extrinsic information vector $\mathbf{A}$ is created with known mutual information $I(X; A)$ and passed into the decoder. The decoding algorithm is run and the mutual information $I(X; E)$ corresponding to the extrinsic information output by the decoder is calculated. This process is repeated for several different extrinsic information input vectors with different mutual information content. The channel noise and hence the received LLRs are kept constant.

For example, for a transmitted vector $\mathbf{x}$ and a BI-AWGN with variance σ^2 the received LLRs are given by

$$R_i = \frac{2}{\sigma^2} y_i = \frac{2}{\sigma^2}(x_i + z_i),$$

where z_i is Gaussian with zero mean and variance σ^2. That is, the vector $\mathbf{R}$ is generated in exactly the same way as for a Monte-Carlo simulation of the decoder BER.

The a priori LLRs for the decoder are, however, generated differently for an EXIT curve than for a BER simulation. Since the whole turbo decoder is not simulated we do not have an extrinsic vector from the other component code decoder. Instead, $\mathbf{A}$ is modeled by a Gaussian random variable:

$$A_i = \mu_A x_i + n_{Ai}, \tag{7.17}$$

where n_{Ai} is Gaussian with zero mean and variance $\sigma_A^2 = 2\mu_A$. Using this model, the mutual information $I(X; A)$ can be calculated easily (see (7.14)):

$$I_A = I(X; A) = 1 - \int \exp\left(\frac{-1}{2\sigma_A{}^2}\left(y - \frac{\sigma_A{}^2}{2}\right)^2\right) \frac{\log_2(1 - e^{-y})}{\sqrt{2\pi}\sigma_A} dy. \tag{7.18}$$

By generating $\mathbf{A}$ in this way we are assuming that $\mathbf{A}$ is an independent, noisy, observation of the source, i.e. that $\mathbf{A}$ is uncorrelated with $\mathbf{R}^{(u)}$. However, for long enough codewords and sufficiently random interleavers, this assumption gives EXIT chart predictions that are very good.

Once **A** and **R** have been generated, the decoder is run in the normal way. The output LLRs **L** are used to calculate the extrinsic LLRs **E** as usual (see (5.2)). The mutual information of **E**, $I(X; E)$, is measured by estimating the pdf $p(E|x)$ from a histogram of the **E** values. A sufficiently large number of **E** values can be created by choosing a long message vector or running the decoder multiple times over many message vectors (which still need to be reasonably long). Given the pdf $p(E|x)$, the mutual information of **E** is calculated using (7.12):

$$I_E = I(X; E) = \frac{1}{2} \int_{e \in E} \sum_{x \in \pm 1} p(e|x) \log_2 \frac{2p(e|x)}{p(e|x=1) + p(e|x=-1)} de. \tag{7.19}$$

Given a variable y with a symmetric pdf, i.e. $p(y|x=1) = p(-y|x=-1)$, if f is the probability density of the log likelihood ratio $L(y)$ then $z = \log(f(z)/f(-z))$ or, equivalently, $f(z) = e^z f(-z)$.

Thus, for a symmetric density $p(E)$, the mutual information between the ith information bit and the ith extrinsic information value E_i is given by

$$I(X; E) = \mathrm{E} \log_2 \frac{2}{1 + e^{-E}},$$

where E is the expectation operator. A plot of $I(X; E)$ versus $I(X; A)$ is called the extrinsic information transfer (EXIT) curve for a component code C. Note that, for each channel signal-to-noise ratio of interest, a new EXIT curve needs to be created.

In summary, the steps to produce an EXIT curve for a component code C are:

(i) Generate an all-ones vector **x**.

(ii) Fix the channel signal-to-noise ratio E_b/N_0 and create the received LLR vectors $\mathbf{R}^{(u)}$, $\mathbf{R}^{(1)}$ and $\mathbf{R}^{(2)}$:

$$R_i = \frac{2}{\sigma^2} y_i = \frac{2}{\sigma^2}(x_i + z_i)$$

where z_i is Gaussian with zero mean and variance $\sigma^2 = (2rE_b/N_0)^{-1}$ and r is the rate of the turbo code.

(iii) For each σ_A value

(a) Create an input extrinsic information vector **A**:

$$A_i = \mu_E x_i + z_i,$$

where z_i is a Gaussian random variable with mean zero and variance $\sigma_A^2 = 2\mu_A$.

(b) Calculate the mutual information between **x** and **A**:

$$I_A = I(X; A) = 1 - \int \exp\left(\frac{-1}{2{\sigma_A}^2} \left(y - \frac{{\sigma_A}^2}{2} \right)^2 \right) \frac{\log_2(1 - e^{-y})}{\sqrt{2\pi}\sigma_A} dy.$$

(c) Run the decoder for C with inputs $\mathbf{R}^{(u)}$, $\mathbf{R}^{(p)}$ and $\mathbf{A}$ and output $\mathbf{L}$. Then calculate $\mathbf{E}$:

$$\mathbf{E} = \mathbf{L} - \mathbf{R}^{(u)} - \mathbf{A}.$$

(d) Estimate the probability distributions $p(E|x = -1)$ and $p(E|x = 1)$ using a histogram of $\mathbf{E}$.

(e) Calculate the mutual information between $\mathbf{x}$ and $\mathbf{E}$:

$$I_E = I(X; E) = \frac{1}{2} \int_{e \in E} \sum_{x \in \pm 1} p(e|x) \log_2 \frac{2p(e|x)}{p(e|x = 1) + p(e|x = -1)} de.$$

(iv) Plot I_E versus I_A.

An EXIT curve for the second component code is calculated in exactly the same way. In fact if the turbo code uses the same component codes then the Code 1 and Code 2 EXIT curves are identical.

Example 7.11 In this example we consider the turbo code presented in Example 5.2. The first component code is the rate-1/2 systematic recursive convolutional code with generator matrix

$$G = \begin{bmatrix} 1 & \dfrac{1 + D^4}{1 + D + D^2 + D^3 + D^4} \end{bmatrix},$$

punctured with the puncturing pattern

$$P = \begin{bmatrix} 1 & 1 \\ 1 & 0 \end{bmatrix}.$$

We will plot the EXIT curve for this code on a BI-AWGN channel with noise variance $\sigma^2 = 0.7942$. A very short message, 10 bits, will be considered and only three points, corresponding to $\sigma_A = 0.01$, 2 and 8, so our results will be completely useless for anything other than as a demonstration of the techniques.

To start we transmit the vector

$$\mathbf{x} = [1\ 1; 1\ 1; 1\ 1; 1\ 1; 1\ 1; 1\ 1; 1\ 1; 1\ 1; 1\ 1; 1\ 1]$$

over an additive Gaussian channel with variance $\sigma^2 = 0.7942$ and obtain the received LLRs

$$\mathbf{R}^{(u)} = [4.1690\ \ 0.8204\ \ 0.6046\ \ 2.9908\ \ -1.9396$$
$$6.9811\ \ -2.3446\ \ 3.3590\ \ 2.0770\ \ 2.9952],$$
$$\mathbf{R}^{(p)} = [1.6099\ \ 0\ \ 1.6181\ \ 0\ \ 2.2966\ \ 0\ \ 2.8677\ \ 0\ \ 3.5903\ \ 0].$$

For our first value, $\sigma_A = 0.01$, we randomly generate a priori LLRs using (7.17):

$$\mathbf{A} = [0.0165 \quad -0.0086 \quad -0.0178 \;\; 0.0111 \;\; 0.0233$$
$$0.0029 \;\; 0.0102 \quad -0.0055 \;\; 0.0096 \quad -0.0151].$$

We now use an adaptive Simpson quadrature to evaluate numerically the integral (7.18) between $-6\sigma_A$ and $6\sigma_A$ to within an error of 1×10^{-6} (using the QUAD function in Matlab), to give $I_A = 1.8 \times 10^{-5}$. Passing $\mathbf{R}$ and $\mathbf{A}$ into the decoder (in this case Algorithm 4.2 since the component code is a convolutional code) gives

$$\mathbf{L} = [4.6848 \;\; 0.0186 \quad -0.7490 \;\; 1.3373 \quad -0.5358$$
$$6.2138 \quad -1.3772 \;\; 3.3292 \;\; 2.0576 \;\; 2.9801].$$

The extrinsic LLRs $\mathbf{E}$ are then given by

$$\mathbf{E} = \mathbf{L} - \mathbf{R}^{(u)} - \mathbf{A} = [0.4993 \quad -0.7932 \quad -1.3359 \quad -1.6646 \;\; 1.3804$$
$$-0.7703 \;\; 0.9571 \quad -0.0243 \quad -0.0290 \;\; 0].$$

Next, a histogram of $\mathbf{E}$ with bins $-5 : 0.5 : 5$ gives non-zero bin values as follows:

$$\begin{array}{lll} H(-1.5) = 2, & H(-1) = 2, & H(0) \;\; = 3, \\ H(0.5) = 1, & H(1) = 1, & H(1.5) = 1. \end{array}$$

So, our estimated pdf is given by

$$\begin{array}{lll} p(E = -1.5|-1) = 0.2, & p(E = -1|-1) = 0.2, & p(E = 0|-1) = 0.3, \\ p(E = 0.5|-1) = 0.1, & p(E = 1|-1) = 0.1, & p(E = 1.5|-1) = 0.1, \end{array}$$
$$p(E|-1) = 0 \;\; \text{otherwise}.$$

Substituting into (7.19) gives $I_E = 0.1490$.

Repeating for $\sigma_A = 2$ gives

$$\mathbf{A} = [1.0095 \;\; 7.3359 \;\; 1.2812 \quad -0.2829 \quad -2.5078$$
$$2.8348 \;\; 0.1499 \quad -0.8421 \;\; 2.5707 \;\; 3.5799],$$

$$I_A = 0.4859,$$

$$\mathbf{E} = [-0.8853 \quad -0.9353 \quad -1.8271 \quad -1.9455 \;\; 3.0121$$
$$-1.1274 \;\; 1.5119 \quad -1.2985 \quad -1.0860 \quad -8.8818e^{-16}],$$

$$I_E = 0.7.$$

Algorithm 7.4 EXIT curve

1: **procedure** EXIT CURVE(K,r,σ,$\mathbf{I}_A$,$\mathbf{X}$)
2:
3: $N = K/r$
4: **for** $i = 1 : K$ **do** ▷ Create the log likelihood ratios from the channel
5: $R_i^{(u)} = \frac{2}{\sigma^2}\left(1 + \text{AWGN}(0, \sigma^2)\right)$
6: **end for**
7: **for** $i = 1 : N - K$ **do**
8: $R_i^{(p)} = \frac{2}{\sigma^2}\left(1 + \text{AWGN}(0, \sigma^2)\right)$
9: **end for**
10:
11: **for** $j = 1 : \text{length}(\mathbf{I}_A)$ **do** ▷ Loop over the input MI values
12: $\sigma_A = J^{-1}(I_{A_j})$
13:
14: **for** $i = 1 : N$ **do** ▷ Create the vector of a priori LLRs
15: $A(i) = \frac{\sigma_A^2}{2} + \text{AWGN}(0, \sigma_A^2)$
16: **end for**
17:
18: $\mathbf{L} = \text{logDecode}(\mathbf{R}^{(u)},\mathbf{R}^{(\mathbf{p})},\mathbf{A})$ ▷ Run the decoder
19: $\mathbf{E} = L - \mathbf{R}^{(u)} - \mathbf{A}$
20:
21: $l = \text{length}(X)$ ▷ Estimate the pdf of $\mathbf{E}$
22: $H = \text{hist}(\mathbf{E},X)$
23: **for** $i = 1 : l$ **do**
24: $p_i = H_i/\text{sum}(H)$
25: **end for**
26:
27: $I_{E_j} = 0$ ▷ Calculate the MI for $\mathbf{E}$
28: **for** $i = 1 : l$ **do**
29: $I_{E_j} = I_{E_j} + \frac{1}{2} p_i \log_2 \frac{2p_i}{p_i + p_{l-i}} + \frac{1}{2} p_{l-i} \log_2 \frac{2p_{l-i}}{p_i + p_{l-i}}$
30: **end for**
31: **end for**
32: **end procedure** output $\mathbf{I}_E$

Repeating for $\sigma_A = 8$ gives

$$\mathbf{A} = [39.6920\ \ 24.1970\ \ 29.2160\ \ 29.3510\ \ 26.6970$$

$$25.4660\ \ 20.0610\ \ 33.9620\ \ 18.2210\ \ 26.3610],$$

$$I_A = 0.9999,$$

$$\mathbf{E} = [6.7742\ \ 1.6181\ \ 8.0761\ \ 2.2966\ \ 5.8869$$

$$2.8677\ \ 2.8677\ \ 3.5903\ \ 3.5903\ \ -0.0001],$$

$$I_E = 0.9.$$

Obviously 10 bits are not enough to obtain very good average mutual information. Indeed, with regard to the last result, the average output mutual information has been calculated to be less than the average input mutual information, which is not actually the case for this channel and this code.

Algorithm 7.4 shows the steps needed to create an EXIT curve for a given component code with rate r and message length K and a BI-AWGN channel with variance σ^2. Rather than inputting a set of σ_A values, the algorithm takes as input a vector $\mathbf{I}_A$ of I_A values and calculates σ_A for each, using (7.16). This enables user control of the spacing of x-axis values on the EXIT chart. The integral (7.19) is estimated using a histogram with bins defined in $\mathbf{X}$.

Example 7.12 Considering again the code from Example 7.11 and repeating the simulation with a message length of 200 000 bits, a channel variance of $\sigma^2 = 0.7942$ and a wider range of input mutual information values gives

$$\mathbf{I}_A = [1.7921e^{-5}\ \ 0.0526\ \ 0.0997\ \ 0.1466\ \ 0.2024\ \ 0.2530\ \ 0.3024$$

$$0.3460\ \ 0.3965\ \ 0.4532\ \ 0.5021\ \ 0.5493\ \ 0.5945\ \ 0.6509$$

$$0.7026\ \ 0.7469\ \ 0.8001\ \ 0.8508\ \ 0.9025\ \ 1.0000],$$

$$\mathbf{I}_E = [0.1760\ \ 0.1963\ \ 0.2287\ \ 0.2688\ \ 0.3118\ \ 0.3636\ \ 0.4073$$

$$0.4537\ \ 0.5114\ \ 0.5599\ \ 0.6085\ \ 0.6595\ \ 0.7128\ \ 0.7685$$

$$0.8101\ \ 0.8484\ \ 0.8879\ \ 0.9152\ \ 0.9485\ \ 1.0000].$$

The corresponding EXIT curve is plotted in Figure 7.10 as the solid curve. Repeating for different channel noise variances gives the second and third curves in Figure 7.10.

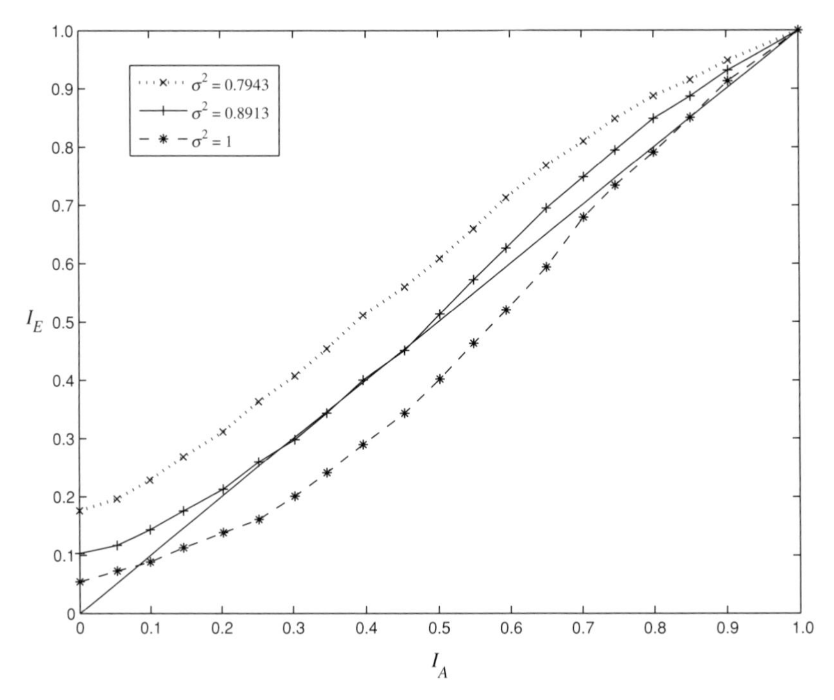

Figure 7.10 EXIT curves for component code of a rate-1/2 turbo code, the punctured convolutional code $[1\ (1 + D^4)/(1 + D + D^2 + D^3 + D^4)]$, on a BI-AWGN channel with varying noise levels.

The final observation required for the EXIT chart is that the output extrinsic information of the first decoder becomes the input extrinsic information for the second decoder ($I^{(2)}_{A,l} = I^{(1)}_{E,l}$) and vice versa ($I^{(1)}_{A,l+1} = I^{(2)}_{E,l}$) (see e.g. Figure 5.1). (Interleaving changes the order of the bits but does not change the average mutual information.) Thus the two EXIT curves can be used to track the change in mutual information over the decoding iterations of the turbo decoder.

An EXIT chart for a turbo code is then simply the plot of EXIT curves for the two component codes. For the first component code, the EXIT curve is plotted in exactly the same way as in Figure 7.10, that is $I^{(1)}_A$ on the x-axis and $I^{(1)}_E$ on the y-axis. For the second component code $I^{(2)}_A$ is plotted on the y-axis and $I^{(2)}_E$ on the x-axis.

Whereas density evolution generates input LLRs for the next component decoder from the pdf of the output LLRs from the previous component decoder, EXIT analysis generates input for the next component decoder with the same mutual information as the output LLRs from the previous component decoder (by assuming a Gaussian pdf). This allows the mutual information transfer function to be simulated separately for each component decoder.

Example 7.13 Consider again the rate-$1/2$ turbo code used in Examples 7.11 and 7.12, simulated on a BI-AWGN channel with noise variance $\sigma^2 = 0.7943$ ($E_b/N_0 = 1$ dB). The code is rate-$1/2$ and so the signal-to-noise ratio is 1 dB. The component codes are the same, so we can re-use the results from Example 7.12 for both. (Although the puncturing pattern is different, the same number of parity bits is punctured for both, so the mutual information is the same.) Figure 7.11 shows the mutual information transfer of the first component code in the upper solid curve, plotted with $I_A^{(1)}$ on the x-axis and $I_E^{(1)}$ on the y-axis, and the mutual information transfer of the second component code in the lower solid curve, plotted with $I_A^{(2)}$ on the y-axis and $I_E^{(2)}$ on the x-axis.

To track the turbo decoder's performance using the EXIT chart, we start on the x-axis with $I_{A,1}^{(1)} = 0$ and read off the Code-1 curve that $I_{E,1}^{(1)} = 0.176$. Thus $I_{A,1}^{(2)} = 0.176$, and reading off the Code-2 curve gives $I_{E,1}^{(2)} = 0.29$. For the second iteration $I_{A,2}^{(2)} = 0.29$, and we can read off the Code-1 curve that $I_{E,2}^{(1)} = 0.39$. Thus $I_{A,2}^{(2)} = 0.39$ and reading off the Code-2 curve gives $I_{E,2}^{(2)} = 0.5$. Repeating this for a further six iterations, the mutual information output reaches 1 and we say that the decoder has successfully converged.

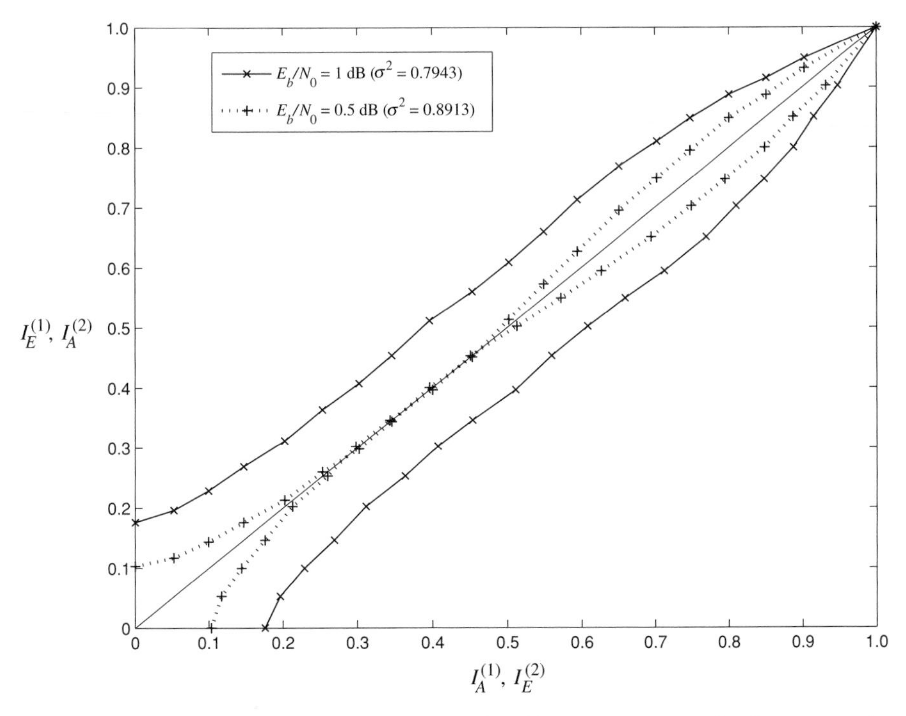

Figure 7.11 EXIT charts for the rate-$1/2$ turbo code with as component codes the punctured convolutional code $[1\ (1 + D^4)/(1 + D + D^2 + D^3 + D^4)]$ on a BI-AWGN channel with signal-to-noise ratios 1 dB and 0.5 dB.

Lastly, we produce a new EXIT curve for the same turbo code but on a BI-AWGN channel with $\sigma^2 = 0.8913$ (signal-to-noise ratio 0.5 dB). The broken curves in Figure 7.11 show this EXIT chart. At this signal-to-noise ratio the EXIT curves cross and the decoder does not converge. We can thus predict that the BER performance for this code will be poor for channels with signal-to-noise ratios around 0.5 dB.

Example 7.13 demonstrates a key benefit of EXIT charts. Since they track the extrinsic information passed between component decoders by a single parameter, the mutual information, a chart of its progress gives an excellent visual representation of the decoder. This makes it easy to see why the iterative decoder in Example 7.13 fails to converge for SNRs lower than 0.5 dB. The EXIT curves show that for some inputs the component codes output extrinsic LLRS with lower mutual information than those given by the a priori input. This causes the two curves to cross and there is no way in which the turbo decoder can improve the mutual information further.

Exit charts for serially concatenated codes

Generating EXIT charts for serially concatenated codes follows the same process as for parallel concatenated codes. The EXIT curves for the component codes are calculated using Algorithm 7.4, and the extrinsic information from the outer decoder becomes the a priori information for the inner decoder and vice versa. The EXIT chart for a serially concatenated code plots $I_E^{(\mathrm{I})}$ versus $I_A^{(\mathrm{I})}$ and $I_A^{(\mathrm{O})}$ versus $I_E^{(\mathrm{O})}$ (see Figure 6.1 for the relationship between $\mathbf{E}^{(\mathrm{I})}$, $\mathbf{A}^{(\mathrm{I})}$, $\mathbf{A}^{(\mathrm{O})}$ and $\mathbf{E}^{(\mathrm{O})}$). Where the serially concatenated codes are convolutional codes using log BCJR decoding, their EXIT curves can be calculated using Algorithm 7.4 for the inner decoder and a slight modification of this algorithm that takes into account the lack of received LLRs and uses different output LLRs, for the outer decoder. The EXIT curves for other decoders are just as straightforward; all that is required is to simulate the decoder using inputs with known mutual information (see (7.14)) and then calculate the mutual information of the output using (7.12).

Example 7.14 Suppose that we have a serially concatenated turbo code with a rate-1 inner convolutional code $1/(1 + D)$ and two alternative rate-1/2 convolutional codes,

$$\begin{bmatrix} 1 & \dfrac{1 + D^4}{1 + D + D^2 + D^3 + D^4} \end{bmatrix} \quad \text{and} \quad \begin{bmatrix} 1 & \dfrac{1 + D}{1 + D + D^2} \end{bmatrix}$$

for the outer code. Figure 7.12 shows the mutual information transfer of the inner component code as the top curve, plotted with $I_A^{(\mathrm{I})}$ on the x-axis and $I_E^{(\mathrm{I})}$ on the y-axis, and the mutual information transfer of the two alternative

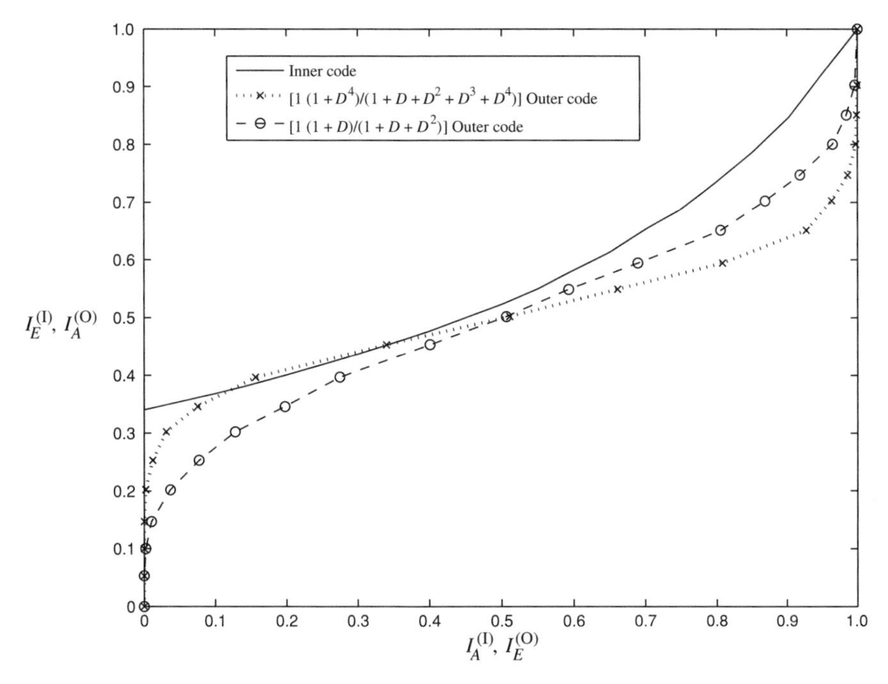

Figure 7.12 The EXIT curves for a serially concatenated turbo code with inner component code $1/(1+D)$ and two possible outer component codes, $[1\ (1+D^4)/(1+D+D^2+D^3+D^4)]$ and $[1\ (1+D)/(1+D+D^2)]$, on a BI-AWGN channel with noise variance $\sigma^2 = 0.7762$.

outer component codes as the middle and bottom curves, plotted with $I_A^{(O)}$ on the y-axis and $I_E^{(O)}$ on the x-axis. The EXIT curves were calculated for a channel noise variance $\sigma^2 = 0.7762$ and, since the code is rate-1/2, the signal-to-noise ratio is 1.1 dB. At this signal-to-noise ratio the EXIT curve for the $[1\ (1+D^4)/(1+D+D^2+D^3+D^4)]$ outer code crosses the EXIT curve of the inner code, suggesting that the decoder will not converge. For this inner code, the better choice of outer component code is thus the $[1\ (1+D)/(1+D+D^2)]$ convolutional code, which does not cross the EXIT curve of the inner code.

7.3.3 EXIT charts for RA codes

The EXIT charts for repeat–accumulate codes are obtained by the same process as those for serially concatenated codes, but now the simulation of the inner decoder requires simulation of the accumulator and parity-check decoder together (see Algorithm 6.2). The EXIT chart for an RA code plots $I_E^{(d)}$ versus $I_A^{(d)}$ and $I_A^{(v)}$ versus $I_E^{(v)}$ (see Figure 6.5 for the relationship between **u**, **d**, **s**, **p**, $\mathbf{E}^{(d)}$, $\mathbf{A}^{(d)}$, $\mathbf{A}^{(v)}$ and $\mathbf{E}^{(v)}$).

An EXIT curve for the accumulator–combiner decoder can be created using Algorithm 7.4 and substituting the accumulator–combiner decoder from Algorithm 6.2 for the decoder. Similarly, an EXIT curve for the repetition code decoder can be created using Algorithm 7.4 and substituting the repetition decoder from Algorithm 6.2 for the decoder.

However, it is not in fact necessary to simulate the repetition decoder to determine its EXIT curve. Recall (see (6.6)) that the repetition code decoder calculates the LLR of a given bit in $\mathbf{v}$ using a priori LLRs in $\mathbf{A}^{(v)}$ for the $q-1$ other bits in $\mathbf{v}$ that are copies of the same message bit. Since each $E_i^{(v)}$ is the sum of $q-1$ LLRs with known variance σ_A^2, its mutual information is, (7.14),

$$I_E^{(v)} = J\left(\sqrt{(q-1)\sigma_A^2}\right). \tag{7.20}$$

For a systematic code each $E_i^{(v)}$ is the sum of $q-1$ LLRs with known variance σ_A^2 and one LLR from the channel. Recall that the channel LLRs are

$$R_i^{(u)} = \frac{2}{\sigma^2}y_i = \frac{2}{\sigma^2}x_i + \frac{2}{\sigma^2}z_i,$$

where z_i is a Gaussian random variable with mean zero and variance σ^2. The received LLRs thus take the form (7.13), with variance

$$\sigma_R^2 = \frac{4}{\sigma^2}.$$

The mutual information of $\mathbf{E}^{(v)}$ for a systematic code is thus

$$I_E^{(v)} = J\left(\sqrt{(q-1)\sigma_A^2 + \sigma_R^2}\right). \tag{7.21}$$

EXIT charts for IRA codes

An IRA code uses multiple repetition codes and combiners with varying rates (equivalently, it has bit and check nodes with varying degrees). In Section 6.2.4 we designated the fraction of bits in $\mathbf{u}$ encoded by a rate-$1/i$ repetition code (equivalently, the fraction of systematic-bit nodes in the Tanner graph with degree i) by v_i and the fraction of interleaved bits in $\mathbf{s}$ encoded by a rate-i accumulator by h_i. Recall that the check nodes connect to i systematic-bit nodes through the interleaver but have total degree $i+2$; the extra two edges connect to the parity-bit nodes (see Figures 6.5 and 6.7). For the edge-degree distribution, the fraction of edges that are connected to degree-i bit nodes (equivalently, the fraction of bits in $\mathbf{v}$ generated by a rate-$1/i$ repetition code) is denoted λ_i, and the fraction of edges that are connected to degree-i check nodes (equivalently, the fraction of bits in $\mathbf{d}$ encoded by a rate-i combiner) is denoted ρ_i. Translating between node degrees and edge degrees, we obtain

$$v_i = \frac{\lambda_i/i}{\sum_j \lambda_j/j},$$

$$h_i = \frac{\rho_i/i}{\sum_j \rho_j/j}.$$

For a systematic IRA code, the output of the repetition decoder corresponding to a degree-d_v bit node (i.e. a rate-$1/d_v$ repetition code) is the sum of $d_v - 1$ LLRs with variance σ_A^2 and one LLR with known variance σ_R^2; thus its mutual information is

$$I_{E,d_v}^{(v)} = J\left(\sqrt{(d_v - 1)\sigma_A^2 + \sigma_R^2}\right).$$

Since a fraction λ_i of the bits in $\mathbf{v}$ were produced by a rate-$1/i$ repetition code, the fraction λ_i of the LLRs in $\mathbf{E}^{(v)}$ will have mutual information $I_{E,i}^{(v)}$. The average mutual information of N variables where λ_i/N of them have mutual information I_i is

$$\sum_i \lambda_i I_i. \tag{7.22}$$

The EXIT curve of a mixture of codes is thus an average of the component EXIT curves. For a repetition code with bit-degree distribution λ the average mutual information of the output extrinsic information is

$$I_E^{(v)} = \sum_i \lambda_i J\left(\sqrt{(i - 1)\sigma_A^2 + \sigma_R^2}\right).$$

Similarly, for non-systematic repetition codes,

$$I_E^{(v)} = \sum_i \lambda_i J\left(\sqrt{(i - 1)\sigma_A^2}\right).$$

A corresponding argument holds for variable-rate combiners, and so

$$I_E^{(d)} = \sum_i \rho_i I_{E,i}^{(d)},$$

where each $I_{E,i}^{(d)}$ must be computed separately. Equation (7.22) also applies equally to irregular turbo codes, which can be described by EXIT curves in the same way.

Example 7.15 The rate-1/2 IRA code in [131, Figure 8] has parameters $h_6 = 1$, $v_2 = 0.063$, $v_3 = 0.631$ and $v_{13} = 0.306$. Translating to edge degrees gives $\rho_6 = 1$, $\lambda_2 = 0.0210$, $\lambda_3 = 0.3157$ and $\lambda_{13} = 0.6633$. Figure 7.13 shows the mutual information transfer on a BI-AWGN channel with signal-to-noise ratio 0.5 dB ($\sigma^2 = 0.8913$) of the combined accumulator and combiner (solid curve), plotted with $I_A^{(d)}$ on the x-axis and $I_E^{(d)}$ on the y-axis, and the mutual information transfer curves of three different repetition codes with $q = 2, q = 3$ and $q = 13$, plotted with $I_A^{(v)}$ on the y-axis and $I_E^{(v)}$ on the x-axis. The irregular-repetition-code EXIT curve, i.e. the weighted sum of the three regular curves, is also shown.

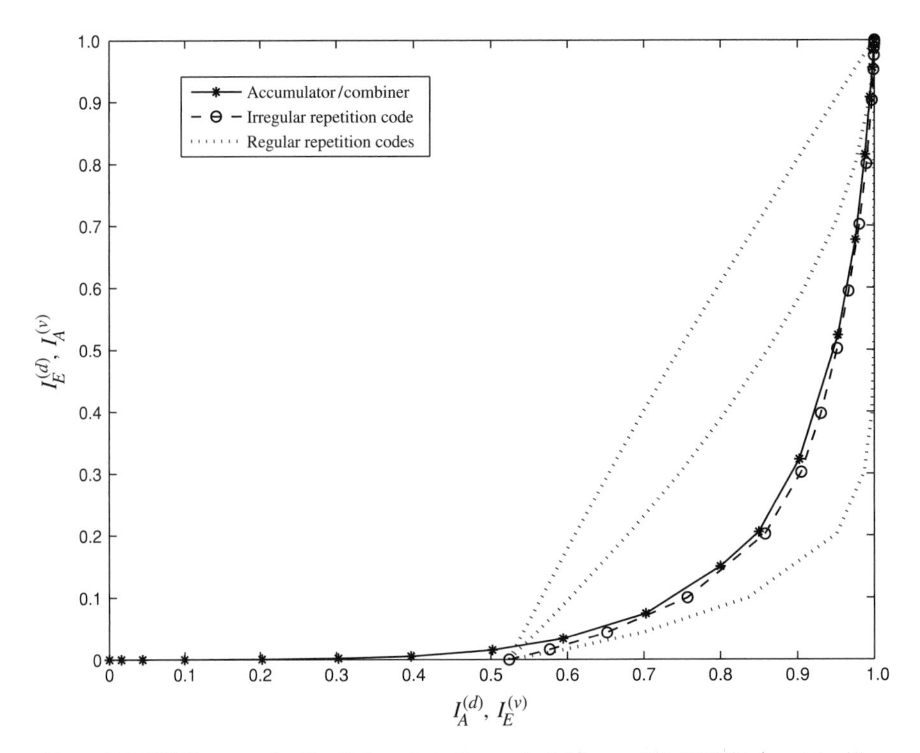

Figure 7.13 EXIT curves for the IRA code in Example 7.15, on a BI-AWGN channel with signal-to-noise ratio 0.5 dB.

This example illustrates how the degrees of an irregular code can be carefully chosen to match the two component curves of the IRA code.

Note that the sum–product decoding and turbo decoding of an RA code are not the same. In an LDPC decoder the parity-bit nodes are decoded in exactly the same way as the systematic-bit nodes (i.e. as for a repetition code). The turbo decoder, however, uses BCJR decoding on a trellis to decode the parity bits; thus the accumulator decoder shows the mutual information transfer of BCJR decoding. Because of this, the EXIT chart derived in Example 7.15 does not show the mutual information transferred when using sum–product decoding for the RA code. The following subsection shows how to generate the EXIT curve for a sum–product decoder.

7.3.4 EXIT charts for LDPC codes

The dual nature of RA codes suggests a close relationship between LDPC codes and turbo codes. Indeed, insights gained from RA codes show how an LDPC code decoder can be considered as a serial concatenation of a repetition-code

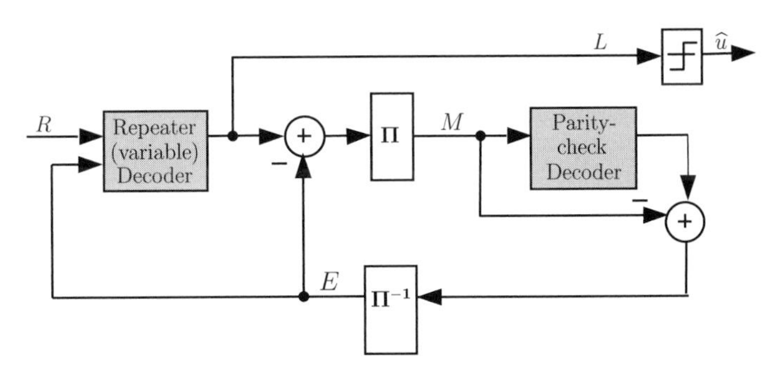

Figure 7.14 Iterative decoding of an LDPC code.

decoder and a parity-check decoder.[1] Figure 7.14 shows the iterative decoding of an LDPC code. Viewed from this perspective the two component decoders are a repetition-code decoder, also called a variable-node decoder (VND) and identical to the repetition decoder for a systematic RA code, and a parity-check decoder, also called a check node decoder (CND) and similar to the parity-check decoder of an RA code.

Generating the EXIT curve for the LDPC repetition decoder is exactly the same as generating the EXIT curve for the repetition decoder of a systematic RA code (i.e. (7.21) is used with q replaced by w_c).

The parity-check decoder in an LDPC code does not have access to channel LLRs. Instead, the jth parity-check decoder receives w_r a priori LLRs $M_{i,j}$ from the repetition decoder, one for each bit involved in that parity-check equation, and outputs w_r extrinsic LLRs, one for each of these bits,

$$E_{j,i} = \log \frac{1 + \prod_{i' \neq i} f(M_{i',j})}{1 - \prod_{i' \neq i} f(M_{i',j})}, \tag{7.23}$$

where the jth parity-check equation includes the ith codeword bit and

$$f(x) = \frac{1 - e^{-x}}{1 + e^{-x}}.$$

The EXIT curve for the LDPC check decoder can be generated using Algorithm 7.4 with (7.23) substituted for the decoder.

Alternatively, the mutual information of the extrinsic information $Ic_E(I_A; w_\mathrm{r})$ output by a parity-check decoder with w_r input LLRs having mutual information I_A can be approximated using the EXIT curve $Ir_E(1 - I_A, w_\mathrm{r})$ of the rate-$1/w_\mathrm{r}$ repetition code decoder having w_r input LLRs with mutual information $1 - I_A$:

$$Ic_E(I_A; w_\mathrm{r}) \approx 1 - Ir_E(1 - I_A, w_\mathrm{r}). \tag{7.24}$$

The extrinsic information $Ic_E(I_A; w_\mathrm{r})$ is thus easily computed using (7.20). This relationship is only exact for the binary erasure channel; nevertheless the result is

[1] Note that this is not true, in general, for the *encoders* of LDPC codes.

still very accurate for Gaussian a priori inputs. Note that this approximation could be used for the parity-check decoder in RA codes. However, the accumulator decoder would still require simulation, reducing the benefits of the approximation in this case.

It has also been shown that the mutual information transfer of the parity-check decoder can be described by

$$I_E = \frac{1}{\log 2} \sum_{i=1}^{\infty} \frac{1}{2i(2i-1)} \left(\Phi_i(J^{-1} I_A) \right)^{w_r - 1}, \tag{7.25}$$

where

$$\Phi_i(x) = \int_{-1}^{+1} \frac{2t^{2i}}{(1-t^2)\sqrt{4\pi x}} \exp \frac{(\log(1+t)/(1-t) - x)^2}{4x} dt.$$

Using (7.25), I_E can be approximated with as much accuracy as required by increasing the number of terms in the sum that are used in the calculation. (See [132] for more details.)

Example 7.16 In Example 7.8 we saw using density evolution that for infinite-length cycle-free regular LDPC codes, a column weight 3 produces the best thresholds. Figure 7.15 shows the EXIT chart for three regular LDPC codes at

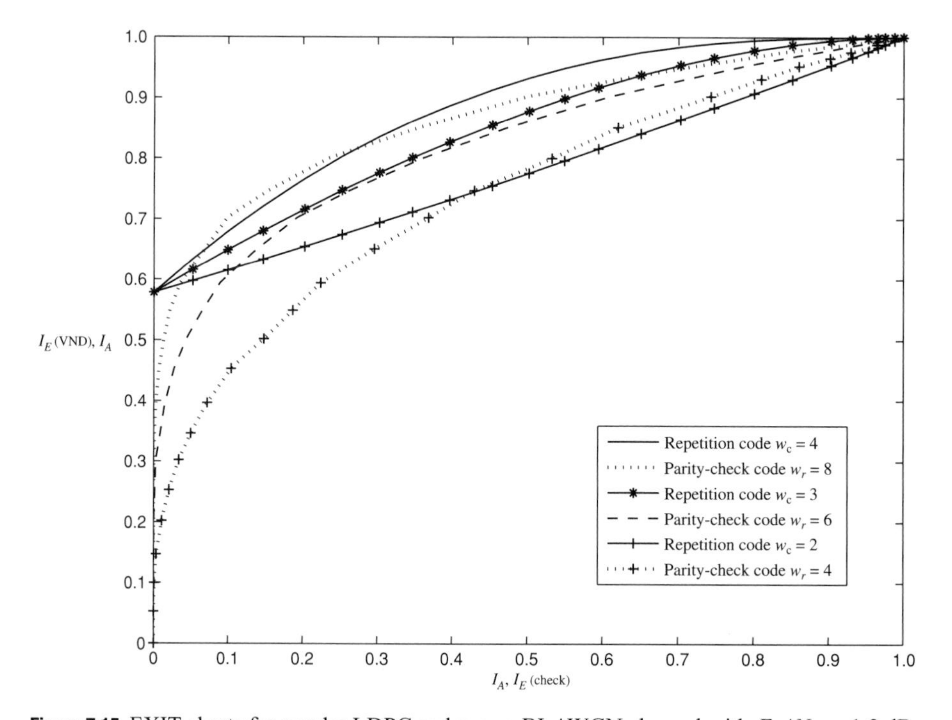

Figure 7.15 EXIT charts for regular LDPC codes on a BI-AWGN channel with $E_b/N_0 = 1.2$ dB.

the same noise variance, $\sigma^2 = 0.75858$ ($E_b/N_0 = 1.2$ dB). The EXIT curves show that the (3, 6)-regular LDPC code has the lowest threshold of the three, owing to a better fit between the EXIT curves of its component codes.

As for IRA codes, irregular LDPC codes can be designed to have high thresholds by choosing the degree distribution to match the EXIT curves of the bit and check nodes closely.

7.3.5 Code design and analysis using EXIT charts

As we have seen, for a given code and channel, EXIT charts can be used to determine whether the turbo decoder is likely to decode correctly, by examination of whether the EXIT curves of the component decoders cross.

By decreasing the signal-to-noise ratio (increasing the noise variance) and finding the point where the EXIT curves cross we can estimate the ensemble threshold. In Example 7.13 we saw that the EXIT curves touched at 0.5 dB, so the ensemble threshold must be at around 0.5 dB.

The EXIT curve also gives an indication of the decoding speed of the turbo decoder. The wider the gap between component codes, the fewer the number of iterations required for convergence. This property can be used to inform code design, since it implies that component codes that have sufficiently separated EXIT curves for a given channel should be chosen. Example 7.14 showed how EXIT curves can be used to design component codes in this way.

Extrinsic information transfer curves can also be used to design capacity-approaching codes. It has been shown that, in order to approach capacity on the binary erasure channel, the component transfer curves must be exactly matched. Empirically, this is also the case for more general channels. A capacity-approaching code will thus require component codes that are as closely matched as possible without actually crossing. Example 7.15 showed how EXIT curves can be used to design component codes that closely match the component EXIT curves and enable convergence at the lowest possible signal-to-noise ratio.

7.4 Bibliographic notes

The type of analysis of message-passing decoding that we now call density evolution (DE) first appeared for regular codes in Gallager's work [10]. For irregular codes, density evolution was first proposed in [14, 133, 134] for the binary erasure channel, applied to hard decision message-passing decoding in [15] and generalized to sum–product decoding on memoryless channels in [127, 135]. Density evolution was first applied to turbo codes in [128]; a good reference is [136]. See [111] for density evolution applied to IRA codes. Online implementations of density evolution can be found at [137] and [138].

Irregular LDPC and RA codes have been shown to asymptotically achieve the capacity of the binary erasure channel [14, 119, 134]. However, the binary erasure channel (BEC) is the only channel for which such a result exists; no such result exists for turbo codes. Nevertheless, using density evolution, thresholds have been found for iterative codes on other binary input memoryless channels that are within a fraction of a decibel of Shannon's capacity limit, earning iterative codes the title capacity-approaching codes.

A number of approaches to the Gaussian approximation of density evolution have been proposed, based on mean LLRs [139], SNRs [140], BERs [141] or mutual information [142]. The advantage of tracking the mutual information is that it gives an easy-to-visualize graphical representation of the decoder, i.e. the extrinsic information transfer (EXIT) chart. Extrinsic information transfer charts were first suggested by ten Brink in 1999 [142] for the analysis of an iterative decoding and demapping scheme. The principle was soon extended to turbo codes [143], RA codes [144] and LDPC codes [131, 132].

In the case of the binary erasure channel, EXIT chart analysis is equivalent to full DE analysis since the densities of iterative decoding for erasures are completely determined by a single parameter. Indeed, analytical expressions are known for the EXIT curves of iterative erasure decoders and, for the binary erasure channel, it has been proved that capacity-approaching codes can be designed by matching the EXIT curves of the component codes [145]. Much less is known about EXIT curves for general binary symmetric memoryless channels. The standard approach is to use a Monte-Carlo simulation (such as Algorithm 7.4) to estimate the EXIT curves, or to approximate the EXIT curves using results from the BEC (as we did for the LDPC check decoder in (7.24)). More recent work has shown that in some cases the EXIT function can be approximated as a series having BEC EXIT functions as its terms [132] (for example (7.25) has been used for the LDPC check decoder). Empirical evidence certainly suggests that designing iterative codes through curve matching also produces capacity-approaching codes on more general memoryless channels.

The density evolution techniques described in this chapter determine, in the case of very long codes, the choice of component codes that gives the best decoding thresholds. Only infinite-length codes have been considered, and the analysis assumed codes that are free of cycles. Nevertheless, density evolution provides the value of the channel-noise standard deviation below which it is possible to decode with zero error using the sum–product algorithm, given enough iterations and a long enough code. Thus, for very long randomly constructed codes, density evolution gives an accurate prediction of the noise value of the waterfall part of the BER curve. Understanding and predicting the performance of iterative decoding at higher noise levels is the topic of the following chapter.

7.5 Exercises

7.1 A code from the (4, 8)-regular LDPC ensemble is to be transmitted on a binary erasure channel with erasure probability $\varepsilon = 0.3$ and decoded with message-passing iterative decoding. What is the probability that a codeword bit will remain erased after five iterations of message-passing decoding if the code's Tanner graph is free of cycles of size 10 or less?

7.2 An irregular LDPC ensemble with degree distribution

$$v_2 = 0.3, \quad v_3 = 0.4, \quad v_4 = 0.3$$

and

$$h_6 = 1$$

is to be transmitted on a binary erasure channel with erasure probability $\varepsilon = 0.3$ and decoded with message-passing iterative decoding. What is the expected probability that a codeword bit will remain erased after four iterations of message-passing decoding if the code Tanner graph is free of cycles of size 8 or less?

7.3 Would you expect a code from the ensemble in Example 7.2 to be able to decode correctly a codeword transmitted on a BEC with erasure probability 0.35, given an unlimited number of iterations and a cycle-free graph?

7.4 Estimate the BEC threshold of the two ensembles

$$\lambda(x) = 0.2068x + 0.6808x^2 + 0.1025x^3 + 0.00988x^8,$$
$$\rho(x) = 0.43217x^4 + 0.46110x^5 + 0.10673x^6$$

and

$$\lambda(x) = 0.0082x + 0.9689x^2 + 0.0077x^3 + 0.0152x^8,$$
$$\rho(x) = 0.35472x^4 + 0.09275x^5 + 0.55252x^6$$

(these ensembles will be considered in Example 8.8).

7.5 Is the BEC threshold of a (3, 6)-regular LDPC ensemble better or worse than the BEC threshold of an RA ensemble with $q = 3$ and $a = 3$?

7.6 Using density evolution, plot the BEC threshold of a rate-$1/4$ regular LDPC ensemble versus the column weight. Is the column weight that produces the best threshold on the BEC the same as that for the BI-AWGN channel?

7.7 Expand the pseudo-code in Algorithm 7.3 to show explicitly how you would generate degree distributions and ensure that they are valid.

7.8 Compare the BI-AWGN thresholds of the two LDPC ensembles considered in Exercise 7.4.

7.9 Show how

$$I(X;Y) = \sum_{x \in X} \sum_{y \in Y} p(x,y) \log_2 \frac{p(x,y)}{p(x)p(y)}$$

can be derived from

$$I(X;Y) = H(X) - H(X|Y),$$

using the definitions of entropy given in Chapter 1.

7.10 Given Figure 7.12, how many decoding iterations would you expect on average will be required to decode SC turbo codes with a rate-1 inner convolutional code $1/(1+D)$ and a rate-1/2 $[1\ (1+D)/(1+D+D^2)]$ convolutional code for the outer code, used on a BI-AWGN channel with $\sigma^2 = 0.7762$?

7.11 Produce an EXIT chart that shows whether iterative sum–product decoding with a (3, 6)-regular LDPC is likely to succeed when $\sigma = 1$. If in fact the decoder is likely to succeed, how many decoding iterations would you expect it to require on average?

7.12 Produce an EXIT chart that shows whether iterative sum–product decoding with a (3, 6)-regular LDPC is likely to succeed when $\sigma = 0.5$. If the decoder is likely to succeed, how many decoding iterations would you expect the decoder to require on average?

7.13 Using EXIT charts show how precoding using an accumulator will affect the threshold of a $q = 3$, $a = 3$ RA code.

7.14 Can precoding using an accumulator improve the threshold of a (3, 6)-regular LDPC code?

8
Error floor analysis

8.1 Introduction

In the previous chapter we analyzed the performance of iterative codes by calculating their threshold and thus comparing their performance in high-noise channels with the channel's capacity. In that analysis we considered the iterative decoding of code ensembles with given component codes, averaging over all possible interleaver–edge permutations. In this chapter we will also use the concept of code ensembles but will turn our focus to low-noise channels and consider the error floor performance of iterative code ensembles. Except for the special case of the binary erasure channel, our analysis will consider the properties of the codes independently of their respective iterative decoding algorithms. In fact, we will assume maximum likelihood (ML) decoding, for which the performance of a code depends only on its codeword weight distribution. Using ML analysis we can

- demonstrate the source of the interleaver gain for iterative codes,
- show why recursive encoders are so important for concatenated codes, and
- show how the error floor performance of iterative codes depends on the chosen component codes.

Lastly, for the special case of the binary erasure channel we will use the concept of stopping sets to analyze the finite-length performance of LDPC ensembles and message-passing decoding.

8.2 Maximum likelihood analysis

Although it is impractical to decode the long, pseudo-random, codes designed for iterative decoding using ML decoding, the ML decoder is the best possible decoder (assuming equiprobable source symbols) and so provides an upper bound on the performance of iterative decoders. Consequently, analyzing the

ML decoding performance of iterative codes can provide beneficial information for code design.

As we are considering the codes independently of their iterative decoding algorithms, we will refer to serially concatenated or parallel concatenated codes rather than turbo, RA or LDPC codes.

In Section 1.3 we saw that the word error rate of an error correction code following ML decoding can be upper bounded using its weight enumerating function (WEF). The weight enumerating function for an error correction code C is

$$A(D) = \sum_{d=d_{\min}}^{N} a_d D^d,$$

where a_d is the number of codewords with weight d. The parameter D allows us to keep track of the weights d.

Similarly, in Section 1.3 we saw that the bit error rate of an error correction code following ML decoding can be upper bounded using its input–output weight enumerating function (IOWEF) or its input–redundancy weight enumerating function (IRWEF). The input–output weight enumerating function for an error correction code C is

$$A(W, D) = \sum_{w=1}^{K} \sum_{d=d_{\min}}^{N} a_{w,d} W^w D^d,$$

where $a_{w,d}$ is the number of codewords with weight d generated by a message with weight w. When the encoder is systematic, the input–redundancy weight enumerating function is defined as

$$A(W, P) = \sum_{w=1}^{K} \sum_{p=1}^{N-K} A_{w,p} W^w P^p,$$

where $A_{w,p}$ is the number of codewords with parity bits having weight p generated by a message with weight w. The parameters W and P perform, for the weights w and p respectively, a similar role to that of D.

Example 8.1 The weight distribution of a rate-$1/q$ repetition code is simple to compute since a weight-w message will produce a weight $d = qw$ codeword. There are $\binom{K}{w}$ possible messages of weight w so there are $\binom{K}{w}$ codewords of weight $d = qw$. Thus

$$a_d = \begin{cases} \binom{K}{d/q}, & d = 0 \pmod{q}, \\ 0 & \text{otherwise}, \end{cases}$$

$$a_{w,d} = \begin{cases} \binom{K}{w}, & d = qw, \\ 0 & \text{otherwise}. \end{cases}$$

8.2.1 Input–output weight enumerating functions for convolutional codes

In this subsection we consider the weight distributions of fixed-length codewords generated by passing a length-K message into a rate-k/n convolutional encoder with free distance d_f. To find the IOWEF of a convolutional code we produce an *augmented state diagram*. The augmented state diagram represents all the paths through the state diagram that begin and end with S_0 but do not otherwise pass through S_0. A series of these paths will be concatenated to form the final codeword, and so we will call them the *candidate paths*.

The augmented state diagram can be formed from the state diagram by splitting the all-zeros state into an initial state S_{in} and a final state S_{out} and removing the loop from S_0 to S_0. In the augmented state diagram S_{in} retains the edges leaving S_0 and S_{out} retains the edges entering S_0.

The purpose of the augmented state diagram is to track the Hamming weight and the number of state transitions associated with each candidate path. To do this, each edge of the augmented state diagram is labeled with $D^d Z$ where d is the output weight associated with that state transition or, if the code is systematic, with $W^w P^p Z$ where w and p are the Hamming weights of the message sequence and parity sequence associated with that state transition. The label Z is the same for every edge and will be used to track the number of state transitions that make up each candidate path.

Example 8.2 The rate-$1/2$ systematic recursive binary convolutional encoder from Figure 4.4 has the state diagram shown in Figure 4.6a. An augmented state diagram for this encoder is shown in Figure 8.1. The shortest path from S_{in} to S_{out} is S_{in}–S_2–S_1–S_{out}. This path has weight $WPZWZWPZ = W^3P^2Z^3$. One length-4 path is S_{in}–S_2–S_3–S_1–S_{out}, with weight $W^2P^4Z^4$. There are two ways to traverse from S_{in} to S_0 via a length-5 path: S_{in}–S_2–S_3–S_3–S_1–S_{out} with weight $W^3P^4Z^5$ or via a length-5 path S_{in}–S_2–S_1–S_2–S_1–S_{out} with weight $W^4P^2Z^5$. There is one further path with input weight 4, S_{in}–S_2–S_3–S_3–S_1–S_{out} with weight $W^4P^4Z^6$, but no further paths with input weight less than 4 or total output weight

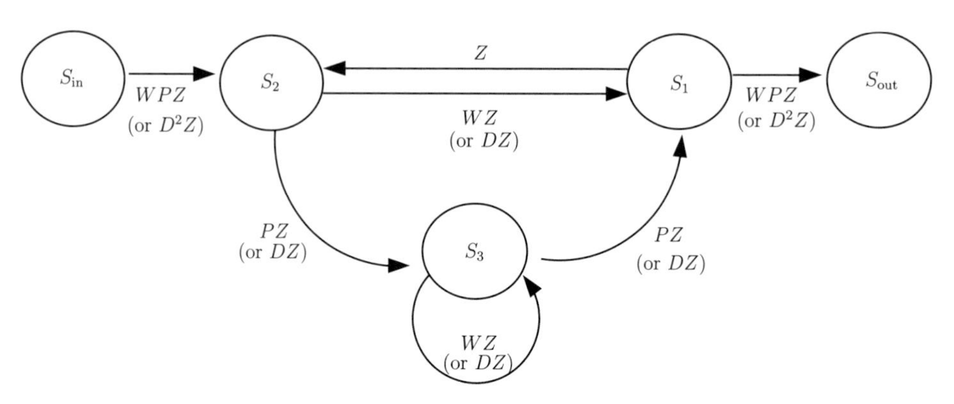

Figure 8.1 The augmented state transition diagram of the rate-1/2 systematic recursive binary convolutional encoder in Figure 4.4.

less than 7. So for this encoder the lowest-weight output is actually produced by a weight-3 input rather than a weight-2 input, and the first few terms of the IRWEF are given by

$$A(W, P) = W^2P^4 + W^3P^2 + W^3P^4 + W^4P^2 + W^4P^4 + \cdots .$$

Translating to the WEF simply requires that the W^wP^p terms be replaced by D^{w+p} terms. For this subset of paths we have one of weight 5, two of weight 6 and one of weight 7. There are in fact three more paths of weight 7 but no more paths with weight less than 7. Thus the first few terms in the WEF function of this code are

$$A(D) = D^5 + 2D^6 + 4D^7 + \cdots$$

Rather than enumerating each path through the trellis (which is completely impractical except for very short message lengths), the IOWEF and IRWEF can found by describing the code's augmented state diagram using a transfer function.

The input–output transfer function of a convolutional code is defined by

$$T(W, D, Z) = \sum_{w,d,l} T_{w,d,l} W^w D^d Z^l,$$

where $T_{w,d,l}$ is the number of candidate paths in the code trellis with an input sequence of weight w, an output sequence of weight d and candidate path length l. Similarly, for a systematic convolutional code the input–redundancy transfer function is defined by

$$T(W, P, Z) = \sum_{w,p,l} T_{w,p,l} W^w P^p Z^l,$$

where $T_{w,p,l}$ is the number of candidate paths in the code trellis with an input sequence of weight w, a parity-check sequence of weight p and candidate path length l.

To describe a transfer function for a given augmented state diagram, the set of edges from from S_{in} to S_{out} is combined into a single edge. The combined weight of this edge gives the transfer function. The edge-combining process requires three rules.

(i) Two edges in series, with weights A and B respectively, form a combined edge with weight AB.
(ii) Two edges in parallel, with weights A and B respectively, form a combined edge with weight $A + B$.
(iii) A forward edge with weight A with a feedback edge (between the same two nodes but going in the opposite direction) of weight B form a combined forward edge with weight $A/(1 - AB)$.

For more complicated encoders, Mason's gain formula [146] may be used to compute their transfer function.

Example 8.3 The augmented state diagram shown in Figure 8.1 has a feedback edge from S_3 to S_3 with weight DZ. Using the above rules it can be replaced by a forward edge with weight $1/(1 - DZ)$. Now the path S_2–S_3–S_1 has three edges in series, with weights DZ, $1/(1 - DZ)$ and DZ respectively. They can be combined to give a single edge from S_2 to S_1 with weight $D^2Z^2/(1 - DZ)$, as shown in Figure 8.2. This new edge is in parallel with the other edge from S_2 to S_1, which has weight DZ. They can be combined to give an edge with weight

$$\frac{D^2Z^2}{1 - DZ} + DZ = \frac{D^2Z^2 + DZ(1 - DZ)}{1 - DZ} = \frac{DZ}{1 - DZ}$$

from S_2 to S_1. There is thus a forward edge from S_2 to S_1 and a feedback edge, with weight Z, from S_1 to S_2; they can be combined to give an edge with weight

$$\frac{DZ/(1 - DZ)}{1 - ZDZ/(1 - DZ)} = \frac{DZ}{1 - DZ - DZ^2}.$$

Finally, considering the three remaining edges in series gives a single edge from S_{in} to S_{out} with weight

$$\frac{DZ}{1 - DZ - DZ^2} D^4Z^2 = \frac{D^5Z^3}{1 - DZ - DZ^2}$$

which is the transfer function for the encoder.

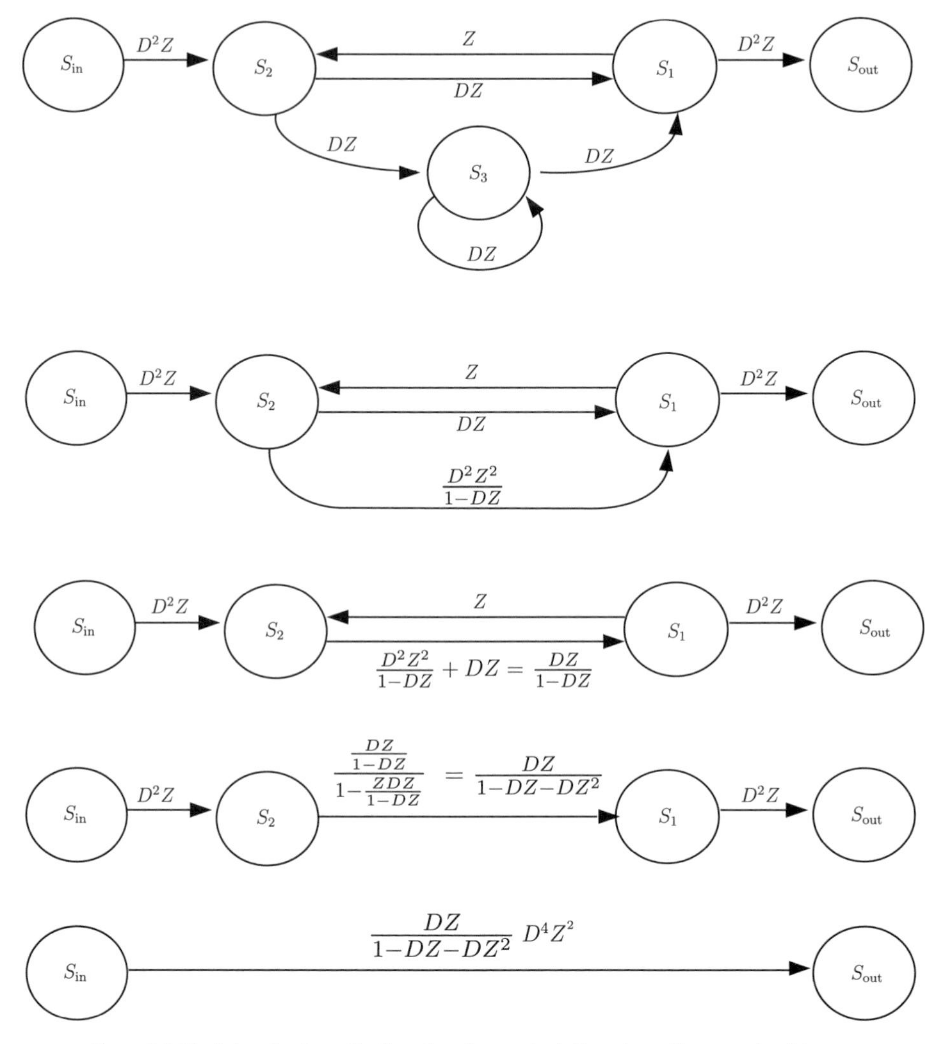

Figure 8.2 Deriving the transfer function for a rate-$1/2$ systematic recursive binary convolutional encoder.

For the WEF the delay terms in Z can be ignored, so that Z becomes 1, and the transfer function then simplifies to

$$T(D) = \frac{D^5}{1 - 2D}.$$

Polynomial division gives the WEF

$$A(D) = D^5 + 2D^6 + 4D^7 + \cdots$$

The input–redundancy transfer function can be found similarly, using the W^w and P^p edge labels, in which case

$$T(W, P, Z) = \frac{W^2Z^2P^2(D^2Z^2 + WZ - W^2Z^2)}{1 - WZ - D^2Z^2 - WZ^2 + W^2Z^3}.$$

The input–output transfer function can be found in the same way, by using the W^w and D^d edge labels.

A message of length K can produce a convolutional codeword made up of a number of concatenated candidate paths, possibly with strings of zeros in between them. The order of the concatenated candidate paths, or the locations of the zeros between them, will make no difference to the Hamming weight of the resulting codeword. What is important are the values w, l and p or d for each path. Assuming k input bits per state transition, there are $T = K/k$ state transitions for a length-K message, and so the sum of the path lengths for the candidate paths in a codeword must be no more than K/k. The sum of the path weights gives the total Hamming weight of the codeword.

For a set of υ candidate paths with total length l there are

$$\binom{K/k - l + \upsilon}{\upsilon}$$

ways in which to arrange them in the available K/k state transitions. Thus the number of messages of weight w that produce a parity sequence of weight p is

$$A_{w,p} = \sum_{\upsilon,l} \binom{K/k - l + \upsilon}{\upsilon} T_{w,p,l,\upsilon},$$

where $T_{w,p,l,\upsilon}$ is the number of sets of υ candidate paths with combined length l, combined message weight w and combined parity weight p that can be generated by a length-K message. For K/k values much larger than the memory of the convolutional code, the low-weight codewords will have values of l that are much less than K/k. Since it is these low-weight codewords that are most likely to cause error events, we can approximate the number of ways of arranging the υ candidate paths by using the inequality

$$\binom{K/k - l + \upsilon}{\upsilon} < \binom{K/k}{\upsilon}.$$

Low-weight codewords will consist of only a small number of candidate paths, so $\binom{K/k}{\upsilon}$ can further be bounded by

$$\binom{K/k}{\upsilon} < \frac{(K/k)^{\upsilon}}{\upsilon!},$$

and we have

$$A_{w,p} < \sum_{\upsilon}^{\upsilon_{\max(w)}} \frac{(K/k)^{\upsilon}}{\upsilon!} \sum_{l} T_{w,p,l,\upsilon}, \tag{8.1}$$

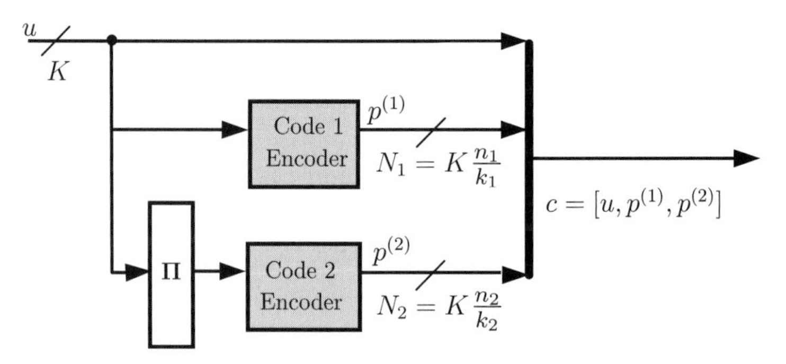

Figure 8.3 Parallel concatenation of two convolutional encoders.

where $\upsilon_{\max(w)}$ is the largest number of candidate paths that can be formed by a weight-w message. Lastly, $A_{w,p}$ can be approximated by its maximum term, for which $\upsilon = \upsilon_{\max(w)}$:

$$A_{w,p} \approx \frac{(K/k)^{\upsilon_{\max(w)}}}{\upsilon_{\max(w)}!} \sum_{l} T_{w,p,l,\upsilon_{\max(w)}}. \tag{8.2}$$

The input–output weight distribution can be found similarly:

$$a_{w,d} < \sum_{\upsilon}^{\upsilon_{\max(w)}} \frac{(K/k)^{\upsilon}}{\upsilon!} \sum_{l} T_{w,d,l,\upsilon}, \tag{8.3}$$

where $T_{w,d,l,\upsilon}$ is the number of sets of υ candidate paths, with combined length l, combined message weight w and combined codeword weight d, that can be generated by a length-K message.

8.2.2 Parallel concatenated code ensembles

In Chapter 7 we derived the threshold performance of turbo codes and saw that thresholds very close to channel capacity could be obtained by turbo code ensembles. As we saw in Chapter 5, the actual bit error correction performance of turbo codes approaches the predicted threshold as the length of the interleaver is increased (the *interleaver gain*). We also saw that turbo codes have a relatively poor bit error floor performance and that the word error rate of turbo codes does not show an interleaver gain. In this section we will explore each of these observations by considering the ML decoding of parallel concatenated codes and applying the union bound, see Section 1.3.2, to model their performance in low-noise channels.

Figure 8.3 shows the parallel concatenation of two component encoders, C_1 and C_2, with rates k_1/n_1 and k_2/n_2 respectively. If a fixed length-K message is encoded then the component codes will produce length $N_1 = Kn_1/k_1$ and length $N_2 = Kn_2/k_2$ outputs respectively, producing an $(N_1 + N_2 - K, K)$ concatenated code.

The weight distribution of codewords formed by concatenated codes is determined by both the choice of component encoders and the choice of interleaver. For the large interleavers used in practice, computing the input–output weight enumerating function (IOWE) of the concatenated encoder is prohibitively complex. A solution is to consider the average IOWE of the encoder over all possible interleavers with a given length. That is, we assume a *uniform interleaver*. A length-K uniform interleaver is a theoretical device that maps a given input block of weight w into all possible $\binom{K}{w}$ permutations with equal probability $1/\binom{K}{w}$. (In practice the performance of this average encoder can be approached by using a pseudo-randomly generated interleaver.)

If the length-K message has weight w, the concatenated parity sequence will have weight p when Code 1 produces a weight-p_1 parity sequence and Code 2 produces a weight-$(p - p_1)$ parity sequence. Thus, assuming a length-K uniform interleaver gives

$$A_{w,p} = \sum_{p_1} A^{(1)}_{w,p_1} A^{(2)}_{w,p-p_1} / \binom{K}{w} \tag{8.4}$$

for the parallel concatenated code, where $A^{(1)}_{w,p_1}$ and $A^{(2)}_{w,p-p_1}$ are the input–redundancy weight distributions of the component encoders.

For rate-k/n convolutional codes as the component codes, the encoders will be terminated and a message of fixed length K encoded. For simplicity we will assume that both component encoders input $k = 1$ bits per state transition. Substituting into (8.1) gives

$$A_{w,p} < \sum_{\upsilon}^{\upsilon_{\max(w)}} \frac{(K/1)^{\upsilon}}{\upsilon!} \sum_{l} T_{w,p,l,\upsilon},$$

for the two convolutional encoders, and bounding the binomial coefficient by

$$\binom{K}{w} > \frac{K^w}{w^w w!},$$

gives

$$\begin{aligned}
A_{w,p} &< \sum_{p_1 = p^{(1)}_{\min}}^{p} \sum_{\upsilon_1}^{\upsilon^{(1)}_{\max(w)}} \sum_{\upsilon_2}^{\upsilon^{(2)}_{\max(w)}} \frac{K^{\upsilon_1}}{\upsilon_1!} \frac{K^{\upsilon_2}}{\upsilon_2!} \frac{w^w w!}{K^w} \sum_{l_1} T^{(1)}_{w,p_1,l_1,\upsilon_1} \sum_{l_2} T^{(2)}_{w,p-p_1,l_2,\upsilon_2} \\
&= \sum_{p_1 = p^{(1)}_{\min}}^{p} \sum_{\upsilon_1}^{\upsilon^{(1)}_{\max(w)}} \sum_{\upsilon_2}^{\upsilon^{(2)}_{\max(w)}} \frac{w^w w!}{\upsilon_1! \upsilon_2!} K^{\upsilon_1 + \upsilon_2 - w} T(w, p_1),
\end{aligned}$$

where $p_{\min}$ is the encoder's minimum-weight parity sequence and

$$T(w, p_1) = \sum_{l_1} T^{(1)}_{w,p_1,l_1,\upsilon_1} \sum_{l_2} T^{(2)}_{w,p-p_1,l_2,\upsilon_2}.$$

In Section 1.3 we saw that the ML decoding message BER of a systematic $(N, K, d_{\min})$ code can be bounded by, (1.43),

$$Pb(C) \leq \sum_{w=1}^{K} \sum_{p=1}^{N-K} \frac{w}{K} A_{w,p} P_{w+p},$$

where $A_{w,p}$ is the number of codewords of weight $d = w + p$ generated by a message of weight w, K is the message length and P_d, which depends on the channel, is the probability of incorrect decoding to a codeword differing in d bit positions from the one sent. Substituting for the $A_{w,p}$ we derived above for a parallel concatenated code gives

$$Pb(C) < \sum_{w,p,p_1,\upsilon_1,\upsilon_2} \frac{w}{K} P_{w+p} \frac{w^w w!}{\upsilon_1!\upsilon_2!} K^{\upsilon_1+\upsilon_2-w} T(w, p_1), \tag{8.5}$$

where

$$\sum_{w,p,p_1,\upsilon_1,\upsilon_2} = \sum_{w=1}^{K} \sum_{p=1}^{N-K} \sum_{p_1=p^{(1)}_{\min}}^{p} \sum_{\upsilon_1}^{\upsilon^{(1)}_{\max(w)}} \sum_{\upsilon_2}^{\upsilon^{(2)}_{\max(w)}}.$$

For a BI-AWGN channel (see (1.34)) we can thus approximate the ML decoding BER of a parallel-concatenated code by

$$Pb(C) < \sum_{w,p,p_1,\upsilon_1,\upsilon_2} \frac{w}{K} Q\left(\sqrt{2(w+p)\mu^2/\sigma^2}\right) \frac{w^w w!}{\upsilon_1!\upsilon_2!} K^{\upsilon_1+\upsilon_2-w} T(w, p_1). \tag{8.6}$$

However, this is an approximate bound, averaged over all interleavers and based on a number of assumptions and thus should be verified using simulations. Nevertheless, it can provide a good indication of the error floor performance of parallel-concatenated codes in low-noise channels, if $T(w, p)$ is known for the component encoders and K is short enough for the calculation of (8.6) to be feasible.

Interleaving gain

We can also use (8.4) to demonstrate the interleaving gain discussed earlier. Substituting (8.2) with $k = 1$, i.e.

$$A^{(i)}_{w,p} \approx \frac{K^{\upsilon^{(i)}_{\max(w)}}}{\upsilon^{(i)}_{\max(w)}!} \sum_{l} T^{(i)}_{w,p,l,\upsilon_{\max(w)}},$$

for the two component encoders and approximating the binomial coefficient by

$$\binom{K}{w} \approx \frac{K^w}{w!}$$

gives

$$\begin{aligned} A_{w,p} &\approx \sum_{p_1=p_{\min}^{(1)}}^{p} \frac{K^{\upsilon^{(1)}_{\max(w)}}}{\upsilon^{(1)}_{\max(w)}!} \frac{K^{\upsilon^{(2)}_{\max(w)}}}{\upsilon^{(2)}_{\max(w)}!} \frac{w!}{K^w} \sum_{l_1} T^{(1)}_{w,p_1,l_1,\upsilon^{(1)}_{\max(w)}} \sum_{l_2} T^{(2)}_{w,p-p_1,l_2,\upsilon^{(2)}_{\max(w)}} \\ &= \sum_{p_1=p_{\min}^{(1)}}^{p} \frac{w!}{\upsilon^{(1)}_{\max(w)}!\upsilon^{(2)}_{\max(w)}!} K^{\upsilon^{(1)}_{\max(w)}+\upsilon^{(2)}_{\max(w)}-w} \widetilde{T}(w, p_1) \end{aligned}$$

for the parallel concatenated code, where

$$\widetilde{T}(w, p_1) = \sum_{l_1} T^{(1)}_{w,p_1,l_1,\upsilon^{(1)}_{\max(w)}} \sum_{l_2} T^{(2)}_{w,p-p_1,l_2,\upsilon^{(2)}_{\max(w)}}.$$

If the same encoders are used for both component codes then

$$A_{w,p} \approx \sum_{p_1} \frac{w!}{(\upsilon_{\max(w)}!)^2} K^{2\upsilon_{\max(w)}-w} \widetilde{T}(w, p_1)$$

and so the BER can be approximated by

$$\begin{aligned} Pb(C) &\approx \sum_{w=1}^{K} \sum_{p=p_{\min}}^{N-K} \sum_{p_1=p_{\min}^{(1)}}^{p} \frac{w}{K} \frac{w!}{(\upsilon_{\max(w)}!)^2} K^{2\upsilon_{\max(w)}-w} \widetilde{T}(w, p_1) P_{w+p} \\ &= \sum_{w=1}^{K} \sum_{p=p_{\min}}^{N-K} \sum_{p_1=p_{\min}^{(1)}}^{p} \frac{ww!}{(\upsilon_{\max(w)}!)^2} K^{2\upsilon_{\max(w)}-w-1} \widetilde{T}(w, p_1) P_{w+p}. \end{aligned} \tag{8.7}$$

The contribution to the error rate of inputs with weight w is

$$Pb(C, w) = \sum_{p=p_{\min}}^{N-K} \sum_{p_1=p_{\min}^{(1)}}^{p} \frac{ww!}{(\upsilon_{\max(w)}!)^2} K^{2\upsilon_{\max(w)}-w-1} \widetilde{T}(w, p_1) P_{w+p}. \tag{8.8}$$

Recall that FIR encoders produce finite-weight outputs for weight $w = 1$ inputs. The number of candidate paths for a weight-1 message is, of course, one, so the contribution to the ML BER of parallel concatenated codes using FIR encoders is

$$\begin{aligned} Pb(C, 1) &= \sum_{p=p_{\min}}^{N-K} \sum_{p_1=p_{\min}^{(1)}}^{p} \frac{1}{(1!)^2} K^{2-1-1} \widetilde{T}(1, p_1) P_{1+p} \\ &= \sum_{p=p_{\min}}^{N-K} \sum_{p_1=p_{\min}^{(1)}}^{p} \widetilde{T}(1, p_1) P_{p+1}. \end{aligned}$$

Meanwhile, for IIR encoders, weight-1 inputs will not produce finite-weight candidate paths. The smallest-weight input to produce finite-weight outputs for IIR encoders is $w = 2$. For an IIR encoder, the number of candidate paths for a weight-2 message must be one (one non-zero entry is required for the candidate path to leave the zero state and one non-zero entry to return the candidate path to the zero state), so (8.8) becomes

$$Pb(C, 2) = \sum_{p=p_{\min}}^{N-K} \sum_{p_1=p_{\min}^{(1)}}^{p} \frac{2 \times 2!}{(1!)^2} K^{2-2-1} \widetilde{T}(2, p_1) P_{p+2}$$

$$= \sum_{p=p_{\min}}^{N-K} \sum_{p_1=p_{\min}^{(1)}}^{p} 4K^{-1} \widetilde{T}(2, p_1) P_{p+2}. \tag{8.9}$$

Comparing the dominant terms in the BER bound for IIR and FIR component encoders we can see clearly that the probability of a bit error is decreased by a factor $1/K$ or, equivalently, the interleaving gain is K^{α} where $\alpha = -1$, when using IIR encoders but not when using FIR encoders (i.e. $\alpha = 0$).

The word error probability, or word error rate (WER), which is essentially given by (8.5) without the w/K term, can be defined similarly. Then the exponent of K in the WER equivalent of (8.7) is $2\upsilon_{\max(w)} - w$. So, even using IIR component codes, where the $w = 2$ term is the first significant term, turbo codes do not provide a WER interleaving gain.

However, generalized turbo codes, i.e. parallel concatenated codes with $J > 2$ component codes, can be proven to have an interleaving gain of K^{α}, where the interleaving gain exponent α is $-J + 1$ for the BER and $-J + 2$ for the WER (see [147] for details). Thus a WER interleaving gain is possible if three or more component codes are concatenated in parallel.

Effective free distance

In Chapter 5 we assumed that the codewords of concatenated IIR codes with large K corresponding to inputs with weight $w = 2$ will dominate the BER performance of the decoding. Considering (8.5) we can see that this assumption is reasonable, since the codewords corresponding to $w = 1$ have very large weight, making the P_{w+p} term very small, while the codewords corresponding to $w > 3$ have large interleaving gain, making the $K^{n_1+n_2-w}$ term very small.

The minimum weight of the codewords corresponding to weight-2 inputs is the effective free distance d_{ef} of the turbo code. Recall from Chapter 5 that the minimum weight parity of a convolutional code for a weight-2 input is p_{ef}. Thus, assuming that both encoders output their minimum-weight outputs (i.e. that the interleaver design is poor), the effective free distance of the turbo code is

$$d_{\text{ef}} = w + p_{\text{ef}}^{(1)} + p_{\text{ef}}^{(2)}.$$

Example 8.4 In Example 5.5 we used Monte-Carlo simulation to plot the BER curve of the original rate-$1/2$ turbo code from [82].

Using a particular pseudo-random interleaver, the number of codewords of weight $d_{\min} = 6$ was found to be 3 (see [86, Section 10.2]) and so, using (1.41),

$$\begin{aligned} Pb(C) &\approx \frac{w}{K} a_{w,d_{\min}} Q(\sqrt{2d_{\text{ef}} r E_b/N_0}) \\ &= \frac{2}{K} 3 Q(\sqrt{6E_b/N_0}). \end{aligned}$$

Substituting $K = 65\,536$, we can now plot this error floor approximation against our simulation results (see Figure 8.4). The relative accuracy of the approximation in the error floor region shows just how much the codewords with minimum free distance dominate the BER performance at these SNRs.

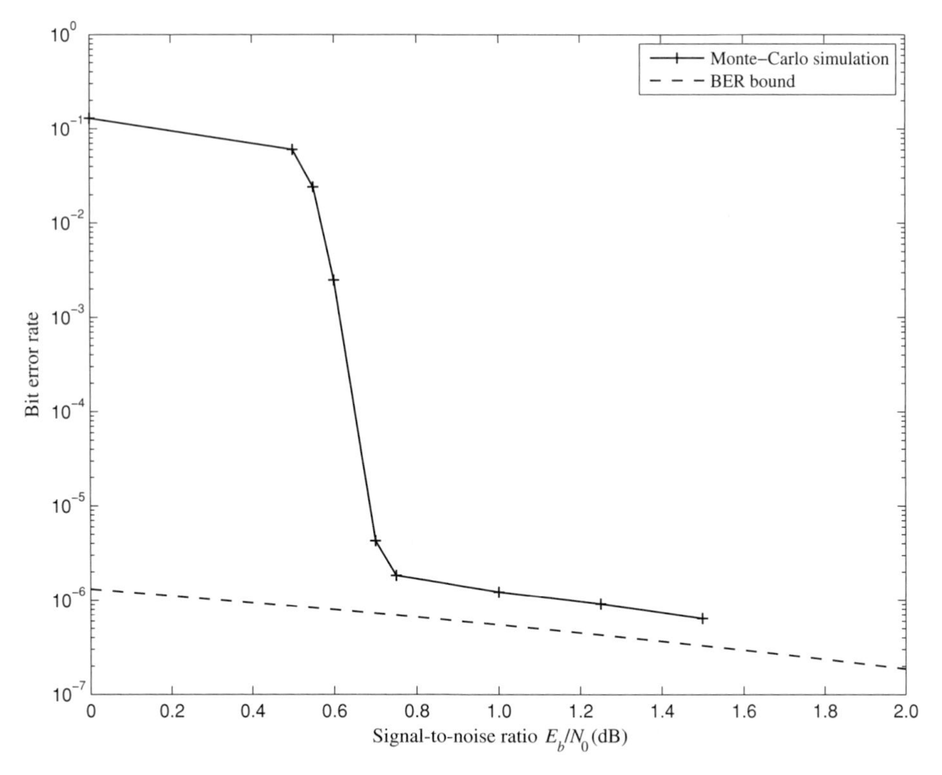

Figure 8.4 The performance of turbo decoding of the code from Example 8.4. The solid curve shows the Monte-Carlo simulation results while the broken curve shows the ML error floor approximation.

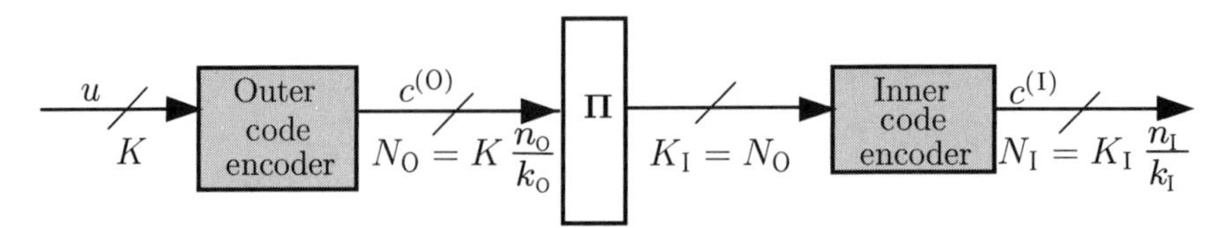

Figure 8.5 Serial concatenation of two convolutional encoders.

8.2.3 Serially concatenated code ensembles

As we saw in Chapter 5, the bit error correction performance of serially concatenated turbo codes approaches the ensemble threshold as the length of the interleaver is increased (the interleaver gain). We also saw that serially concatenated turbo codes can have improved bit error floor performances compared with parallel concatenated codes and that the WER of serially concatenated turbo codes does show an interleaver gain. In this subsection we obtain insights into each of these observations by considering the ML decoding performance of serially concatenated codes and applying the union bound to model their performance in low-noise channels.

Figure 8.5 shows the serial concatenation of two component encoders C_{O} and C_{I} with rates $k_{\mathrm{O}}/n_{\mathrm{O}}$ and $k_{\mathrm{I}}/n_{\mathrm{I}}$ respectively. If a fixed length-K message is encoded then the component codes will produce length $N_{\mathrm{O}} = Kn_{\mathrm{O}}/k_{\mathrm{O}}$ and length $N_{\mathrm{I}} = K_{\mathrm{I}}n_{\mathrm{I}}/k_{\mathrm{I}} = Kn_{\mathrm{O}}n_{\mathrm{I}}/(k_{\mathrm{O}}k_{\mathrm{I}})$ outputs respectively to produce an (N_{I}, K) concatenated code.

For serial concatenation with an $(N_{\mathrm{O}}, K_{\mathrm{O}}, d_{\mathrm{min}}^{(\mathrm{O})})$ outer code C_{O}, an inner code C_{I} with parameters $(N_{\mathrm{I}}, K_{\mathrm{I}}, d_{\mathrm{min}}^{(\mathrm{I})})$ and a length $N_{\mathrm{O}} = K_{\mathrm{I}}$ interleaver, the concatenated code is an $(N = N_{\mathrm{I}}, K = K_{\mathrm{O}}, d_{\mathrm{min}})$ code.[1] Note that d_{min} is not necessarily the same as $d_{\mathrm{min}}^{(\mathrm{I})}$ since the message sequences that produce minimum-weight inner codewords may not be codewords of the outer code.

In general, a weight-w message will produce a weight-d concatenated codeword for every i where the outer code produces a weight-i output for a weight-w input and the inner code produces a weight-d output for a weight-i input. Thus, assuming a uniform interleaver, for a serially concatenated code we have

$$a_{w,d} = \sum_i a_{w,i}^{(1)} a_{i,d}^{(2)} \Big/ \binom{K_{\mathrm{I}}}{i}, \tag{8.10}$$

where $a_{w,i}^{(\mathrm{O})}$ and $a_{i,d}^{(\mathrm{I})}$ are the input–output weight distributions of the outer and inner component encoders respectively.

[1] Although it is straightforward to concatenate several inner messages or outer codewords together to fit the interleaver length, or vice versa, it is simpler for our presentation to assume that $K_{\mathrm{I}} = N_{\mathrm{O}}$ is the interleaver length.

The IOWE for the component convolutional codes, (8.3), are

$$a^{(\mathrm{O})}_{w,i} < \sum_{\upsilon=1}^{\upsilon^{(\mathrm{O})}_{\max(w)}} \frac{(K_\mathrm{O}/k_\mathrm{O})^\upsilon}{\upsilon!} \sum_l T^{(\mathrm{O})}_{w,i,l,\upsilon}$$

and

$$a^{(\mathrm{I})}_{i,d} < \sum_{\upsilon=1}^{\upsilon^{(\mathrm{I})}_{\max(w)}} \frac{(K_\mathrm{I}/k_\mathrm{I})^\upsilon}{\upsilon!} \sum_l T^{(\mathrm{I})}_{i,d,l,\upsilon}.$$

Now bounding $\binom{K_\mathrm{I}}{i}$ by a binomial coefficient,

$$\binom{K_\mathrm{I}}{i} > \frac{K_\mathrm{I}^i}{i^i i!},$$

and substituting into (8.10) we obtain

$$a_{w,d} < \sum_{i=d^{(\mathrm{O})}_{\min}}^{N_\mathrm{O}} \sum_{\upsilon_1=1}^{\upsilon^{(\mathrm{O})}_{\max(w)}} \sum_{\upsilon_2=1}^{\upsilon^{(\mathrm{I})}_{\max(w)}} \frac{(K_\mathrm{O}/k_\mathrm{O})^{\upsilon_1}}{\upsilon_1!} \frac{(K_\mathrm{I}/k_\mathrm{I})^{\upsilon_2}}{\upsilon_2!} \frac{i^i i!}{K_\mathrm{I}^i} f(i, w, d) \qquad (8.11)$$

for a serially concatenated code, where

$$f(i, w, d) = \sum_{l_1} T^{(\mathrm{O})}_{w,i,l_1,\upsilon_1} \sum_{l_2} T^{(\mathrm{I})}_{i,d,l_2,\upsilon_2}.$$

From Section 1.3 we see that the message BER of an $(N, K, d_{\min})$ code can be bounded by, cf. (1.41),

$$Pb(C) \leq \sum_{w=1}^{K} \sum_{d=d_{\min}}^{N} \frac{w}{K} a_{w,d} P_d,$$

where $a_{w,d}$ is the number of codewords of weight d generated by a message of weight w and P_d, which depends on the channel, is the probability of incorrect decoding to a codeword differing in d bit positions from the one sent.

Substituting (8.11) into the above BER equation, we can thus approximate the ML decoding performance of serially concatenated codes by

$$Pb(C) < \sum_{w,d,i,\upsilon_1,\upsilon_2} \frac{w}{K} P_d \frac{(K_\mathrm{O}/k_\mathrm{O})^{\upsilon_1}}{\upsilon_1!} \frac{(K_\mathrm{I}/k_\mathrm{I})^{\upsilon_2}}{\upsilon_2!} \frac{i^i i!}{K_\mathrm{I}^i} f(i, w, d),$$

where

$$\sum_{w,d,i,\upsilon_1,\upsilon_2} = \sum_{w=1}^{K} \sum_{d=d_{\min}}^{N} \sum_{i=d^{(\mathrm{O})}_{\min}}^{N_\mathrm{O}} \sum_{\upsilon_1=1}^{\upsilon^{(\mathrm{O})}_{\max(w)}} \sum_{\upsilon_2=1}^{\upsilon^{(\mathrm{I})}_{\max(w)}}.$$

Noting that $K_\mathrm{O} = N_\mathrm{O}k_\mathrm{O}/n_\mathrm{O} = K_\mathrm{I}k_\mathrm{O}/n_\mathrm{O}$ and $K = K_\mathrm{O}$ gives

$$Pb(C) < \sum_{w,d,i,\upsilon_1,\upsilon_2} P_d \frac{w i^i i!}{\upsilon_1!\upsilon_2! k_\mathrm{O} n_\mathrm{O}^{\upsilon_1-1} k_\mathrm{I}^{\upsilon_2}} K_\mathrm{I}^{\upsilon_1+\upsilon_2-i-1} f(i,w,d). \quad (8.12)$$

Again this is an approximate bound, averaged over all interleavers and based on a number of assumptions, and thus should be verified using simulations. Nevertheless, it can provide a good indication of the error floor performance of serially concatenated codes in low-noise channels if $T_{w,d,l,\upsilon}$ is known for the component encoders and K is short enough for the calculation of (8.12) to be feasible.

Interleaving gain

From (8.12) we can see that the exponent of K_I is

$$\alpha = \upsilon_1 + \upsilon_2 - i - 1.$$

To give an interleaving gain, α needs to be negative for all w, d and i.

Firstly, the maximum number of candidate paths in a weight-i output from the outer code is

$$\upsilon_1 = \left\lfloor \frac{i}{d_\mathrm{f}^{(\mathrm{O})}} \right\rfloor,$$

where $d_\mathrm{f}^{(\mathrm{O})}$ is the minimum free distance of the outer code. Since the outer code need not be IIR a block code may used, in which case $d_\mathrm{min}^{(\mathrm{O})}$ is substituted for $d_\mathrm{f}^{(\mathrm{O})}$ and υ_1 becomes the maximum number of non-zero codewords in the length-N_O block of codewords.

For an FIR inner code we note that a message sequence of weight-1 will produce a finite-weight candidate path, and so a weight-i message can produce an inner-code codeword formed from up to $\upsilon_2 = i$ candidate paths. Thus

$$\max \alpha = \upsilon_1 + i - i - 1 = \upsilon_1 - 1,$$

so α is non-negative and there is no interleaving gain.

For an IIR inner code we know that a message sequence of weight at least 2 is required for each candidate path, and so a weight-i input can produce up to $\upsilon_2 = \lfloor i/2 \rfloor$ candidate paths in the inner-code codeword. Thus

$$\max \alpha = \left\lfloor \frac{i}{d_\mathrm{f}^{(\mathrm{O})}} \right\rfloor + \left\lfloor \frac{i}{2} \right\rfloor - i - 1 = \left\lfloor \frac{i}{d_\mathrm{f}^{(\mathrm{O})}} \right\rfloor - \left\lfloor \frac{i+1}{2} \right\rfloor - 1.$$

For $d_f^{(O)} \geq 2$,

$$\left\lfloor \frac{i}{d_f^{(O)}} \right\rfloor < \left\lfloor \frac{i+1}{2} \right\rfloor$$

for all i and thus α is always negative; thus the largest α is

$$\max \alpha = -\left\lfloor \frac{d_f^{(O)}+1}{2} \right\rfloor, \tag{8.13}$$

corresponding to $i = d_f^{(O)}$, i.e. $v_l = 1$. Consequently, serially concatenated codes with inner IIR convolutional encoders always have a BER interleaving gain, and this gain can be increased by choosing an outer code with large minimum free distance.

The interleaving gain for the word error probability, or word error rate, can be found by dropping the w/K factor from (1.41) and following the same arguments as above:

$$\max \alpha = -\left\lfloor \frac{d_f^{(O)}-1}{2} \right\rfloor. \tag{8.14}$$

Thus, serially concatenated codes have a WER interleaving gain if $d_f^{(O)} \geq 3$.

Example 8.5 Non-systematic repeat–accumulate codes (without a combiner) are serially concatenated codes with an outer repetition code and an inner accumulator code. The outer (repetition) code of RA and IRA codes has minimum distance $d_{\min}^{(O)} = q_{\min}$, where $q_{\min}$ is the minimum-repetition parameter (equivalently, the minimum degree of all systematic-bit nodes). The inner (accumulator) code has $d_{\text{ef}} = 1$, since a weight-2 input with consecutive ones will produce a weight-1 output. Thus the BER interleaving gain exponent of non-systematic RA and IRA codes is

$$\alpha = -\left\lfloor \frac{q_{\min}+1}{2} \right\rfloor.$$

Effective free distance

For concatenated codes with large K_I we can assume that the term in (8.12) with smallest interleaver gain (i.e. maximum α) will dominate the sum. (This is just as for parallel concatenated codes, where the codewords corresponding to $w = 2$ resulted in the minimum interleaving gain and so were assumed to dominate the BER performance.) In the case of SC codes the weight $d = d_{\max(\alpha)}$ of the codewords corresponding to $\max \alpha$ will play an important role in the

BER performance through P_d. Assuming an IIR inner code, if we consider the lowest-weight d for the inner-code codeword associated with $\max \alpha$ then we have

$$v_2 = \left\lfloor \frac{i}{2} \right\rfloor = \left\lfloor \frac{d_\mathrm{f}^{(\mathrm{O})}}{2} \right\rfloor$$

candidate paths in the codeword from the inner code. If $d_\mathrm{f}^{(\mathrm{O})}$ is even then we have at most $d_\mathrm{f}^{(\mathrm{O})}/2$ candidate paths in the codeword, each with input weight 2, and if $d_\mathrm{f}^{(\mathrm{O})}$ is odd then we have $(d_\mathrm{f}^{(\mathrm{O})} - 3)/2$ candidate paths in the codeword, each with input weight 2, and one candidate path with input weight 3. Since the inner code has effective free distance $d_\mathrm{ef}^{(\mathrm{I})}$, the weight-2 input paths must each produce outputs of weight at least $d_\mathrm{ef}^{(\mathrm{I})}$. Thus, for $d_\mathrm{f}^{(\mathrm{O})}$ even,

$$d_{\max \alpha} = \frac{d_\mathrm{f}^{(\mathrm{O})}}{2} d_\mathrm{ef}^{(\mathrm{I})}, \tag{8.15}$$

and, for $d_\mathrm{f}^{(\mathrm{O})}$ odd,

$$d_{\max \alpha} = \frac{d_\mathrm{f}^{(\mathrm{O})} - 3}{2} d_\mathrm{ef}^{(\mathrm{I})} + d_{\min(w=3)}^{(\mathrm{I})}, \tag{8.16}$$

where $d_{\min(w=3)}^{(\mathrm{I})}$ is the minimum-weight output of the inner encoder for a weight-3 input. We can see then that, for a serially concatenated code with large interleaver length, the important component code parameters are $d_\mathrm{f}^{(\mathrm{O})}$ for the outer code and $d_\mathrm{ef}^{(\mathrm{I})}$, and possibly $d_{\min(w=3)}^{(\mathrm{I})}$, for the inner code.

Example 8.6 We remind the reader that non-systematic repeat–accumulate codes (without a combiner) are serially concatenated codes with an outer repetition code and an inner accumulator code. The outer (repetition) code of RA and IRA codes has minimum distance $d_{\min}^{(\mathrm{O})} = q_{\min}$, where $q_{\min}$ is the minimum-repetition parameter. The inner (accumulator) code has $d_\mathrm{ef} = 1$ since a weight-2 input with consecutive ones will produce a weight-1 output. Now we calculate the effective free distance of such codes; it is given by

$$d_{\max \alpha} = \frac{q_{\min}}{2}$$

for $q_{\min}$ even and

$$d_{\max \alpha} = \frac{q_{\min} - 3}{2} + d_{\min(w=3)}^{(\mathrm{I})}$$

for $q_{\min}$ odd.

For an accumulator, a weight-3 input will not produce a finite-weight output and so $d^{(\mathrm{I})}_{\min(w=3)}$ grows with N. Consequently, for RA and IRA codes with $q_{\min}$ odd the significant term in the error performance may correspond to source messages with weight 2. Further, an alternative inner code, such as the $1/(1 + D + D^{\psi})$ encoder proposed for w3IRA codes, can be chosen to improve $d^{(\mathrm{I})}_{\mathrm{ef}}$ and hence the error floor performance when q is even.

8.2.4 The error floor of irregular block codes

As we saw in Chapter 3, LDPC codes that are capacity-approaching tend to have poor error floor performances while codes with extremely low error floors have thresholds far from capacity. We observed that the large number of weight-2 nodes returned by optimizing the degree distribution for the code threshold can limit the decoding performance of LDPC codes in low-noise channels. Here we will consider the ML decoding performance of LDPC codes to see how degree-2 codes can result in a reduced minimum distance for the ensemble.

To determine the contribution of degree-2 bit nodes to the error floor performance of irregular codes, we define the subgraph $\mathcal{T}_2$ of an irregular Tanner graph as the set of all bit nodes with degree 2, all edges emanating from a degree-2 bit node and all check nodes connected to those edges. To find the low-weight codewords in a code with Tanner graph $\mathcal{T}_2$, we note that a codeword of weight d between degree-2 nodes is also a cycle of size $2d$.

Assuming that the edges from the degree-2 nodes are evenly distributed to the check nodes, the approximate number of cycles of size $2d$ in $\mathcal{T}_2$ (see [37]) is

$$\frac{(\lambda_2 \rho'(1))^d}{2d} + O(N^{-1/3}), \tag{8.17}$$

where λ_2 is the fraction of edges connected to degree-2 bit nodes and $\rho(x)$, (3.4), describes the edge-degree distribution of the check nodes.

Since a codeword of weight d is formed by every cycle of weight $2d$, we can approximate the first few terms of the expected weight enumerating function of an irregular LDPC code from the ensemble with degree distribution λ, ρ as

$$A(D) \approx \sum_d \frac{(\lambda_2 \rho'(1))^d}{2d} D^d. \tag{8.18}$$

(This result is derived rigorously in [37].)

Using (1.37),

$$Pe(C) \leq \sum_d a_d P_d, \tag{8.19}$$

we can approximate the WER performance of an irregular LDPC ensemble with

$$a_d = \frac{(\lambda_2 \rho'(1))^d}{2d}$$

on a binary input memoryless symmetric channel with pairwise error probability P_d.

For a BI-AWGN channel we have from (1.34)

$$P_d = Q(\sqrt{2d\mu^2/\sigma^2}) = Q(\sqrt{2dr E_b/N_0}), \tag{8.20}$$

where

$$Q(x) = \frac{1}{\sqrt{2\pi}} \int_x^{\infty} e^{-t^2/2}\, dt.$$

Thus, the contribution to the ML WER on the BI-AWGN channel with signal-to-noise ratio E_b/N_0 of errors within bits corresponding to the $v_2 N$ degree-2 bit nodes in a rate-r ensemble is

$$Pe(C) \approx \sum_{d=2}^{v_2 N} \frac{(\lambda_2 \rho'(1))^d}{2d} Q(\sqrt{2dr E_b/N_0}).$$

The BER is

$$Pc(C) \approx \sum_{d=2}^{v_2 N} \frac{d}{N} \frac{(\lambda_2 \rho'(1))^d}{2d} Q(\sqrt{2dr E_b/N_0}).$$

Note that we are calculating the *codeword* BER, as we have not defined an encoder for the LDPC ensemble so do not know the mapping of messages to codewords.

Approximating the BER by just the first term, $d = 2$, gives

$$Pc(C) \approx \frac{2}{N} \frac{(\lambda_2 \rho'(1))^2}{4} Q(\sqrt{4r E_b/N_0}). \tag{8.21}$$

Example 8.7 Figure 8.6 shows the Monte-Carlo simulated ensemble-average error correction performance and error floor approximation (8.21) of codes with varying degree distributions. Code 1 is from Example 3.2, and the Code-2 degree distribution from [19] is

$$\lambda(x) = 0.0082x + 0.9689x^2 + 0.0077x^3 + 0.0152x^8,$$
$$\rho(x) = 0.35472x^4 + 0.09275x^5 + 0.55252x^6.$$

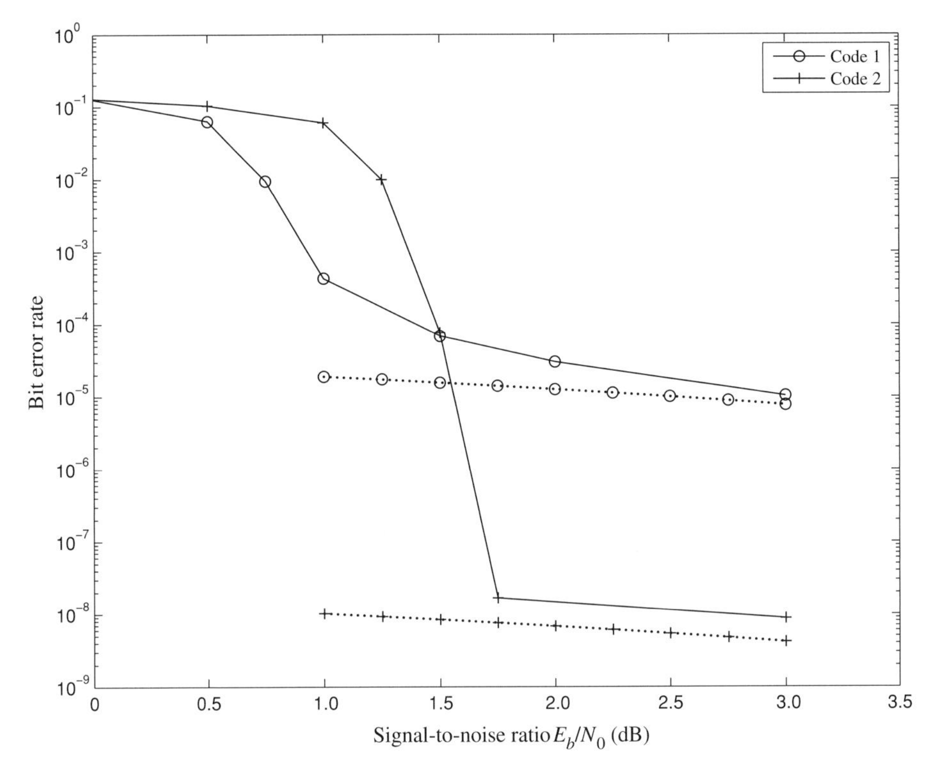

Figure 8.6 Average BER decoding performances (solid curves) and ML error floor approximations (dotted curves) of the LDPC ensembles from Example 8.7. After [19], © IEEE 2006.

Often LDPC codes are constructed from expurgated ensembles. For example, codes free of 4-cycles are usually considered. For these codes, considering the impact of codewords of size 2 would not make sense since the subgraph induced by the weight-2 nodes cannot have codewords of size 2 in a girth-6 graph. For ensembles with girth g the WER floor of LDPC ensembles on the BI-AWGN channel can be modeled by setting $d = g/2$ as the smallest term in the sum (8.21). Approximating the WER of ensembles with girth g by just the smallest term gives

$$Pe(C) \approx Q\left(\sqrt{2\frac{g}{2}r\frac{E_b}{N_0}}\right)\frac{(\lambda_2\rho'(1))^{g/2}}{g}, \tag{8.22}$$

and similarly for the BER.

Threshold versus error floor

As a consequence of approximating the error floor performance of irregular codes using only the degree-2 bit nodes, the resulting expression can be used to constrain the optimization process to find ensembles with the best possible threshold subject to some minimum bound on the WER floor.

Approximating the error function by

$$Q(x) = \frac{1}{\sqrt{2\pi}} \int_x^\infty e^{-t^2/2}\, dt \le e^{-x^2/2},$$

(8.22) can be rewritten as

$$W_g = (e^{-\mu^2/\sigma^2})^{g/2} \frac{(\lambda_2 \rho'(1))^{g/2}}{g}. \tag{8.23}$$

Recall that differential evolution is constrained by the stability condition for the binary input BI-AWGN channel with variance σ:

$$\lambda_2 \rho'(1) < e^{1/2\sigma^2}. \tag{8.24}$$

Using (8.23) we can explicitly trade off threshold for error floor by adjusting the condition in (8.24) to give

$$\lambda_2 \rho'(1) < (1 - (1 - E)^2) e^{1/2(\sigma^*)^2}, \tag{8.25}$$

where E is defined to provide the required WER and σ^* is the threshold value of the erasure probability returned by density evolution. Thus $\sigma > \sigma^*$ corresponds to the error floor region of the WER curve, making σ^* an ideal value at which to evaluate (8.22).

In practice we apply this condition as a constraint on λ_2:

$$\lambda_2 < \frac{(1 - (1 - E)^2)}{\sum_i (i-1)\rho_i} e^{1/2\sigma^2}. \tag{8.26}$$

For $E = 1$ the optimal (for threshold) irregular code distribution is returned, while $E = 0$ gives the optimal degree distribution having no weight-2 degrees. As E is varied between 0 and 1, the increase in E provides an improved threshold at the expense of a reduction in error floor performance.

The constraint (8.26) can be implemented quite simply by scaling the relevant density-evolution stability condition by the required value of E, where E is a function of both the desired constraint on the WER floor and the girth of the code to be constructed.

Example 8.8 Figure 8.7 shows the Monte-Carlo simulated ensemble-average error correction performance of codes with varying degree distributions. The unconstrained distribution is given in Example 3.2 and the constrained degree distributions from [19] are

$$\lambda(x) = 0.2068x + 0.6808x^2 + 0.1025x^3 + 0.00988x^8,$$

$$\rho(x) = 0.43217x^4 + 0.46110x^5 + 0.10673x^6$$

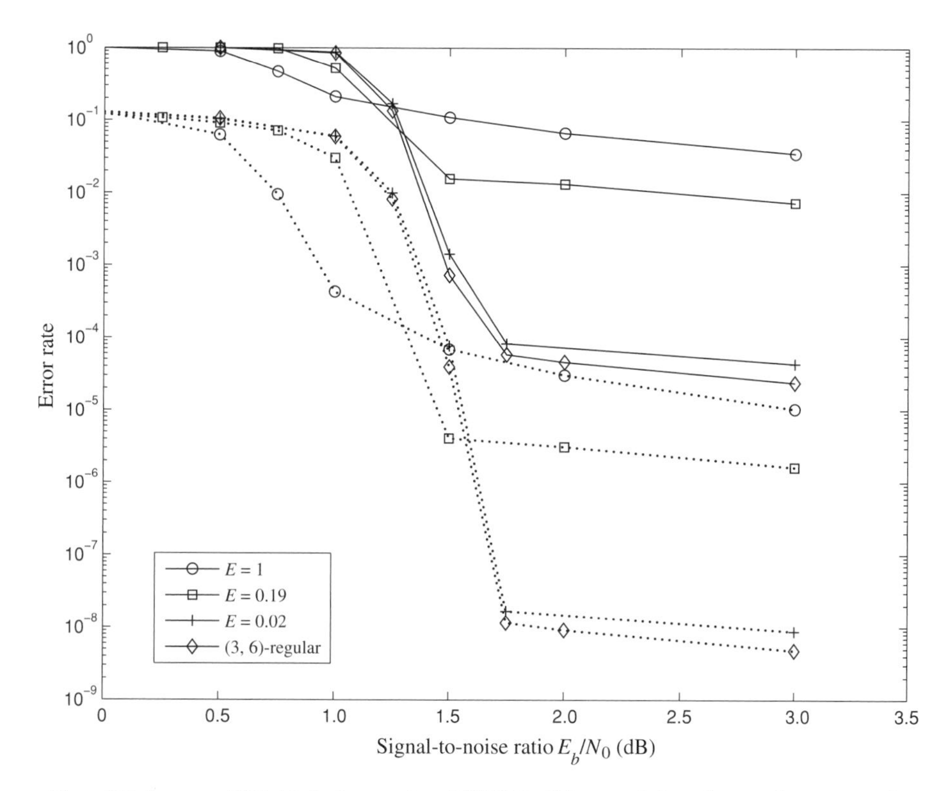

Figure 8.7 Average BER (dotted curves) and WER (solid curves) decoding performances of LDPC ensembles with constrained degree distributions from Example 8.8. After [19], © IEEE 2006.

for $E = 0.19$ and

$$\lambda(x) = 0.0082x + 0.9689x^2 + 0.0077x^3 + 0.0152x^8,$$

$$\rho(x) = 0.35472x^4 + 0.09275x^5 + 0.55252x^6$$

for $E = 0.02$. The constrained degree distributions allow a trade-off between threshold and error floor performance.

8.2.5 Designing iterative codes to improve the interleaving gain and error floor

The ML analysis of iterative codes leads to several observations on how to design iterative codes so as to improve their interleaving gain and error floor.

For parallel concatenated turbo codes:

- The interleaving gain for length-K interleavers scales the bit error probability by a factor K^{-1} for IIR component codes and by a factor K^0 for FIR component codes; consequently

  - the component codes need to be IIR to ensure an interleaving gain, and
  - longer interleavers provide larger interleaving gains.
- To achieve a WER interleaving gain requires generalized turbo codes with more than two parallel concatenated component codes.
- Choosing component codes with large p_{ef} values (equivalently, large d_{ef} values, since $d_{\mathrm{ef}} = p_{\mathrm{ef}} + 2$) will improve the error floor performance of the code.

For serially concatenated turbo codes and non-systematic RA codes, if the inner code is FIR then no interleaving gain is possible, and if the inner code is IIR then the interleaving gain for length-K_{I} interleavers scales the bit error probability by a factor K_{I}^{α} where $\alpha = \left\lfloor (d_{\mathrm{f}}^{(\mathrm{O})} + 1)/2 \right\rfloor$ and $d_{\mathrm{f}}^{(\mathrm{O})}$ is the free distance of the outer code. Thus:

- The inner code needs to be IIR to ensure an interleaving gain.
- The outer code should have a large $d_{\mathrm{f}}^{(\mathrm{O})}$ to improve the interleaving gain (but it need not be IIR).
- Longer interleavers provide larger interleaving gains.
- The error floor performance can be improved by choosing an outer code with large d_{f}.

Comparing the interleaver gain exponents of serially concatenated and parallel concatenated codes having two component codes, $\alpha = -\left\lfloor (d_{\mathrm{f}}^{(\mathrm{O})} + 1)/2 \right\rfloor$ and $\alpha = -1$ respectively, shows that serially concatenated codes can yield higher interleaving gains than parallel concatenated codes. Furthermore, unlike parallel concatenation, the serial concatenation of two component codes can also provide a WER interleaving gain. These results explain the difference in the WER performance of parallel and serially concatenated codes that we saw in Chapters 5 and 6.

It is of interest to note that, with the exception of the case $\upsilon = 2$, the component convolutional encoders found, using density evolution, to produce rate-1/2 turbo codes with the best thresholds (see Example 7.9) are not the encoders with the best effective free distance (see the table in Section 5.4.2). Similarly, the error floor performance of irregular LDPC codes is poorer for those degree distributions with the best thresholds. Whether in general an ensemble of iteratively decodable codes cannot simultaneously be capacity-approaching and have a low error floor has yet to be proved or disproved.

8.2.6 Asymptotic (in the code length) performance of iterative ensembles

The performance of error correction codes in low-noise channels is dominated by errors due to low-weight codewords. This makes the minimum distance of the codes particularly important for determining their error floor performance. Finding the minimum distance of an arbitrary code requires an exhaustive search over all codewords. However, again the concept of code ensembles is useful: we can talk about the expected minimum distance of an ensemble of codes. In particular an ensemble is considered asymptotically "good" if its minimum distance grows linearly as the code length N is increased, i.e.

$$d_{\min} = N\delta$$

for some constant δ.

Firstly, for regular and irregular LDPC ensembles without any degree-2 bit nodes it has been proven that there exists a value δ such that all but a fraction of codes from the ensemble will have minimum distance

$$d_{\min} \geq \lfloor N\delta \rfloor.$$

For irregular codes with weight-2 bit nodes (i.e. $\lambda'(0) = \lambda_2 \neq 0$), the minimum distance of a code with degree distribution λ, ρ can increase linearly with the code length if

$$\lambda_2 \rho'(1) < 1,$$

but it is very unlikely to do so otherwise. Indeed, for

$$\lambda_2 \rho'(1) > 1,$$

all except possibly a small fraction of codes with degree distribution λ, ρ will have codewords that grow at most with the logarithm of the minimum distance. (For a precise definition and details on how to find δ see [10] for regular codes and [37] for irregular codes.)

For regular LDPC codes, the probability that a randomly chosen code from the ensemble (w_c, w_r, N) has minimum distance at least $\lfloor N\delta \rfloor$ is at least

$$1 - O(N^{-w_c+2}),$$

while for irregular codes the probability that a randomly chosen code from the ensemble (λ, ρ, N) has minimum distance at least $\lfloor N\delta \rfloor$ is at least

$$\max\{1 - \log \sqrt{1/(1 - \lambda_2 \rho'(1))} + O(N^{-1/3}), 0\}.$$

Thus, for large enough N a reasonable proportion of all codes with a given set of parameters can achieve a minimum distance $d_{\min} = \lfloor N\delta \rfloor$. The smaller the code length, the larger the proportion of codes from the ensemble with a minimum distance less than $\lfloor N\delta \rfloor$; however, for small N a larger number of attempts to construct a "good" code is more feasible and so we can assume that it is possible, without too much difficulty, to construct a random parity-check matrix with $d_{\min}$ at least $\lfloor N\delta \rfloor$ when $\lambda_2\rho'(1) < 1$. Indeed, for small N this bound is not tight and we can do much better.

Unfortunately, however, it is conjectured that it is exactly the codes with

$$\lambda_2\rho'(1) > 1$$

that are required for capacity-approaching performances (this is known as the flatness condition [129]).

The minimum distance of a code randomly chosen from the ensemble of parallel concatenated convolutional codes with J parallel encoders grows as

$$K^{(J-2)/J},$$

where the interleaver length is K (see [148] for details). Standard turbo codes (i.e. those with $J = 2$) can be shown to have minimum distances that increase with approximately the base-3 logarithm of the message length (see [149] and references therein for the details). Meanwhile, the minimum distance of a randomly chosen code from the ensemble of serially concatenated codes with a recursive inner convolutional code grows as

$$K^{(d_{\mathrm{f}}^{(\mathrm{O})}-2)/d_{\mathrm{f}}^{(\mathrm{O})}}$$

with interleaver length K and where $d_{\mathrm{f}}^{(\mathrm{O})}$ is the free distance of the outer code (see [148]). Similarly, RA, IRA and ARA ensembles have sub-linear minimum distances.

Attempts to design iterative codes with both capacity-approaching thresholds and linear minimum distances have focused on improving the thresholds of codes that already have the required distance properties, for example by using precoding. For example, the serial concatenation of an accumulator (as the precoder) can be used with a regular LDPC code as the inner code. The accumulator does not change the overall code rate (it is rate 1) but it can improve both the threshold of the ensemble and the asymptotic growth of the ensemble's minimum distance.

Asymptotically vanishing word error probability

While the focus of analysis of iteratively decoded codes is usually the bit error rate, it can be equally important that the word error probability of an ensemble approaches zero as the code length goes to infinity. Using the linear growth in minimum distance of randomly chosen ensembles with union bounds on ML decoding, it can be shown that, for turbo codes with more than three component encoders, there exists a noise threshold below which the word error probability approaches zero (see [147]). Bounds showing that the word error rate of ML decoding asymptotically approaches zero have been found for LDPC codes (see [150]) and these results have been extended to iterative decoding algorithms (see [151]).

8.3 Finite-length analysis

Using the binary erasure channel (BEC) we can go much further in understanding the performance of finite-length (FL) LDPC codes with iterative decoding algorithms. On the BEC, a transmitted symbol is either received correctly or completely erased, the latter with probability ε. As we saw in Chapter 2, the iterative decoding of an LDPC code transmitted over a binary erasure channel consists of finding parity-check equations that check on only one erased bit and of choosing the value of that bit to satisfy the parity-check equation. Owing to the correction of these bits, new parity-check equations that check on only one erased bit may be created and these are then used to correct further erased bits in the subsequent iteration. The decoder will continue until all the erased bits are corrected or until every parity-check equation that includes one remaining erased bit also includes at least one other erased bit. When this occurs no further erased bits can be corrected; the set of remaining erased bits is called a stopping set.

Recall that a stopping set $\mathcal{S}$ is a subset of bit nodes in a Tanner graph for which every check node connected by a graph edge to a bit node in $\mathcal{S}$ is connected to at least one other bit node in $\mathcal{S}$. If all the bits in a stopping set are erased, the message-passing decoder will be unable to correct any erased stopping-set bit. Indeed, the key to FL analysis is that the collection of stopping sets in an LDPC Tanner graph determines exactly the erasure patterns for which the message-passing decoding algorithm will fail. A stopping set containing v bit nodes is said to be a *size-v* stopping set.

In general, FL analysis calculates the word erasure probability for an LDPC ensemble, given that v bits are erased, by calculating the total number $T(v)$ of Tanner graph subgraphs on a fixed set of v bit nodes and the number $B(v)$ of those subgraphs that contain stopping sets. Then $B(v)/T(v)$ gives the conditional probability of unsuccessful decoding given that v erasures occur. The functions $T(v)$ and $B(v)$ are calculated combinatorially; $B(v)$ is found by iteratively building up graphs one bit node at a time and counting the total number of ways in which this can be done.

The LDPC ensembles that we consider first are the set of regular Tanner graphs with m check nodes of degree at most w_{r} and N bit nodes of degree l. We will say that the check nodes have w_{r} "sockets" to which graph edges may connect. Note that this ensemble definition allows a single bit node to be connected to a single check node by more than one edge. (We will present a modification to avoid this below.)

Firstly, a v-bit subgraph $\mathcal{T}$ of a Tanner graph from this ensemble contains v bit nodes, all the vl edges emanating from them and all the check nodes connected to one or more of these edges. The total number $T(v)$ of all possible subgraphs on a fixed set of v bit nodes is given by

$$T(v) = (vl)!\binom{mw_\text{r}}{vl}, \tag{8.27}$$

as there are in total mw_r available sockets to choose for the vl graph edges, which can be done in $\binom{mw_\text{r}}{vl}$ ways, and $(vl)!$ possible permutations of these edges.

The total number of possible subgraphs on a fixed set of v bit nodes that form a stopping set is given by (see [152])

$$\text{coef}(g(x, w_\text{r}, t), vl)(vl)!, \tag{8.28}$$

where

$$g(x, w_\text{r}, t) = ((1 + x)^{w_\text{r}} - 1 - w_\text{r}x)^t$$

and $\text{coef}(g(x), i)$ denotes the coefficient of x^i in $g(x)$; $\text{coef}(g(x, w_\text{r}, t), vl)$ gives the number of ways in which vl edges can be allocated to t check nodes with w_r sockets such that each check node receives two or more edges. The factor $(vl)!$ counts all the possible permutations of these edges.

The number of degree-1 check nodes in $\mathcal{T}$ is denoted s, and the number of degree ≥ 2 check nodes in $\mathcal{T}$ is denoted t. Using this notation, the subgraphs $\mathcal{T}$ that are stopping sets are easily seen to be the subgraphs with $s = 0$.

Erasing the v bits corresponding to a stopping set will cause the message-passing decoder to fail, but so too will erasing v bits such that some subset of the v bits erases a stopping set. To incorporate these potential decoding failures, $B(v)$ is found by starting with the known stopping sets in (8.28) and iteratively building up larger subgraphs by adding one bit node at a time and counting the total number of ways in which this can be done.

When adding a new degree-l bit node, the l new edges can either: (a) create Δs new degree-1 check nodes, increasing s by Δs; (b) add edges to σ of the existing degree-1 check nodes, decreasing s by σ and increasing t by σ (this is called *covering* the degree-1 check nodes); (c) add Δt new degree-2 check nodes, increasing t by Δt; (d) add τ edges to cover free slots in the existing check nodes with degree ≥ 2; (e) add ϕ extra edges to free slots in the $\Delta t + \sigma$ newly created check nodes with degree ≥ 2; or (f) some combination of the above. Here

$$l = \Delta s + \sigma + 2\Delta t + \tau + \phi.$$

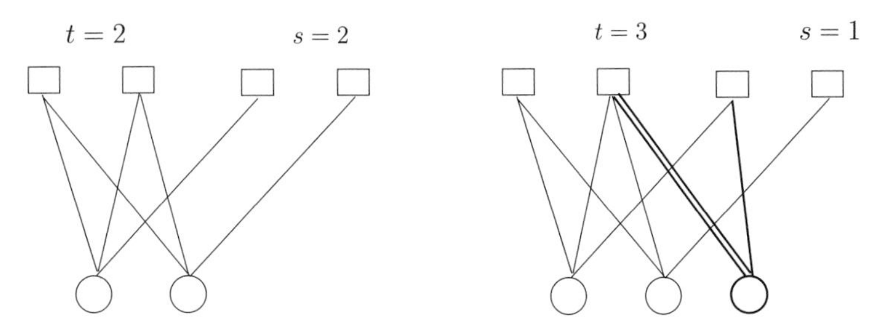

Figure 8.8 Addition of a degree $l = 3$ node (bold) to a subgraph that originally contained $v' = 2$ bit nodes, $s' = 2$ degree-1 check nodes and $t' = 2$ degree ≥ 2 check nodes. The three extra edges of the additional bit node are divided so that $\tau = 2$ and $\sigma = 1$, to give a final subgraph with $t = 3$ and $s = 1$.

As an example, Figure 8.8 shows a possible addition of bit nodes to an existing subgraph. Note that v, s and t denote the final number of each node type after the extra bit node has been added. The numbers of each node type in the original subgraph, to which a bit node is now being added, are thus $v' = v - 1$, $s' = s + \sigma - \Delta s$ and $t' = t - \Delta t - \sigma$.

If $\Delta s + \Delta t$ new check nodes are created then there are $\binom{t+s}{\Delta t + \Delta s}\binom{\Delta t + \Delta s}{\Delta t}$ choices of location for them. For each of the Δs new degree-1 check nodes there are w_r ways in which the check node socket may be chosen; in total, therefore, this can be done in $w_r^{\Delta s}$ ways. Which of the $s + \sigma - \Delta s$ original degree-1 nodes to cover can be chosen in $\binom{s + \sigma - \Delta s}{\sigma}$ ways. Then the $\sigma + 2\Delta t + \phi = l - \Delta s - \tau$ edges allocated to the $\Delta t + \sigma$ new degree-2 nodes must be added, so that each check node has at least two sockets filled. This can be done in

$$\operatorname{coef}(f(x), l - \Delta s - \tau)$$

ways, where

$$f(x) = ((1 + x)^{w_r - 1} - 1)^{\sigma}((1 + x)^{w_r} - 1 - w_r x)^{\Delta t}.$$

Lastly, the extra τ edges are allocated to the remaining sockets in the original degree ≥ 2 check nodes. There are $(t - \Delta t - \sigma)w_r$ sockets in the $t - \Delta t - \sigma$ original check nodes with degree ≥ 2; however, the existing $v - 1$ bit nodes are using

$$\omega = (v - 1)l - (s + \sigma - \Delta s)$$

of these sockets. Thus the free sockets can be chosen in

$$N_\tau = \binom{w_r(t - \Delta t - \sigma) - \omega}{\tau}$$

ways. Finally, a subgraph with α bit nodes connected to degree-1 check nodes can be constructed in α different ways. Scaling by $1/\alpha = \Delta s/s$ ensures that each subgraph is only counted once.

Putting all this together and recursively adding one bit node at a time calculates (for a given v, s and t) the contribution of all the possible stopping-set subgraphs including those stopping sets on some subset of the v bits.

The quantity $A(v, t, s, X)$ denotes the number of stopping sets of size $\leq v$ within a subgraph $\mathcal{T}$ containing v unordered degree-l bit nodes, t check nodes with degree ≥ 2 and s degree-1 check nodes. Here X is a variable used to differentiate between the bit error rate ($X = 1$) and the word error rate ($X = 0$).

The function $A(v, t, s, X)$ is initialized by the total number of possible stopping sets with size exactly v, from (8.28) (see [152]):

$$A(v, t, 0, X) = \operatorname{coef}(g(x, w_{\mathrm{r}}, t), vl)\frac{(vl)!}{v!(l!)^v}\left(\frac{v}{N}\right)^X, \tag{8.29}$$

where the factor $(v/N)^X$ calculates the fraction of codeword bits in the stopping set, which is necessary if the bit error rate ($X = 1$) rather than the word error rate ($X = 0$) is required.

Next, $A(v, t, s, X)$ can be calculated for all values of s and t by examining all the ways in which an extra bit node can be added to each size-($v - 1$) subgraph and tracking the values for s and t for each of the newly formed size-v subgraphs:

$$\begin{aligned} &A(v, t, s, X) \\ &\quad = \sum_{\Delta s=1}^{l} \sum_{\sigma=0}^{l-\Delta s} \sum_{\Delta t=0}^{\lfloor (l-\Delta s-\sigma)/2 \rfloor} \sum_{\tau=0}^{l-\Delta s-\sigma-2\Delta t} \binom{\Delta t + \Delta s}{\Delta t} \\ &\qquad \times \operatorname{coef}(f(x), l - \Delta s - \tau)\binom{w_{\mathrm{r}}(t - \Delta t - \sigma) - \omega}{\tau}\binom{s + \sigma - \Delta s}{\sigma} \\ &\qquad \times w_{\mathrm{r}}^{\Delta s}\frac{\Delta s}{s} A(v - 1, t - \Delta t - \sigma, s + \sigma - \Delta s, X)\binom{t + s}{\Delta t + \Delta s} \end{aligned} \tag{8.30}$$

where $\omega = (v - 1)l - (s + \sigma - \Delta s)$.

The number of subgraphs on v bit nodes that contain stopping sets, $B(v, X)$, is found by summing the contribution for a given v over every possible s and t: thus $B(v, 0)$ is the number of stopping sets of size v or less on a fixed set of v degree-l bit nodes. We have

$$B(v, X) = v!(l!)^v \sum_{t=0}^{m} \sum_{s=0}^{m-t} \binom{M}{t + s} A(v, t, s, X). \tag{8.31}$$

The ensemble-average bit erasure rate $P(\varepsilon, X = 1)$ and word erasure rate $P(\varepsilon, X = 0)$ following decoding are each given by the probability that an erasure of e bits will occur multiplied by the probability that such an erasure will include a stopping set:

$$P(\varepsilon, X) = \sum_{e=0}^{N} \binom{N}{e} \varepsilon^e (1-\varepsilon)^{N-e} \frac{B(e, X)}{T(e)}. \tag{8.32}$$

Expurgated ensembles

Some ensembles are required to be "expurgated" so that they do not contain stopping sets of size smaller than $S_{\min}$ or greater than $S_{\max}$. Such stopping sets are discounted by setting $A(v, t, 0, X) = 0$ if $v < S_{\min}$ or $v > S_{\max}$. However, the expression (8.28) may include graphs of size $v \geq S_{\min}$ which are themselves stopping sets but which also contain as subgraphs stopping sets of size less than $S_{\min}$. To avoid this, $A(v, t, s, X)$ and $T(v)$ are initialized recursively to avoid the inclusion of these expurgated stopping sets.

We define a new function $C(v, t, 0)$ in (8.33) below to count the number of possible stopping sets on v bits recursively, so that stopping sets on v bits which contain subgraphs that are stopping sets with size $v' < S_{\min}$ are not included. The function $C(v, t, s)$ is initialized with a single empty graph, $C(0, 0, 0) = 1$, and graphs are built up one bit node at a time in a similar manner to (8.30), with the exception that subgraphs with both $s = 0$ and $v \leq S_{\min}$ or $\geq S_{\max}$, i.e. the expurgated stopping sets, are not included. Thus we have (see [160] for more details)

$$\begin{aligned} \text{Initialization:}\quad & C(v, t, s) = 0 \ \text{ except for } \ C(0, 0, 0) = 1, \\ \text{Then:}\quad & \forall v \in \{0, \ldots, N\}, t \in \{0, \ldots, m\} \text{ and} \\ & s \in \{0, \ldots, m-t\} \ : \ s > 0 \ \text{and/or} \ S_{\min} \leq v \leq S_{\max}, \end{aligned}$$

$$\begin{aligned} & C(v, t, s) \\ & = \sum_{\Delta s=0}^{l} \sum_{\sigma=0}^{l-\Delta s} \sum_{\Delta t=0}^{\lfloor (l-\Delta s-\sigma)/2 \rfloor} \sum_{\tau=0}^{l-\Delta s-\sigma-2\Delta t} \binom{t+s}{\Delta t + \Delta s} \\ & \times \binom{\Delta t + \Delta s}{\Delta t} \operatorname{coef}(f(x), l - \Delta s - \tau) \binom{w_{\mathrm{r}}(t - \Delta t - \sigma) - \omega}{\tau} \\ & \times \binom{s + \sigma - \Delta s}{\sigma} w_{\mathrm{r}}^{\Delta s} C(v - 1, t - \Delta t - \sigma, s + \sigma - \Delta s). \end{aligned} \tag{8.33}$$

Next $C(v, t, 0)$ is used to initialize $A(v, t, s, X)$:

$$A(v, t, 0, X) = \frac{C(v, t, 0)}{v!} \left(\frac{v}{n}\right)^X, \tag{8.34}$$

and we have

$$A(v,t,s,X) = \sum_{\Delta s=1}^{l}\sum_{\sigma=0}^{l-\Delta s}\sum_{\Delta t=0}^{\lfloor (l-\Delta s-\sigma)/2\rfloor}\sum_{\tau=0}^{l-\Delta s-\sigma-2\Delta t}\binom{t+s}{\Delta t+\Delta s}$$
$$\times\binom{\Delta t+\Delta s}{\Delta t}\operatorname{coef}(f(x),l-\Delta s-\tau)\binom{w_{\mathrm{r}}(t-\Delta t-\sigma)-\omega}{\tau}$$
$$\times\binom{s+\sigma-\Delta s}{\sigma}w_{\mathrm{r}}^{\Delta s}\frac{\Delta s}{s}A(v-1,t-\Delta t-\sigma,s+\sigma-\Delta_s), \tag{8.35}$$

where $\omega = (v-1)l - (s+\sigma-\Delta s)$.

Similarly, $T(v)$ counts the total number of ways in which a subgraph on v message nodes can be constructed:

$$T(v) = v!(l!)^v\sum_{t=0}^{m}\sum_{s=0}^{m-t}\binom{m}{t+s}\frac{C(v,t,s)}{v!}. \tag{8.36}$$

Equations (8.31) and (8.32) are unchanged for expurgated ensembles.

Ensembles without repeated edges

The ensembles considered here are the regular ensembles of the previous section, with m check nodes of degree at most w_{r} and N bit nodes of degree l, with the extra condition that a single bit node can be connected to a single check node by at most one edge. This ensemble definition more closely maps to the LDPC codes commonly used in practice, which are constructed without repeated edges.

Recall that Δt is the number of new degree-2 nodes added by a single bit node, and ϕ is the number of extra edges added to degree-1 nodes that have already had an extra edge added to them by the current bit node. Thus the explicit addition of repeated edges in finite-length analysis can be avoided by setting Δt and ϕ to zero. However, repeated edges may still be added even if $w_{\mathrm{r}} > 3$ when τ is greater than 1 (consider the node added in Figure 8.8). The reason is that the number of ways in which these τ edges can be added is the number of ways of choosing τ free sockets in the degree ≥ 2 check nodes, allowing two free sockets to be contained in the same node. Upper and lower bounds on the number of ways in which these τ edges can be added, without allowing repeated edges, can be used to bound the finite-length performance of these ensembles.

The upper bound is derived by assuming that the existing edges are evenly distributed amongst the $t-\sigma$ check nodes and then counting the number of ways in which τ edges can be added to τ different check nodes. An uneven distribution of free sockets in the check nodes will give fewer options for allocating the τ edges.

In the existing graph on $v-1$ bit nodes there are $(v-1)l$ edges; however, $s+\sigma-\Delta s$ of them are allocated to the $s+\sigma-\Delta s$ degree-1 check nodes, and so there are

$$\omega = (v-1)l - s - \sigma + \Delta s$$

edges to the existing $t-\sigma$ degree ≥ 2 check nodes. Assuming an even distribution of the w edges into the existing $t-\sigma$ check nodes, each node has $w_r - \omega/(t-\sigma)$ free sockets. Which τ check nodes to cover can be chosen in $\binom{t-\sigma}{\tau}$ ways, and the socket to use for each check node can be chosen in $(w_r - \omega/(t-\sigma))^\tau$ ways, giving

$$N_\tau \leq \binom{t-\sigma}{\tau}\left(w_r - \frac{\omega}{t-\sigma}\right)^\tau, \tag{8.37}$$

where N_τ is the number of ways of choosing the free sockets; it can be lower bounded by taking the most uneven distribution of existing edges to the $t-\sigma$ check nodes. Given ω edges to $t-\sigma$ check nodes, and at least two edges to each check node, the maximum number a of check nodes that can be full is given by

$$a = \left\lfloor \frac{\omega - 2(t-\sigma)}{w_r - 2} \right\rfloor.$$

The remaining check nodes have only two edges each, except for a single check node that has any remaining edges allocated to it, so that it has

$$b = 2 + (\omega - 2(t-\sigma) - (w_r-2)a)$$

of its slots taken. Thus in total there are a full check nodes, one half-full check node and the remaining $t-\sigma-a-1$ check nodes, which have only two sockets taken.

Choosing τ of the $t-\sigma-a$ check nodes with only two sockets taken in order to add one edge to each can be done in

$$\binom{t-\sigma-a-1}{\tau}$$

ways, while choosing one free socket in each chosen node can be done in $(w_r-2)^\tau$ ways. However, one of the τ edges could also be allocated to the half-empty node. In this case that socket can be chosen in $w_r - b$ ways and the remaining $\tau-1$ edges allocated to the degree-2 check nodes in

$$\binom{t-\sigma-a-1}{\tau-1}(w_r-2)^{\tau-1}$$

ways. Thus, in total,

$$N_\tau \geq \binom{t-\sigma-a-1}{\tau-1}(w_r-2)^{\tau-1}(w_r-b) + \binom{t-\sigma-a-1}{\tau}(w_r-2)^{\tau}. \tag{8.38}$$

As before we calculate the number of stopping sets on v bits, $C(v, t, s)$.

Initialization: $C(v,t,s)=0$ except for $C(0,0,0)=1$,

Then: $\forall v \in \{0,\ldots,N\}, t \in \{0,\ldots,m\}$ and

$s \in \{0,\ldots,m-t\}$: $s > 0$ and/or $S_{\min} \leq v \leq S_{\max}$,

$$C(v,t,s) = \sum_{\Delta s=0}^{l} \sum_{\sigma=0}^{l-\Delta s} C(v-1, t-\sigma, s+\sigma-\Delta s)\binom{t+s}{\Delta s} \times w_r^{\Delta s}\binom{s+\sigma-\Delta s}{\sigma}(w_r-1)^{\sigma} N_\tau, \tag{8.39}$$

$$A(v,t,0,X) = \frac{C(v,t,0)}{v!}\left(\frac{v}{n}\right)^X, \quad \text{then } \forall v \in \{0,\ldots,N\},$$

$t \in \{0,\ldots,m\}$ and $s \in \{1,\ldots,m-t\}$,

$$A(v,t,s,X) = \sum_{\Delta s=1}^{l} \sum_{\sigma=0}^{l-\Delta s} A(v-1, t-\sigma, s+\sigma-\Delta s, X)\binom{t+s}{\Delta s} \times w_r^{\Delta s}\binom{s+\sigma-\Delta s}{\sigma}(w_r-1)^{\sigma} N_\tau \frac{\Delta s}{s}, \tag{8.40}$$

where $\tau = l - \sigma - \Delta s$.

For an upper bound on $C(v,t,s)$, N_τ from (8.37) is substituted into (8.39) while, for a lower bound on $C(v,t,s)$, N_τ from (8.38) is substituted into (8.39). Similarly, for an upper bound on $A(v,t,s,X)$, N_τ from (8.37) is substituted into (8.39) and (8.40) and, for a lower bound on $A(v,t,s,X)$, N_τ from (8.38) is substituted into (8.39) and (8.40). An upper bound on $P(\varepsilon, X)$ in (8.32) is found by using the upper bound on N_τ when calculating $B(v,X)$ and the lower bound when calculating $T(v)$. Similarly a lower bound on $P(\varepsilon, X)$ in (8.32) is found by using the lower bound on N_τ when calculating $B(v,X)$ and the upper bound when calculating $T(v)$.

Using finite-length analysis we can find the expected performance on the binary erasure channel of the ensemble of LDPC codes over all erasure probabilities. This is particularly useful in channels with low erasure rates, where Monte-Carlo simulation is not feasible.

Example 8.9 In Example 7.8 we saw, using density evolution, that for infinite-length cycle-free regular LDPC codes, a column weight 3 produces the best thresholds. In considering finite-length codes, however, we saw using simulation, in Figure 3.1, that it is often codes with large column weight that perform best in low-noise channels.

Figure 8.9 shows the expected performance of ensembles of length-100 rate-1/2 regular LDPC codes on a binary erasure channel calculated using finite-length analysis. We can see that ensembles of column-weight-3 codes do not always perform better than ensembles of codes with larger column weights. In fact, at smaller erasure probabilities codes with longer column weights give the best performance.

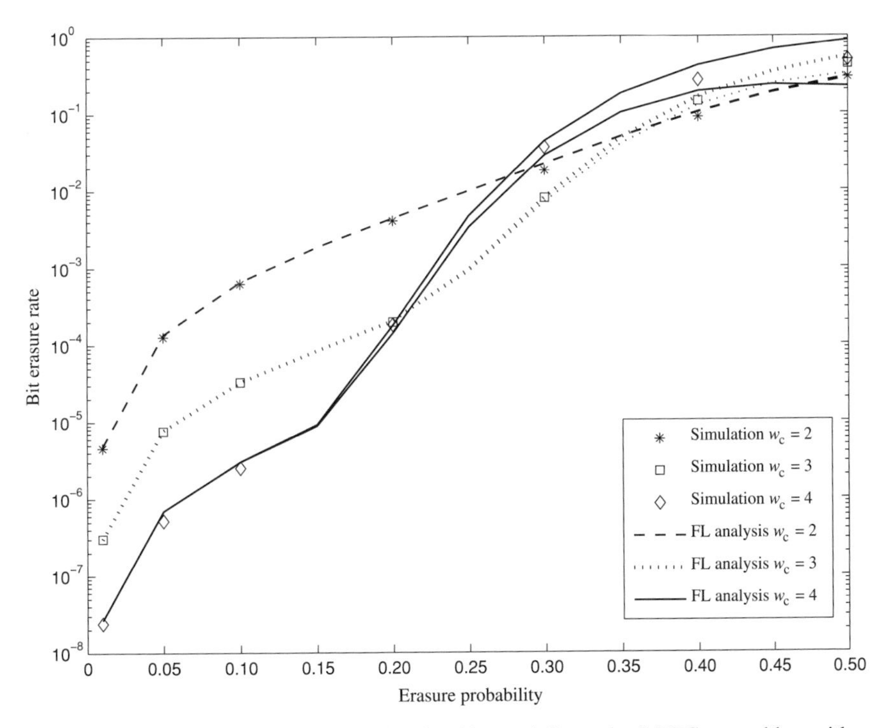

Figure 8.9 The expected performance of length-100 rate-1/2 regular LDPC ensembles, with no repeated edges, on a BEC calculated using finite-length analysis.

Pseudo-codewords

The role of stopping sets in predicting the performance of message-passing decoding on the BEC tells us that for message-passing decoding, unlike for ML decoding, properties other than the codeword set influence decoder performance. The same is true of message-passing decoding on more general channels, where stopping sets can be generalized to *pseudo-codewords* or the related concepts of near-codewords, trapping sets, instantons and absorbing sets. However, defining the configurations that lead to decoder failure in the general case is less straightforward.

The low complexity of iterative message-passing decoding stems from the fact that the algorithm operates locally on the Tanner graph representing the matrix H (i.e. each node in the decoding algorithm acts only on the messages that it can see, not on the other messages in the graph). This same operation also leads to a fundamental weakness of the algorithm: because nodes act locally, any given node does not know the whole structure of the Tanner graph.

An alternative graph that would produce, locally, the same LLRs as an actual Tanner graph is called a finite lift or *cover* of the Tanner graph. A pseudo-codeword can be thought of as a valid codeword in one of these lifts of the Tanner graph. Since the message-passing decoder cannot distinguish locally between the actual Tanner graph and the lifted graphs, any codeword in any lifted graph is equally as valid as a codeword in the Tanner graph to the iterative decoder. Thus when we say that the decoder has failed to converge it has actually converged to one of these pseudo-codewords. Using the concept of pseudo-codewords to analyze and design new iterative codes is an area of current research.

8.4 Bibliographic notes

We began this chapter with a presentation of a maximum likelihood union bound for concatenated codes developed by Benedeto and Montorsi for parallel concatenation [153–155], by Benedeto, Divsalar, Montorsi and Pollara for serially concatenated codes [109, 116, 117] and proved formally in [147].

The main obstacle to applying the ML bounds of (8.5) and (8.12) is computing the weight spectra of the component codes. However, ML union bounds aid understanding of the role of each component code in a concatenated code and have been used extensively to develop design criteria and to search for good codes (see e.g. [93, 96, 109]). A result of this ML analysis is the observation that, to optimize the error floor performance of the resulting turbo code, both component encoders in parallel concatenated codes, and the inner component encoder of serially concatenated codes, should be IIR and chosen with the best possible effective free distance. The results of an extensive search for convolutional encoders with the best effective free distance for varying rates and memory order are published in [90].

Finding lower bounds on the minimum distance of an ensemble can also give a good indication of the performance of iterative codes in the error floor region; a "good" ensemble is one with minimum distance increasing linearly with the code length [37, 149, 156]. While some iterative code ensembles can be shown to have linear minimum distance, these ensembles are unfortunately not the same as those shown to be capacity-achieving. Current research is focused on defining iterative code ensembles with both capacity-approaching thresholds and linear minimum distance (see e.g. [157]).

In general, union bounds are only useful at medium-to-high SNRs, and current research effort in this area is centered on finding improved ML bounds for concatenated codes (see e.g. [158]) that extend the range of valid SNRs. Just as significantly, union bounds are based on ML decoding while iterative error correction codes are decoded with sub-optimal iterative algorithms. Results on the performance of concatenated codes with iterative decoding still require simulation, whether using the EXIT chart or DE analysis of Chapter 7 or full Monte-Carlo simulation of the decoder.

For the special case of the binary erasure channel, a combinatorial characterization of decoding failures can be used to derive expressions for the average bit and block erasure probabilities of some LDPC ensembles. This finite-length analysis was first presented in [39] for regular LDPC ensembles and extended in [159] to allow for irregular bit degrees, in [152] to allow for expurgated ensembles and in [160] to allow for LDPC ensembles without repeated edges. Stopping sets can also be defined for turbo code ensembles [161] and are useful for understanding the performance of turbo decoders. Unfortunately, a combinatorial characterization of the stopping set distribution of turbo ensembles is more difficult.

A topic of current research is the search for a finite-length analysis for more general memoryless channels based on extending the concept of stopping sets to the more general pseudo-codewords (see e.g. [162–166]). A good source for more information on pseudo-codewords and related topics is [167].

In general, though, when we are considering the iterative decoding of turbo, LDPC and RA codes we can either assume that the code is infinitely long (so that the graph is cycle-free) and consider its threshold or assume that the iterative decoder is actually performing as an ML decoder (which can be reasonable in the error floor region). What is lacking is a complete analysis of iterative decoding on graphs with cycles. A step in that direction is recent work considering the iterative decoder as a non-linear dynamical feedback system [168–171]. Analysis of the fixed point solutions of the system is directed towards understanding the effect of cycles in the graphs on the convergence and stability of the decoding as well as towards considering the relationship between the fixed points of iterative and ML decoding.

8.5 Exercises

8.1 Find the transfer function of an accumulator and give the first few terms of its IOWE function.

8.2 Find the transfer functions of the encoder in Example 4.1 (Figure 4.1). Give the first few terms of its input–redundancy weight enumerating function (IRWEF).

8.3 Find the transfer function of the $1/(1 + D + D^6)$ convolutional code and give the first few terms of its input–output weight enumerating function (IOWEF).

8.4 The UMTS standard turbo encoder is formed using the concatenation of two copies of the convolutional encoder

$$\left[1 \quad \frac{1 + D + D^3}{1 + D^2 + D^3}\right].$$

Find the effective free distance of the turbo code.

8.5 The CDMA2000 standard turbo encoder is formed using the concatenation of two copies of the convolutional encoder

$$\left[1 \quad \frac{1 + D + D^3}{1 + D^2 + D^3} \quad \frac{1 + D + D^2 + D^3}{1 + D^2 + D^3}\right].$$

Find the effective free distance of the turbo code.

8.6 For each of the first three convolutional codes in the table from Example 7.9 find the effective free distance of a turbo code using parallel concatenation of the same encoder twice.

8.7 What is the effective free distance of a w3IRA code with $q = 2$ repetition outer code and $1/(1 + D + D^6)$ inner convolutional code ($a = 1$)?

8.8 For each of the first three convolutional codes in Table 5.1, find a codeword with the effective free distance given in the turbo code, using parallel concatenation of the same encoder twice. Use any interleaver you like.

8.9 Can you constrain the interleaver of an RA code with $q = 4$ and $a = 1$ so as to ensure that codewords with the effective free distance are never produced?

8.10 Given the graph on the left in Figure 8.8, in how many ways can you add a degree-3 bit node so that the resulting graph has
(a) $s = 2$, (b) $s = 3$, (c) $t = 3$, (d) $t = 4$.

Assume that repeated edges and stopping sets are allowed.

8.11 Repeat Exercise 8.10 but assume that stopping sets of size 2 or less are not allowed.

8.12 Repeat Exercise 8.10 but assume that repeated edges are not allowed.

8.13 For a length-6 (2, 3)-regular LDPC ensemble calculate
(a) $T(v)$ for $v = 1, 2, 3$,
(b) $B(v, X)$ for $v = 1, 2, 3$ and $X = 1, 0$,
(c) the first three terms of (8.32) (i.e. $e = 1, 2, 3$) for the average bit and word erasure probabilities using this ensemble on a BEC with erasure probability $\varepsilon = 0.1$.

References

[1] T. J. Richardson and R. L. Urbanke, *Modern Coding Theory*. New York: Cambridge University Press, 2008.

[2] T. M. Cover and J. A. Thomas, *Elements of Information Theory*. New York: John Wiley & Sons, 1991.

[3] J. R. Pierce, *An Introduction to Information Theory, Symbols, Signals and Noise*, 2nd edn. Mineola, NY: Dover, 1980.

[4] S. Roman, *Introduction to Coding and Information Theory*, Undergraduate Texts in Mathematics series. New York: Springer-Verlag, 1997.

[5] S. B. Wicker, *Error Control Systems for Digital Communication and Storage*. Upper Saddle River, NJ: Prentice Hall, 1995.

[6] T. K. Moon, *Error Correction Coding: Mathematical Methods and Algorithms*. Hoboken, NJ: Wiley-Interscience, 2005.

[7] S. Lin and D. J. Costello, Jr, *Error Control Coding: Fundamentals and Applications*, 2nd edn. Upper Saddle River, NJ: Pearson Prentice Hall, 2004.

[8] F. J. MacWilliams and N. J. A. Sloane, *The Theory of Error-Correcting Codes*. Amsterdam: North-Holland, 1977.

[9] W. C. Huffman and V. Pless, *Fundamentals of Error Correcting Codes*. Cambridge, UK: Cambridge University Press, 2003.

[10] R. G. Gallager, *Low-Density Parity-Check Codes*. Cambridge, MA: MIT Press, 1963.

[11] R. M. Tanner, "A recursive approach to low complexity codes," *IEEE Trans. Inform. Theory*, vol. IT-27, no. 5, pp. 533–547, September 1981.

[12] D. J. C. MacKay, "Good error-correcting codes based on very sparse matrices," *IEEE Trans. Inform. Theory*, vol. 45, no. 2, pp. 399–431, March 1999.

[13] C. Berrou and A. Glavieux, "Near optimum error correcting coding and decoding: turbo codes," *IEEE Trans. Commun.*, vol. 44, no. 10, pp. 1261–1271, October 1996.

[14] M. G. Luby, M. Mitzenmacher, M. A. Shokrollahi, D. A. Spielman and V. Stemann, "Practical loss-resilient codes," in *Proc. 30th ACM Symp. on the Theory of Computing*, pp. 249–258, 1998.

[15] M. G. Luby, M. Mitzenmacher, M. A. Shokrollahi and D. A. Spielman, "Improved low-density parity-check codes using irregular graphs," *IEEE Trans. Inform. Theory*, vol. 47, no. 2, pp. 585–598, February 2001.

[16] S.-Y. Chung, G. D. Forney, Jr, T. J. Richardson and R. L. Urbanke, "On the design of low-density parity-check codes within 0.0045 dB of the Shannon limit," *IEEE Commun. Letters*, vol. 5, no. 2, pp. 58–60, February 2001.

[17] T. J. Richardson and R. L. Urbanke, "Efficient encoding of low-density parity-check codes," *IEEE Trans. Inform. Theory*, vol. 47, no. 2, pp. 638–656, February 2001.

[18] D. J. MacKay, ⟨http://wol.ra.phy.cam.ac.uk/mackay/⟩.

[19] S. Johnson and S. Weller, "Constraining LDPC degree distributions for improved error floor performance," *IEEE Commun. Lett.*, vol. 10, no. 2, pp. 103–105, February 2006.

[20] X.-Y. Hu, M. Fossorier and E. Eleftheriou, "Approximate algorithms for computing the minimum distance of low-density parity-check codes," in *Proc. Int. Symp on Information Theory (ISIT 2004)*, pp. 475ff, 2004.

[21] A. Orlitsky, K. Viswanathan and J. Zhang, "Stopping set distribution of LDPC code ensembles," *IEEE Trans. Inform. Theory*, vol. 51, no. 3, pp. 929–953, March 2005.

[22] S. J. Johnson and S. R. Weller, "A family of irregular LDPC codes with low encoding complexity," *IEEE Commun. Lett.*, vol. 7, no. 2, pp. 79–81, February 2003.

[23] A. J. Blanksby and C. J. Howland, "A 690-mW 1-Gb/s 1024-b, rate-1/2 low-density parity-check code decoder," *IEEE J. Solid-State Circuits*, vol. 37, no. 3, pp. 404–412, March 2002.

[24] L. Fanucci, P. Ciao and G. Colavolpe, "VLSI design of a fully-parallel high-throughput decoder for turbo Gallager codes," *IEICE Trans. Fund. Electronics, Communications and Computer Sci.*, vol. E89-A, no. 7, pp. 1976–1986, July 2006.

[25] V. Nagarajan, N. Jayakumar, S. Khatri and G. Milenkovic, "High-throughput VLSI implementations of iterative decoders and related code construction problems," in *Proc. IEEE Global Telecommunications Conf. (GLOBECOM 2004)*, vol. 1, pp. 361–365, 2004.

[26] L. Zhou, C. Wakayama and C.-J. R. Shi, "CASCADE: a standard super-cell design methodology with congestion-driven placement for three-dimensional interconnect-heavy very large scale integrated circuits," *IEEE Trans. Computer-Aided Design*, no. 7, July 2007.

[27] G. Masera, F. Quaglio and F. Vacca, "Implementation of a flexible LDPC decoder," *IEEE Trans. Circuits and Systems – II: Express Briefs*, vol. 54, no. 6, pp. 542–546, June 2007.

[28] J.-Y. Lee and H.-J. Ryu, "A 1-Gb/s flexible LDPC decoder supporting multiple code rates and block lengths," *IEEE Trans. Consumer Electronics*, vol. 54, no. 2, pp. 417–424, May 2008.

[29] A. Tarable, S. Benedetto and G. Montorsi, "Mapping interleaving laws to parallel turbo and LDPC decoder architectures," *IEEE Trans. Inform. Theory*, vol. 50, no. 9, pp. 2002–2009, September 2004.

[30] K. Andrews, S. Dolinar and J. Thorpe, "Encoders for block-circulant LDPC codes," in *Proc. Int. Symp. on Information Theory (ISIT 2005)*, pp. 2300–2304, 2005.

[31] S. J. Johnson and S. R. Weller, "Codes for iterative decoding from partial geometries," *IEEE Trans. Comms*, vol. 52, no. 2, pp. 236–243, February 2004.

[32] P. O. Vontobel and R. M. Tanner, "Construction of codes based on finite generalized quadrangles for iterative decoding," in *Proc. Internat. Symp. on Information Theory (ISIT 2001)*, Washington, DC, p. 223, 2001.

[33] D. J. C. MacKay and R. M. Neal, "Near Shannon limit performance of low density parity check codes," *Electron. Lett.*, vol. 32, no. 18, pp. 1645–1646, March 1996, reprinted in *Electron. Lett*, vol. 33(6), pp. 457–458, March 1997.

[34] R. J. McEliece, D. J. C. MacKay and J.-F. Cheng, "Turbo decoding as an instance of Pearl's 'belief propagation' algorithm," *IEEE J. Selected Areas Commun.*, vol. 16, no. 2, pp. 140–152, February 1998.

[35] T. Etzion, A. Trachtenberg and A. Vardy, "Which codes have cycle-free Tanner graphs?" *IEEE Trans. Inform. Theory*, vol. 45, no. 6, pp. 2173–2181, September 1999.

[36] C. Di, T. J. Richardson and R. L. Urbanke, "Weight distributions: how deviant can you be?" in *Proc. Int. Symp. on Information Theory (ISIT 2001)*, Washington, DC, p. 50, 2001.

[37] C. Di, T. Richardson and R. Urbanke, "Weight distribution of low-density parity-check codes," *IEEE Trans. Inform. Theory*, vol. 52, no. 11, pp. 4839–4855, November 2006.

[38] M. Sipser and D. A. Spielman, "Expander codes," *IEEE Trans. Inform. Theory*, vol. 42, no. 6, pp. 1710–1722, November 1996.

[39] C. Di, D. Proietti, I. E. Telatar, T. J. Richardson and R. L. Urbanke, "Finite-length analysis of low-density parity-check codes on the binary erasure channel," *IEEE Trans. Inform. Theory*, vol. 48, no. 6, pp. 1570–1579, June 2002.

[40] D. Hösli, E. Svensson and D. Arnold, "High-rate low-density parity-check codes: construction and application," in *Proc. 2nd Int. Symp. on Turbo Codes, Brest, France*, pp. 447–450, 2000.

[41] V. Sorokine, F. R. Kschischang and S. Pasupathy, "Gallager codes for CDMA applications – Part I: Generalizations, constructions, and performance bounds," *IEEE Trans. Commun.*, vol. 48, no. 10, pp. 1660–1668, October 2000.

[42] J. A. McGowan and R. C. Williamson, "Removing loops from LDPC codes," in *Proc. 4th Australian Communications Theory Workshop (AusCTW'03)*, Melbourne, pp. 140–143, 2003.

[43] X.-Y. Hu, E. Eleftheriou and D.-M. Arnold, "Regular and irregular progressive edge-growth Tanner graphs," *IEEE Trans. Inform. Theory*, vol. 51, no. 1, pp. 386–398, January 2005.

[44] T. Richardson and R. Urbanke, "The renaissance of Gallager's low-density parity-check codes," *IEEE Commun. Mag.*, vol. 41, no. 8, pp. 126–131, August 2003.

[45] R. L. Townsend and E. J. Weldon, "Self-orthogonal quasi-cyclic codes," *IEEE Trans. Inform. Theory*, vol. IT-13, no. 2, pp. 183–195, April 1967.

[46] M. Karlin, "New binary coding results by circulants," *IEEE Trans. Inform. Theory*, vol. IT-15, no. 1, pp. 81–92, 1969.

[47] S. Lin and D. Costello, Jr, *Error Control Coding: Fundamentals and Applications*, Prentice-Hall Series in Computer Applications in Electrical Engineering. Englewood Cliffs, NJ: Prentice-Hall, 1983.

[48] R. M. Tanner, "Spectral graphs for quasi-cyclic LDPC codes," in *Proc. Int. Symp. on Information Theory (ISIT 2001)*, Washington, DC, p. 226, 2001.

[49] M. P. C. Fossorier, "Quasi-cyclic low-density parity-check codes from circulant permutation matrices," *IEEE Trans. Inform. Theory*, vol. 50, no. 8, pp. 1788–1793, August 2004.

[50] Z. Li, L. Chen, L. Zeng, S. Lin and W. Fong, "Efficient encoding of quasi-cyclic low-density parity-check codes," *IEEE Trans. Commun.*, vol. 54, no. 1, pp. 71–81, January 2006.

[51] Y. Xu and G. Wei, "On the construction of quasi-systematic block-circulant LDPC codes," *IEEE Commun. Letters*, vol. 11, no. 11, pp. 886–888, November 2007.

[52] J. Xu and S. Lin, "A combinatoric superposition method for constructing LDPC codes," in *Proc. Int. Symp. on Information Theory (ISIT 2003)*, Yokohama, Japan, p. 30, 2003.

[53] J. Thorpe, "Low-density parity-check codes constructed from protographs," IPN Progress Report 42-154, JPL, Tech. Rep., August 2003.

[54] J. Thorpe, K. Andrews and S. Dolinar, "Methodologies for designing LDPC codes using protographs and circulants," in *Proc. Int. Symp. on Information Theory 2004 (ISIT 2004)*, pp. 238ff, 2004.

[55] S. J. Johnson, "Burst erasure correcting LDPC codes," *IEEE Trans. Commun.*, vol. 57, no. 3, pp. 641–652, March 2009.

[56] D. Divsalar, S. Dolinar and C. Jones, "Protograph LDPC codes over burst erasure channels," in *Proc. IEEE Military Communications Conf. (MILCOM) 2006*, pp. 1–7, 2006.
[57] K. Andrews, D. Divsalar, S. Dolinar, J. Hamkins, C. Jones and F. Pollara, "The development of turbo and LDPC codes for deep-space applications," *Proc. IEEE*, vol. 95, no. 11, pp. 2142–2156, November 2007.
[58] K. Karplus and H. Krit, "A semi-systolic decoder for the PDSC-73 error-correcting code," *Discrete Applied Math.*, vol. 33, no. 1–3, pp. 109–128, November 1991.
[59] J. L. Fan, *Constrained Coding and Soft Iterative Decoding*, Kluwer International Series in Engineering and Computer Science. Kluwer Academic Publishers, 2001.
[60] R. Lucas, M. P. C. Fossorier, Y. Kou and S. Lin, "Iterative decoding of one-step majority logic decodable codes based on belief propagation," *IEEE Trans. Commun.*, vol. 48, no. 6, pp. 931–937, June 2000.
[61] Y. Kou, S. Lin and M. P. C. Fossorier, "Low-density parity-check codes based on finite geometries: a rediscovery and new results," *IEEE Trans. Inform. Theory*, vol. 47, no. 7, pp. 2711–2736, November 2001.
[62] H. Tang, J. Xu, S. Lin and K. Abdel-Ghaffar, "Codes on finite geometries," *IEEE Trans. Inform. Theory*, vol. 51, no. 2, pp. 572–596, February 2005.
[63] B. Vasic and O. Milenkovic, "Combinatorial constructions of low-density parity-check codes for iterative decoding," *IEEE Trans. Inform. Theory*, vol. 50, no. 6, pp. 1156–1176, June 2004.
[64] S. Lin, H. Tang, Y. Kou, J. Xu and K. Abdel-Ghaffar, "Codes on finite geometries," in *Proc. IEEE Information Theory Workshop (ITW 2001)*, Cairns, Australia, pp. 14–16, 2001.
[65] H. Tang, J. Xu, Y. Kou, S. Lin and K. Abdel-Ghaffar, "On algebraic construction of Gallager low density parity check codes," in *Proc. Int. Symp. on Information Theory (ISIT 2002)*, Lausanne, Switzerland, p. 482, 2002.
[66] S. J. Johnson and S. R. Weller, "Resolvable 2-designs for regular low-density parity-check codes," *IEEE Trans. Commun.*, vol. 51, no. 9, pp. 1413–1419, September 2003.
[67] B. Vasic, "Structured iteratively decodable codes based on Steiner systems and their application in magnetic recording," in *Proc. IEEE Globecom Conf.*, San Antonio, TX, pp. 2954–2960, 2001.
[68] G. A. Margulis, "Explicit constructions for graphs without short cycles and low density codes," *Combinatorica*, vol. 2, no. 1, pp. 71–78, 1982.
[69] J. Rosenthal and P. O. Vontobel, "Constructions of LDPC codes using Ramanujan graphs and ideas from Margulis," in *Proc. 38th Annual Allerton Conf. on Communications, Control and Computing*, Monticello House, IL, 2000.
[70] J. Lafferty and D. Rockmore, "Codes and iterative decoding on algebraic expander graphs," in *Proc. Int. Symp. on Information Theory and its applications (ISITA 2000)*, Hawaii, USA, 2000.
[71] J. W. Bond, S. Hui, and H. Schmidt, "Constructing low-density parity-check codes with circulant matrices," in *Proc. IEEE Information Theory Workshop (ITW 1999)*, Metsovo, Greece, p. 52, 1999.
[72] E. F. Assmus, Jr and J. D. Key, *Designs and their Codes*, Cambridge Tracts in Mathematics series, vol. 103. Cambridge, UK: Cambridge University Press, 1993.
[73] P. J. Cameron and J. H. van Lint, *Graphs, Codes and Designs*, London Mathematical Society Lecture Note Series, no. 43. Cambridge, UK: Cambridge University Press, 1980.
[74] I. Anderson, *Combinatorial Designs: Construction Methods*, Mathematics and its Applications series. Chichester: Ellis Horwood, 1990.

[75] L. R. Bahl, J. Cocke, F. Jelinek and J. Raviv, "Optimal decoding of linear codes for minimizing symbol error rate," *IEEE Trans. Inform. Theory*, vol. IT-20, no. 2, pp. 284–287, March 1974.

[76] A. J. Viterbi, "Error bounds for convolutional codes and an asymptotically optimal decoding algorithm," *IEEE Trans. Inform. Theory*, vol. IT-13, pp. 260–269, April 1967.

[77] P. Elias, "Coding for noisy channels," in *IRE Conv. Rept.*, Pt. 4, pp. 37–47, 1955.

[78] J. M. Wozencraft and B. Reiffen, *Sequential Decoding*. Cambridge, MA: MIT Press, 1961.

[79] J. L. Massey, *Threshold Decoding*. Cambridge, MA: MIT Press, 1963.

[80] G. D. Forney Jr, "The Viterbi algorithm," *Proc. IEEE*, vol. 61, pp. 268–278, March 1973.

[81] G. D. Forney Jr, "The Viterbi algorithm: A personal history," April 2005, available on the arXiv preprint server.

[82] C. Berrou, A. Glavieux and P. Thitimajshima, "Near Shannon limit error-correcting coding and decoding: turbo-codes," in *Proc. IEEE Int. Conf. on Communications (ICC 93)*, Geneva, Switzerland, pp. 1064–1070, 1993.

[83] A. Dholakia, *Introduction to Convolutional Codes*. Dordrecht, The Netherlands: Kluwer Academic, 1994.

[84] L. C. H. Lee, *Convolutional Coding: Fundamentals and Applications*. Norwood, MA: Artec House, 1997.

[85] R. Johannesson and K. S. Zigangirov, *Fundamentals of Convolutional Coding*. Piscataway, NJ: IEEE Press, 1999.

[86] C. B. Schlegel and L. C. Pérez, *Trellis and Turbo Coding*. Piscataway, NJ: IEEE Press and Wiley Interscience, 2004.

[87] D. J. Costello Jr, J. Hagenauer, I. Hideki and S. B. Wicker, "Applications of error-control coding," *IEEE Trans. Inform. Theory*, vol. 44, no. 6, pp. 2531–2560, 1998.

[88] S. Dolinar and D. Divsalar, "Weight distributions for turbo codes using random and non-random permutations," JPL TDA Progress Report pp. 42–144, August 1995.

[89] R. Garello, P. Pierleoni and S. Benedetto, "Computing the free distance of turbo codes and serially concatenated codes with interleavers: algorithms and applications," *IEEE J. Selected Areas Commun.*, vol. 19, no. 5, pp. 800–812, May 2001.

[90] S. Benedetto, R. Garello and G. Montorsi, "A search for good convolutional codes to be used in the construction of turbo codes," *IEEE Trans. Commun.*, vol. 46, no. 9, pp. 1101–1105, September 1998.

[91] A. Abbasfar, D. Divsalar and K. Yao, "A class of turbo-like codes with efficient and practical high-speed decoders," in *Proc. IEEE Military Communications Conf. 2004 (MILCOM 2004)* pp. 245–250, 2004.

[92] C. Heegard and S. Wicker, *Turbo Coding*, Kluwer International Series in Engineering and Computer Science. Boston: Kluwer Academic, 1999.

[93] S. Benedetto and G. Montorsi, "Design of parallel concatenated convolutional codes," *IEEE Trans. Commun.*, vol. 44, no. 5, pp. 591–600, May 1996.

[94] D. Divsalar and R. McEliece, "Effective free distance of turbo codes," *Electron. Lett.*, vol. 32, no. 5, pp. 445, February 1996.

[95] A. Graelli Amat, G. Montorsi and S. Benedetto, "A new approach to the construction of high-rate convolutional codes," *IEEE Commun. Lett.*, vol. 5, no. 11, pp. 453–455, November 2001.

[96] S. Benedetto, R. Garello and G. Montorsi, "A search for good convolutional codes to be used in the construction of turbo codes," *IEEE Trans. Commun.*, vol. 46, no. 9, pp. 1101–1105, September 1998.

[97] L. Dinoi and S. Benedetto, "Design of fast-prunable s-random interleavers," *IEEE Trans. Wireless Commun.*, vol. 4, no. 5, pp. 2540–2548, September 2005.
[98] J. Sun and O. Takeshita, "Interleavers for turbo codes using permutation polynomials over integer rings," *IEEE Trans. Inform. Theory*, vol. 51, no. 1, pp. 101–119, January 2005.
[99] N. Wiberg, H.-A. Loeliger and R. Kötter, "Codes and iterative decoding on general graphs," *Eur. Trans. Telecommun.*, vol. 6, pp. 513–525, September/October 1995.
[100] N. Wiberg, "Codes and decoding on general graphs," Ph.D. dissertation, Linköping University, Sweden, Department of Electrical Engineering, 1996.
[101] F. Kschischang and B. Frey, "Iterative decoding of compound codes by probability propagation in graphical models," *IEEE J. Select. Areas Commun.*, vol. 16, no. 2, pp. 219–230, February 1998.
[102] F. Kschischang, "Codes defined on graphs," *IEEE Commun. Mag.*, vol. 41, no. 8, pp. 118–125, August 2003.
[103] F. R. Kschischang, B. J. Frey and H.-A. Loeliger, "Factor graphs and the sum–product algorithm," *IEEE Trans. Inform. Theory*, vol. 47, no. 2, pp. 498–519, February 2001.
[104] B. Vucetic and J. Yuan, *Turbo Codes Principles and Applications*, Kluwer International Series in Engineering and Computer Science. Kluwer Academic, 2000.
[105] W. E. Ryan, "Concatenated codes and iterative decoding," *Wiley Encyclopedia of Telecommunications*. Wiley and Sons, 2003.
[106] S. Benedetto, G. Montorsi and D. Divsalar, "Concatenated convolutional codes with interleavers," *IEEE Commun. Mag.*, vol. 41, no. 8, pp. 102–109, August 2003.
[107] E. Yeo and V. Anantharam, "Iterative decoder architectures," *IEEE Commun. Mag.*, vol. 41, no. 8, pp. 132–140, August 2003.
[108] M. Tchler and J. Hagenauer, "EXIT charts of irregular codes," in *Proc. Conf. on Information Sciences and Systems (CISS 2002), Princeton*, 2002.
[109] S. Benedetto, D. Divsalar, G. Montorsi and F. Pollara, "Serial concatenation of interleaved codes: performance analysis, design, and iterative decoding," *IEEE Trans. Inform. Theory*, vol. 44, no. 3, pp. 909–926, May 1998.
[110] H. Pfister, I. Sason and R. Urbanke, "Capacity-achieving ensembles for the binary erasure channel with bounded complexity," *IEEE Trans. Inform. Theory*, vol. 51, no. 7, pp. 2352–2379, July 2005.
[111] A. Roumy, S. Guemghar, G. Caire and S. Verdú, "Design methods for irregular repeat–accumulate codes," *IEEE Trans. Inform. Theory*, vol. 50, no. 8, pp. 1711–1727, August 2004.
[112] S. Johnson and S. Weller, "Constructions for irregular repeat–accumulate codes," in *Proc. Int. Symp. on Information Theory (ISIT 2005)*, pp. 179–183, 2005.
[113] A. Abbasfar, D. Divsalar and K. Yao, "Accumulate–repeat–accumulate codes," *IEEE Trans. Commun.*, vol. 55, no. 4, pp. 692–702, April 2007.
[114] S. J. Johnson and S. R. Weller, "Combinatorial interleavers for systematic regular repeat–accumulate codes," *IEEE Trans. Commun.*, vol. 56, no. 8, pp. 1201–1206, August 2008.
[115] S. J. Johnson and S. R. Weller, "Practical interleavers for repeat–accumulate codes," *IEEE Trans. Commun.*, vol 57, no. 5, pp. 1225–1228, May 2009.
[116] S. Benedetto and G. Montorsi, "Serial concatenation of block and convolutional codes," *Electron. Lett.*, vol. 32, no. 10, pp. 887–888, May 1996.
[117] S. Benedetto and G. Montorsi, "Iterative decoding of serially concatenated convolutional codes," *Electron. Lett.*, vol. 32, no. 13, pp. 1186–1188, June 1996.

[118] D. Divsalar, H. Jin and R. J. McEliece, "Coding theorems for 'turbo-like' codes," in *Proc. 36th Allerton Conf. on Communications, Control and Computing*, Allerton, IL, pp. 201–210, 1998.

[119] H. Jin, D. Khandekar and R. J. McEliece, "Irregular repeat–accumulate codes," in *Proc. 2nd Int. Symp. on Turbo Codes and Related Topics*, Brest, France, pp. 1–8, 2000.

[120] M. Yang, W. E. Ryan and Y. Li, "Design of efficiently encodable moderate-length high-rate irregular LDPC codes," *IEEE Trans. Commun.*, vol. 52, no. 4, pp. 564–571, April 2004.

[121] R. Echard and S.-C. Chang, "The π-rotation low-density parity-check codes," in *Proc. Global Telecommunications Conf.*, San Antonio, TX, pp. 980–984, 2001.

[122] R. M. Tanner, "On quasi-cyclic repeat-accumulate codes," in *Proc. 37th Annual Allerton Conf. on Communication, Control and Computing*, Monticello, IL, 1999.

[123] I. Sason and R. Urbanke, "Complexity versus performance of capacity-achieving irregular repeat–accumulate codes on the binary erasure channel," *IEEE Trans. Inform. Theory*, vol. 50, no. 6, pp. 1247–1256, June 2004.

[124] A. Abbasfar, D. Divsalar and K. Yao, "Accumulate repeat accumulate codes," in *Proc. Global Telecommunications Conf. (GLOBECOM '04)*, vol. 1, pp. 509–513, 2004.

[125] H. Pfister and I. Sason, "Accumulate–repeat–accumulate codes: capacity-achieving ensembles of systematic codes for the erasure channel with bounded complexity," *IEEE Trans. Inform. Theory*, vol. 53, no. 6, pp. 2088–2115, June 2007.

[126] A. Papoulis, *Probability, Random Variables and Stochastic Processes*, 2nd edn. Singapore: McGraw-Hill, 1984.

[127] T. J. Richardson and R. L. Urbanke, "The capacity of low-density parity-check codes under message-passing decoding," *IEEE Trans. Inform. Theory*, vol. 47, no. 2, pp. 599–618, February 2001.

[128] T. Richardson and R. Urbanke, "Thresholds for turbo codes," in *Proc. IEEE Int. Symp. on Information Theory*, pp. 317ff, 2000.

[129] A. Shokrollahi, "Capacity-achieving sequences," 2000 (online). Available at citeseer.ist.psu.edu/shokrollahi00capacityachieving.html.

[130] I. Sason and R. Urbanke, "Parity-check density versus performance of binary linear block codes over memoryless symmetric channels," *IEEE Trans. Inform. Theory*, vol. 49, no. 7, pp. 1611–1635, July 2003.

[131] S. ten Brink, G. Kramer and A. Ashikhmin, "Design of low-density parity-check codes for modulation and detection," *IEEE Trans. Commun.*, vol. 52, no. 4, pp. 670–678, April 2004.

[132] E. Sharon, A. Ashikhmin and S. Litsyn, "Analysis of low-density parity-check codes based on EXIT functions," *IEEE Trans. Commun.*, vol. 54, no. 8, pp. 1407–1414, August 2006.

[133] M. G. Luby, M. Mitzenmacher and M. A. Shokrollahi, "Analysis of random processes via and–or tree evaluation," in *Proc. 9th Annual ACM-SIAM Symp. on Discrete Algorithms*, pp. 364–373, 1998.

[134] M. G. Luby, M. Mitzenmacher, M. A. Shokrollahi and D. A. Spielman, "Efficient erasure correcting codes," *IEEE Trans. Inform. Theory*, vol. 47, no. 2, pp. 569–584, February 2001.

[135] T. J. Richardson, M. A. Shokrollahi and R. L. Urbanke, "Design of capacity-approaching irregular low-density parity-check codes," *IEEE Trans. Inform. Theory*, vol. 47, no. 2, pp. 619–637, February 2001.

[136] D. Divsalar, S. Dolinar and F. Pollara, "Iterative turbo decoder analysis based on density evolution," *IEEE J. Select. Areas Commun.*, vol. 19, no. 5, pp. 891–907, May 2001.

[137] A. Amraoui and R. L. Urbanke, *LdpcOpt:* ⟨http://lthcwww.epfl.ch/research/ldpcopt/⟩.

[138] D. Hayes, S. Weller and S. Johnson, *LODE:* ⟨http://sigpromu.org/ldpc/DE/⟩.

[139] S.-Y. Chung, T. Richardson and R. Urbanke, "Analysis of sum–product decoding of low-density parity-check codes using a Gaussian approximation," *IEEE Trans. Inform. Theory*, vol. 47, no. 2, pp. 657–670, February 2001.

[140] H. E. Gamal and A. R. Hammons, "Analyzing the turbo decoder using the Gaussian approximation," *IEEE Trans. Inform. Theory*, vol. 47, no. 2, pp. 671–686, February 2001.

[141] F. Lehmann and G. M. Maggio, "Analysis of the iterative decoding of LDPC and product codes using the Gaussian approximation," *IEEE Trans. Inform. Theory*, vol. 49, no. 11, pp. 2993–3000, November 2003.

[142] S. ten Brink, "Convergence of iterative decoding," *Electron. Lett.*, vol. 35, no. 13, pp. 1117–1118, June 1999.

[143] S. ten Brink, "Convergence behaviour of iteratively decoded parallel concatenated codes," *IEEE Trans. Commun.*, vol. 49, no. 10, pp. 1727–1737, October 2001.

[144] S. ten Brink, "Design of repeat–accumulate codes for iterative detection and decoding," *IEEE Trans. Signal Processing*, vol. 51, no. 11, pp. 2764–2772, November 2003.

[145] A. Ashikhmin, G. Kramer and S. ten Brink, "Extrinsic information transfer functions: model and erasure channel properties," *IEEE Trans. Inform. Theory*, vol. 50, no. 11, pp. 2657–2673, November 2004.

[146] S. Mason and H. Zimmermann, *Electronic Circuits, Signals and Systems*. New York: John Wiley, 1960.

[147] H. Jin and R. J. McEliece, "Coding theorems for turbo code ensembles," *IEEE Trans. Inform. Theory*, vol. 48, no. 6, pp. 1451–1461, June 2002.

[148] N. Kahale and R. Urbanke, "On the minimum distance of parallel and serially concatenated codes," in *Proc. IEEE Int. Symp. on Information Theory 1998*, pp. 31ff, 1998.

[149] A. Perotti and S. Benedetto, "A new upper bound on the minimum distance of turbo codes," *IEEE Trans. Inform. Theory*, vol. 50, no. 12, pp. 2985–2997, December 2004.

[150] G. Miller and D. Burshtein, "Bounds on the maximum-likelihood decoding error probability of low-density parity-check codes," *IEEE Trans. Inform. Theory*, vol. 47, no. 7, pp. 2696–2710, November 2001.

[151] M. Lentmaier, D. Truhachev, K. Zigangirov and D. Costello, "An analysis of the block error probability performance of iterative decoding," *IEEE Trans. Inform. Theory*, vol. 51, no. 11, pp. 3834–3855, November 2005.

[152] T. J. Richardson and R. L. Urbanke, "Finite-length density evolution and the distribution of the number of iterations on the binary erasure channel," unpublished manuscript, available at ⟨http://lthcwww.epfl.ch/papers/RiU02.ps⟩.

[153] S. Benedetto and G. Montorsi, "Average performance of parallel concatenated block codes," *Electron. Lett.*, vol. 31, no. 3, pp. 156–158, February 1995.

[154] S. Benedetto and G. Montorsi, "Performance evaluation of turbo-codes," *Electron. Lett.*, vol. 31, no. 3, pp. 163–165, February 1995.

[155] S. Benedetto and G. Montorsi, "Unveiling turbo codes: some results on parallel concatenated coding schemes," *IEEE Trans. Inform. Theory*, vol. 42, no. 2, pp. 409–428, March 1996.

[156] M. Breiling, "A logarithmic upper bound on the minimum distance of turbo codes," *IEEE Trans. Inform. Theory*, vol. 50, no. 8, pp. 1692–1710, August 2004.

[157] D. Divsalar, S. Dolinar and C. Jones, "Construction of protograph LDPC codes with linear minimum distance," in *Proc. IEEE Int. Symp. on Information Theory, 2006*, pp. 664–668, 2006.

[158] D. Divsalar and E. Biglieri, "Upper bounds to error probabilities of coded systems beyond the cutoff rate," *IEEE Trans. Commun.*, vol. 51, no. 12, pp. 2011–2018, December 2003.

[159] H. Zhang and A. Orlitsky, "Finite-length analysis of LDPC codes with large left degrees," in *Proc. Int. Symp. on Information Theory (ISIT 2002)*, Lausanne, Switzerland, p. 3, 2002.

[160] S. J. Johnson, "A finite-length algorithm for LDPC codes without repeated edges on the binary erasure channel," *IEEE Trans. Inform. Theory*, vol. 55, no. 1, pp. 27–32, January 2009.

[161] J. Lee, R. Urbanke and R. Blahut, "Turbo codes in binary erasure channel," *IEEE Trans. Inform. Theory*, vol. 54, no. 4, pp. 1765–1773, April 2008.

[162] B. Frey, R. Koetter and A. Vardy, "Signal-space characterization of iterative decoding," *IEEE Trans. Inform. Theory*, vol. 47, no. 2, pp. 766–781, February 2001.

[163] R. Smarandache and P. Vontobel, "Pseudo-codeword analysis of Tanner graphs from projective and Euclidean planes," *IEEE Trans. Inform. Theory*, vol. 53, no. 7, pp. 2376–2393, July 2007.

[164] C. Kelley and D. Sridhara, "Pseudocodewords of Tanner graphs," *IEEE Trans. Inform. Theory*, vol. 53, no. 11, pp. 4013–4038, November 2007.

[165] M. Ivkovic, S. Chilappagari and B. Vasic, "Eliminating trapping sets in low-density parity-check codes by using Tanner graph covers," *IEEE Trans. Inform. Theory*, vol. 54, no. 8, pp. 3763–3768, August 2008.

[166] S.-T. Xia and F.-W. Fu, "Minimum pseudoweight and minimum pseudocodewords of LDPC codes," *IEEE Trans. Inform. Theory*, vol. 54, no. 1, pp. 480–485, January 2008.

[167] P. Votobel, ⟨http://www.hpl.hp.com/personal/Pascal Vontobel/pseudocodewords/⟩.

[168] T. Richardson, "The geometry of turbo-decoding dynamics," *IEEE Trans. Inform. Theory*, vol. 46, no. 1, pp. 9–23, January 2000.

[169] D. Agrawal and A. Vardy, "The turbo decoding algorithm and its phase trajectories," *IEEE Trans. Inform. Theory*, vol. 47, no. 2, pp. 699–722, February 2001.

[170] M. Fu, "Stochastic analysis of turbo decoding," *IEEE Trans. Inform. Theory*, vol. 51, no. 1, pp. 81–100, January 2005.

[171] B. Rüffer, C. Kellett, P. Dower and S. Weller, "Belief propagation as a dynamical system: the linear case and open problems," submitted to *IET Control Theory Appl.*

Index

FEB 23 2010